Nels[obscured by label]

BIOLOGY

BOB RITTER
University of Alberta/Austin O'Brien High School
Edmonton, Alberta

DR. WALLIE SAMIRODEN
Professor
Dept. of Secondary Education
University of Alberta
Edmonton, Alberta

Reviewer:
MONICA QUINLAN
Science Teacher
Father Lacombe High School
Calgary Catholic School Board
Calgary, Alberta

Teacher's Resource

Nelson Canada

Published in 1993 by
Nelson Canada,
A Division of Thomson Canada Limited
1120 Birchmount Road
Scarborough, Ontario M1K 5G4

This book is printed on acid-free paper. The choice of paper reflects Nelson Canada's goal of using, within the publishing process, the available resources, technology, and suppliers that are as environment friendly as possible.

All student investigations have been designed to be as safe as possible, and have been reviewed by professionals specifically for that purpose. As well, appropriate warnings concerning potential safety hazards are included where applicable to particular investigations. However, responsibility remains with the student, the classroom teacher, the school principal, and the school board.

ISBN 0-17-603861-2

Canadian Cataloguing in Publication Data

Ritter, Robert John, 1950 –
Nelson biology. Teacher's resource

Supplement to: Nelson Biology
ISBN 0-17-603861-2

1. Biology - Study and teaching (Secondary).
I. Samiroden, Wallie II. Title.

QH308.7.N452 1993 574 C93-093227-7

Cover photo: The cover shows a stand of yellow cedars in British Columbia. The inset shows a cougar, a common inhabitant of western Canada. Tim Fitzharris

Executive Editor: Lynn Fisher
Project Manager: Ruta Demery
Project Co-ordinator: Jennifer Dewey
Supervising Production Editor: Cecilia Chan
Illustrations and Typesetting: VISU*TronX* – J. Loates/S. Calverley

Printed and Bound in Canada
1 2 3 4 5 6 7 8 9 WC 9 8 7 6 5 4 3

C O N T E N T S

Textbook Features

Demonstrations

It is not uncommon for a particular laboratory group working in isolation to miss the entire focus of a laboratory activity; the demonstration is designed to ensure a basic understanding of concepts within a dynamic learning approach, prior to work being done in smaller groups. The demonstration is designed to initiate large-group discussion, where each member of the class benefits from a commonality of questions and discussed answers. Demonstrations provide a common focus for all class members; they address the immediate application of written material with an emphasis on discovery. The demonstration is an integral part of the program because it ensures each student gains a minimum grounding in essential concepts. Imbedded observation questions make the activity interactive by encouraging students to question observations. The *Teacher's Resource* gives the estimated time for the procedure and offers suggested methods for organizing the activity.

Laboratory Activities

Laboratory activities are designed to encourage small-group cooperation and organization. While small groups collect and record data throughout the activity, the application questions can be done individually. Laboratory activities can become an important component of the evaluation process. The laboratory activities are integrated within the student book, thereby ensuring proper background information. All too often laboratory activities are conducted at the end of a chapter, when students have already acquired the necessary information to predict outcomes. Consequently, the laboratory activities lose their appeal as they become predictable exercises. Conversely, when laboratory activities are conducted too early, without the appropriate background information, students are often frustrated by the experience.

Case Studies

Case studies are intended to provide students with experience in investigative research. Although organized in much the same manner as demonstrations and laboratory activities, they emphasize data analysis and interpretation rather than observation. The case studies take advantage, where appropriate, of Canadian examples to introduce a concept.

Questions

Three types of questions are found in each chapter of *Nelson Biology.*

- Review questions are lower-level reading comprehension questions. These are found in two or three locations throughout each chapter.
- Applying the concepts, higher-level application questions, are presented at the end of each chapter. These questions stress the integration of everyday knowledge with the concepts learned in the chapter. A technological perspective is introduced.
- Critical-thinking questions are also found at the end of each chapter and are designed partly to introduce societal issues that arise through the use of science by technology. This section is designed to encourage lateral thinking skills. Questions are more open-ended, often not calling for just a single answer. Students are encouraged to explore different pathways and to consider alternatives.

Social Issues

A point-counterpoint format is used to introduce social issues and to set the stage for student debates. By participating in debate students are provided with an opportunity to share scientific information; thus, *Nelson Biology* addresses the call for enhancing students' "scientific literacy." The debate format encourages scientific discourse and promotes the notion that peers can contribute to the learning of science. The debates, then, are intended to present divergent perspectives; they promote a mutual respect and understanding of different worldviews.

Career Investigation

Each unit in *Nelson Biology* presents a career investigation section, designed to encourage career exploration. Special attention is paid to divergent abilities and interests. For example, a stockbroker's skills can be investigated when considering biotechnology as a growth industry. This reinforces the notion that the need for scientific knowledge in building a career is not restricted to researchers and science teachers.

Research in Canada

This section highlights Canadian contributions to science developments, both presently and in the past. By describing ongoing research, this feature of *Nelson Biology* demonstrates that science is not a finite set of answers to be learned, but an ongoing process. The profiles can be used to uncover the dynamic nature of scientific paradigms, and to demonstrate a dialectic of viewpoints.

Concept Maps and Summary Tables

Students are provided with concept maps throughout the textbook. Concept maps are designed to provide students with an opportunity to organize cause-and-effect relationships. Summary tables are designed to provide students with a quick reference or an overview of a complicated concept. In the *Teacher's Resource* students are encouraged to develop their own concept maps.

Teacher's Resource Features

Suggestions for Introducing a New Unit and Initiating the Study of a New Chapter

Beginning a new unit of study or chapter of the text provides the opportunity for a teacher to gain the attention and interest of the students. These sections of the *Teacher's Resource* will provide some suggestions for initiating the study of each unit and chapter. These suggestions can help the teacher learn of the conceptions that students bring to the class. The suggestions offered are very limited, and teachers are encouraged to expand on any of them and/or develop others as they see being appropriate. Another source of ideas for these pre-unit activities are the questions that are given at the ends of chapters under the headings of Applying the Concepts and Critical-Thinking Questions.

Addressing Alternate Conceptions

As each new chapter of the text is initiated, the teacher can learn of the beliefs and understandings that the students hold about the subject prior to the study of the material. These understandings may be at odds with reality and/or accurate scientific explanations. Such alternate conceptions frequently stand in the way of learning. But, when a teacher knows something of the alternate conceptions that students hold, lessons and learning activities can be planned in ways that challenge such conceptions and lead students to gaining more accurate understandings.

Making Connections

This section attempts to integrate information from a variety of disciplines. Students come to recognize that physiological systems are influenced by genetics which in turn influences adaptation and diversity.

Possible Journal Entries

The journal is designed to provide students with an opportunity to begin thinking about the manner in which they construct knowledge and develop personal meaning. Suggestions in this section focus students on metacognitive processes. The journal should provide the students and the teacher with a record of how the students' thinking has changed and how their attitudes about biology have evolved.

Using the Videodisc

The *Living Textbook Principles of Biology* videodisc package has been correlated to *Nelson Biology*. The videodisc was produced by Optical Data Corporation and can be ordered from Perceptix Inc. (1-800-267-7788).

The barcodes and frame numbers appear in a chart for each chapter, which briefly provides a description of the still or movie clip, as well as the page reference in the student text where the information is presented. You can access the videodisc by using the barcodes or frame numbers that appear in the chart. The videodisc program contains a wealth of visual images, for only some of which we have provided a reference.

In order to use the videodisc with the frame numbers, you will need a videodisc player with a remote control device. To use the barcode correlations, you will need a videodisc player, remote control device, and barcode scanner. For a sequence of still frames, we have provided only the barcode for the initial frame. Use your remote control to step through the sequence.

Ideas for Initiating a Discussion

This section lists suggestions for the use of tables, diagrams, and photographs to encourage students to talk about their learning. The purpose of these discussions is to transform the students' learning from a passive to an active process.

Constructivist Approach

The text is written in a manner that attempts to engage students in a conversation with the author. As students read, they construct their own meanings. Laboratory activities, case studies, application questions, critical-thinking questions, and debates will cause students to challenge some of their learning frameworks. The approach also acknowledges that prior learning has occurred. Previous formalized learning and everyday lived (experiential) learning are woven into a fabric of understanding.

As students construct a knowledge base, gaps are found in their understanding. The traditional approach provides a solid basis prior to application and assumes that student learning parallels teaching. A constructivist approach requires the re-examination of beliefs and assumptions as knowledge is formalized. The foundations for understanding are either reinforced or changed as knowledge is constructed.

This approach also recognizes that some flaws will be encountered as students begin building a basis of understanding. Students using this constructivist approach will identify difficulties or imperfections in the structural basis of their theories as they begin to assemble higher levels of understanding. The constructivist approach necessitates that existing assumptions are challenged and scientific knowledge is reorganized to provide a more solid basis. Unlike a traditional view of science that presupposes that scientific knowledge is constructed on irrefutable axioms, this view presupposes that knowledge is tentative. Assumptions must be continuously addressed and questioned.

The tendency to avoid potential misconceptions, by providing a complete answer, usually amplifies the separation of "science as taught" from "lived experiences." Students tend to accept scientific definitions of things such as "organ systems" or "energy" as having meaning in science classes, but often reject their meanings or fail to acknowledge them when dealing with technological applications beyond the domain of the classroom. An approach which stresses the linear, incremental building of scientific knowledge and fosters the idea that an understanding of science begins with a sequential linking of details only magnifies the problem. This view of scientific knowledge is analogous to building a pyramid, in that a solid base must be established before students can begin adding progressive levels of understanding. At the base of the pyramid are irrefutable truths or axioms that have been tested by time. The axioms become the factual basis on which an understanding is built. The recognition of linkages is often relegated to the teacher, and teacher-directed lessons dominate curriculum discussions and the organization of knowledge.

Social Issues

Social issues are presented in a pro/con format (identified as point/counterpoint) at the end of every chapter. The point/counterpoint structure is designed to highlight a dialectic of opposing views, but does not assume that there are only two views, or that students who accept one of the arguments raised by one side of the dialectic will necessarily reject all of the aspects made by the other viewpoint. The issues can be organized into a debate that investigates divergent points of view or the questions can be addressed by individuals as they reflect upon the social implication of technological applications of science. A debate format provides the added advantage of a sharing of ideas. Whatever strategy is applied, the emphasis of the social issues is founded upon critical thinking which stems from further research and the reflection on scientific concepts within a social context.

Should the debating strategy be employed, students are provided with an opportunity to share scientific information. The term "scientific literacy" is often used, but little provision is made for its development in current textbooks. Students are often encouraged to discuss themes of poetry in English classes, or to provide their opinions about current events in Social Studies classes, yet they are rarely given the same opportunity in science classes. Some science classes tend to promote a didactic presentation of material. The debate encourages scientific discourse and promotes the notion that peers can contribute to the learning of science. No group is expected to win the debate. Although one group may garner more popular support than another, the objective is not to promote consensus. Viewpoints founded upon divergent philosophical or theological tenets are to be understood, not changed. Ideally, students should learn that the views of the majority are not inherently correct because they reflect opinions held by a greater populace. Those who hold the most common opinions must be sensitive to the religious, cultural, and moral beliefs of others, and must acknowledge that opinions, like scientific facts, may change over time. The debate serves as a means of demonstrating that both sides of the dialectic can be founded upon rational arguments, which hold divergent worldviews and priorities. All too often controversial issues are presented within a *good guy—bad guy* image. The industrialist and environmentalist perspectives often come into conflict; however, both groups usually present sound and logical rationales for their opinions. Although both groups evaluate a problem and its potential solutions from different frameworks, both attempt to provide society with a better quality of life. Their opinions about what constitutes a better quality of life are formulated on different priorities. The social issues are intended to present divergent solutions and perspectives that demonstrate different priorities. It is hoped that an understanding of the reasons for divergent points of view will help create a respect for different worldviews, and promote forbearance and mutual respect.

Selecting Controversial Issues for the Classroom

Although there is no single answer to this potential dilemma, the following points may provide some guidance:

- What is the maturity and intellectual level of the students? The parameters of an issue may have to be defined to focus discussion on the issue. Some issues may not be understood by younger students.
- Does the issue place any group or individual student at risk? Many perspectives are founded upon divergent worldviews which may be influenced by religion or cultural mores. The social issues are not designed to reach consensus, but promote an appreciation and respect for different worldviews.
- Do students consider the issue to be important?
- Does the issue support the curriculum? Issues that do not have a direct relationship with the knowledge presented in the science course will confuse rather than enhance student learning.

- Is the issue one that can be treated within the time allotted? Simplistic overviews of issues can be avoided by allowing sufficient time for research and brainstorming.
- Can the issue be properly researched? Students should be encouraged to move beyond emotional responses. The point/counterpoint format presented in the textbook is not designed to summarize the points to be debated, but serves as an outline for beginning to understand the question. Research and reflection are required before beginning the debate.
- Does the teacher feel comfortable talking about the issue?
- Will the issue clash with community standards or subject a particular subgroup of students to ridicule?

Decision-Making Process

Students are often asked to suspend judgment until all of the facts are gathered; however, most people tend to personalize controversial issues rather quickly. In that most technology is directed toward specific interest groups, judgments are often linked to the benefits supplied to a specific group. Therefore, it is recommended that students be allowed to evaluate an issue as they begin to gather data. First impressions are not always negative, but students should be open to changing their first impressions.

The following may help students gain a better understanding of the issue and the problems presented by different resolutions.

1. The issue is identified and individual research begins.
   - Students seek resources which help them understand the complexity of the issue.
   - Students consider possible resolutions suggested by experts.
   - Students evaluate the resolution from a personal point of view.
   - Trade-offs are considered.
2. Students are organized into small groups composed of individuals who hold similar opinions (homogeneous groups). The issue is discussed by group members.
   - The group attempts to identify assumptions that they have used in formulating their opinions.
   - The group identifies its own set of values and priorities.
   - The group attempts to identify political influences, economic influences, cultural influences, and religious influences.
   - The group attempts to identify subgroups that will be compromised.
   - The group speculates about the reasons for opposition.
3. Students are organized into larger groups composed of individuals who hold divergent opinions (heterogeneous groups). The issue is discussed by group members.
   - Divergent viewpoints are identified.
   - The consequences of each resolution and the rationale for accepting each viewpoint are discussed.
   - Personal viewpoints are questioned and, possibly, defended.
4. Students reflect personally on the issue.

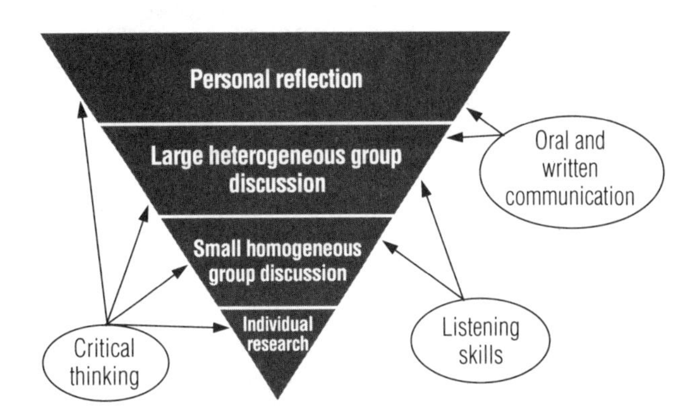

Implementing Controversial Issues

Plan to make social issues a regular part of your science program. If you include at least one controversial issue in every unit, students will expect and prepare for controversy. Skills and attitudes developed during one controversy can be refined and enhanced in others. Students soon come to recognize that a deeper understanding of science-related social issues is inextricably tied to an understanding of scientific knowledge. We hope that the opportunity to experience science and scientific innovation within a social context will encourage students to maintain a lifelong interest in science.

The following recommendations may help your debate:

Suggestion	Reason
• Select an independent chairperson to monitor the time of presentation and response to questioning. Do not partake in questioning until the summary, if at all.	• During the debate, students often look to the teacher for confirmation of their worldview. A teacher who generally has a greater information base, makes a worthy ally.
• Provide some initial background information and ask probing questions before beginning the debate or role-playing scenario.	• Students should understand the scientific and technological underpinnings that provide the framework for understanding the issue. The boundaries of the issue must be defined.
• Do not expect to find a winner of the debate. Discourage the notion that one viewpoint is correct and the other is wrong.	• The importance of the debate should not be measured by one side dominating another. Most debates don't have clear winners and losers. They do, however, provide an opportunity to consider the opposing viewpoint and critically analyze one's own viewpoint.
• In selecting group members for the interactive debate or role-playing scenario consider gender balance. It is to be hoped that all groups will have a balance of motivated students who are willing to express themselves. The discussion should not be approached from the idea that the group that dominates the conversation wins the argument.	• As much as possible, viewpoints should not be aligned with easily identified subgroups. The focus should be placed on the argument, not upon who delivers the argument. Opinions should not be viewed as female responses or male opposition to a viewpoint. Similarly, a point of view held by an honors student is not necessarily more just than that of a student who finds the work difficult.
• Students preparing for debates should coordinate their presentations with other group members. A list of important points should be written out for quick reference.	• Redundant arguments should be avoided. Students involved in the debate must be accountable. Coordination ensures that each student makes an important contribution. Evidence of research should be present.
• During the debate or forum, students should be directed to listen to and respect divergent points of view.	• Ideally, controversial issues should promote tolerance and understanding. A greater understanding of and appreciation for the reasons that certain groups hold certain values can be promoted.

Possible structure for implementation

Select a small group of students who agree with one response to the social issue and another group of students, preferably of equal numbers, who support divergent or opposing points of view. Students will be given an opportunity to provide arguments that support and defend their viewpoint. Each student will also be provided with an opportunity to challenge opposing viewpoints.

Outline of Social Issues in the Text

	Chapter	Social Issue
1	Equilibrium in the Biosphere	The Greenhouse Effect
2	Energy and Ecosystems	Economics and the Environment
3	Aquatic and Snow Ecosystems	Northern Development
4	Adaptation and Change	Pesticides and Evolution
5	Development of the Cell Theory	Limits to Cell Technology
6	Chemistry of Life	Athletes and Performance-Enhancing Drugs
7	Energy within the Cell	Interdependence of Cellular Respiration and Photosynthesis
8	Organs and Organ Systems	Artificial Organs
9	Digestion	Fad Diets
10	Circulation	Heart Care
11	Blood and Immunity	Compulsory AIDS Testing
12	Breathing	Smoking
13	Kidneys and Excretion	Fillings and Kidney Disease
14	The Endocrine System and Homeostasis	Growth Hormone: The Anti-aging Drug
15	The Nervous System and Homeostasis	A New View of Drug Addiction
16	Special Senses	Rock Concerts and Hearing Damage
17	The Reproductive System	Fetal Alcohol Syndrome
18	Asexual Cell Reproduction	Biotechnology and Agriculture
19	Sexual Cell Reproduction	Limits on Reproductive Technology
20	Genes and Heredity	Genetic Screening
21	The Source of Heredity	The Potential of Gene Therapy
22	DNA: The Molecule of Life	Nonhuman Life Forms as Property
23	Protein Synthesis	Biological Warfare
24	Population Genetics	Interbreeding of Plains and Woodland Bison
25	Populations and Communities	Forest Fires and Ecology

Cooperative Learning

Cooperative learning is one of the most thoroughly researched of all instructional methods and consensus has been reached on a number of results. There is agreement that cooperative learning methods can and usually do have a positive effect on student achievement. Other outcomes seen in many studies of cooperative learning include gains in self-esteem, liking for school and of the subject being studied, reduction of time-on-task, and better attendance. Extended experiences with cooperative learning can increase the ability to work effectively with others. Cooperative learning also improves the social acceptance of mainstreamed academically handicapped students by their classmates.

Cooperative learning is organizing and structuring a lesson so that students work and learn together to accomplish a goal. A number of components within cooperative learning facilitate the transition from the more traditional lecture or teacher-directed style to that of group learning. We recognize this approach as another way by which teachers can meet different learning styles and maintain a high level of student involvement. Through cooperative learning approaches, students work with and teach other students. It promotes:

- Discussion and argumentation among students
- The elevation of self-esteem among group members when they agree
- Shared and individual responsibility, since each member has a responsibility to the group and the group assists each member
- Interaction among all class members through flexible groupings
- Active learning in which students are engaged in teaching each other

Comparison of Traditional Teacher-Directed Class and Cooperative-Learning Class

Whole-Class Question-Answer	Numbered Heads Together
The teacher asks questions.	The teacher numbers students from 1 to 4 within each group.
Students who wish to respond raise their hands.	The teacher asks questions.
The teacher calls on one student—maybe a student who didn't have his hand raised.	Teacher tells students to "put their heads together" to ensure that each group member knows the answer.
The student attempts to state the correct answer.	Teacher calls a number from 1 to 4 and students with that number can raise their hands to respond.

As can be seen from the comparison in the table above, the traditional classroom teacher-student interaction tends to be one of competition, in which a student's self-esteem can be frequently at risk. In a cooperative learning situation where a group gains some consensus on the answer to a teacher's question, students are put into less threatening situations, and even those students who might not become involved in answering questions in the traditional classroom can become more actively involved.

Groupings

It is worthwhile to survey your students and get to know them in order to form heterogeneous groups based on differing abilities, sex, ethnic backgrounds, and personalities. Groups may last a few minutes, half an hour, a week, a month, or an entire school year, depending on the topic and the task. At the beginning, pairs are easiest for students. It is advisable for groups to stay together until they experience some success. Knowing that the group either "sinks" or "swims" together motivates the members to work together as a team.

Structure	Brief Description	Academic and Social Functions
Some Possible Structures and Their Functions*		
Team Building		
Round-robin	Each student in turn shares something with his/her teammates	Expressing ideas and opinions, creating stories. Equal participation, getting acquainted with each other.
Class Building		
Corners	Each student moves to a group in a corner or location as determined by the teacher through specified alternatives. Students discuss within groups, then listen to and paraphrase ideas from other groups.	Seeing alternative hypotheses, values, and problem-solving approaches. Knowing and respecting differing points of view.
Mastery		
Numbered heads together	The teacher asks a question, students consult within their groups to make sure that each member knows the answer. Then one student answers for the group in response to the number called out by the teacher.	Review, checking for knowledge, comprehension, analysis, and divergent thinking. Tutoring.
Color-coded co-op cards	Students memorize facts using a flash-card game or an adaptation. The game is structured so that there is a maximum probability for success at each step, moving from short- to long-term memory. Scoring is based on improvement.	Memorizing facts. Helping, praising.
Pairs check	Students work in pairs within groups of four. Within pairs students alternate—one solves a problem while the other coaches. After every problem or so, the pair checks to see if they have the same answer as the other pair.	Practicing skills. Helping, praising.
Concept Development		
Three-step interview	Students interview each other in pairs, first one way, then the other. Each student shares information learned during interviews with the group.	Sharing personal information such as hypotheses, views on an issue, or conclusions from a unit. Participation, involvement.
Think-pair-share	Students think to themselves on a topic provided by the teacher; they pair up with another student to discuss it; and then share their thoughts with the class.	Generating and revising hypotheses, inductive and deductive reasoning, and application. Participation and involvement.
Team word-webbing	Students write simultaneously on a piece of paper, drawing main concepts, supporting elements, and bridges representing the relation of concepts/ideas.	Analysis of concepts into components, understanding multiple relations among ideas, and differentiating concepts. Role taking.
Multifunctional		
Roundtable	Each student in turn writes one answer as a paper and a pencil are passed around the group. With simultaneous roundtable, more than one pencil and paper are used at once.	Assessing prior knowledge, practicing skills, recalling information, and creating designs. Team building, participation of all.
Partners	Students work in pairs to create or master content. They consult with partners from other teams. Then, they share their products or understandings with the other partner pair in their team.	Mastery and presentation of new material, concept development. Presentation and communication skills.
Jigsaw	Each student from each team becomes an "expert" on one topic area by working with members from other teams assigned to the same topic area. On returning to their own teams, each one teaches the other members of the group, and students are assessed on all aspects of the topic.	Acquisition and presentation of new material, review, and informed debate. Interdependence, status equalization.

*Adapted from Spencer Kagan (1990), "The Structural Approach to Cooperative Learning," *Educational Leadership*, December 1989/January 1990.

Roles for Cooperative learning

Roles for Group Members

Leader
- Designates roles for group
- Gets group started
- Keeps group on task
- Encourages participation
- Promotes quality work

Recorder
- Records all group work
- Takes part in discussion

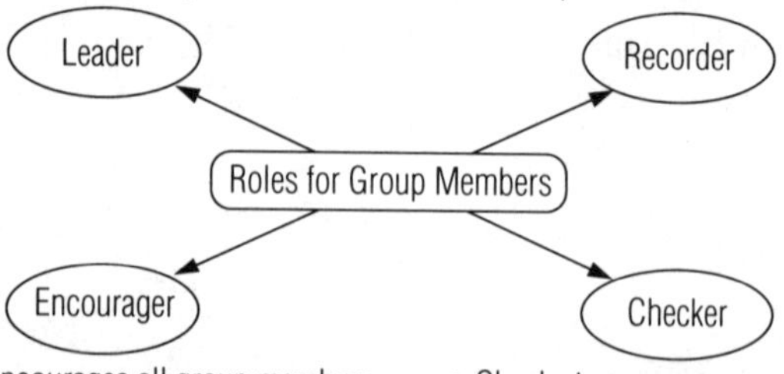

Encourager
- Encourages all group members
- Supports involvement
- Values contributions

Checker
- Checks to ensure group answers have addressed the questions
- Ensures that all group members agree and understand answers

More Cooperative Learning Participant Roles

During cooperative learning activities, students can and should have the opportunity of assuming a variety of roles in relation to the particular activity being used. Some of these roles are outlined.

Active listener: Repeats or paraphrases what has been said by the different members of the group.
Idea giver: Contributes ideas, information, opinions.
Materials manager: Collects and distributes all necessary material for the group.
Observer: Completes checklists for the group, e.g., social skills exhibited by the individual members.
Questioner: Seeks information, opinions, explanations, and justifications from other members of the group.
Reader: Reads any textual materials to the group.
*Reporter:*Prepares and/or makes a report on behalf of the group.
*Summarizer:*Summarizes the work, conclusions, or results of the group so that they can be presented coherently.
Timekeeper: Keeps the group members focused on the task and keeps time.

There may be other roles that can be identified in relation to specific activities. For example, in a laboratory situation or when different kinds of equipment and materials are being used, someone can be responsible for ensuring that safety measures are being followed, or the equipment is clean prior to and at the end of the activity.

Behaviors of Group Members

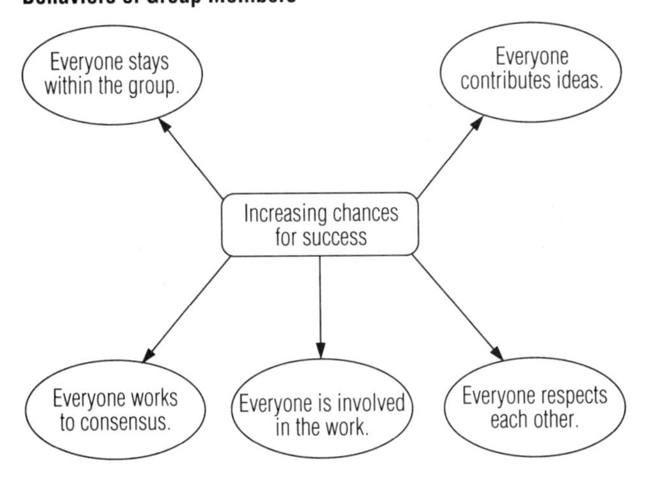

Four Phases of Cooperative Learning Lessons

I Organizational Prelesson Decisions

a What academic and social objectives will be emphasized? In other words, what content and skills are to be learned and what interaction skills are to be emphasized or practiced?

b What will be the group size? Or, what is the most appropriate group size to facilitate the achievement of the academic and social objectives? This will depend on the amount of individual involvement expected (small groups promote more individual involvement), the task (diverse thinking is promoted by larger groups), nature of task or materials available, and the time available (shorter times demand smaller groupings to promote involvement).

c Who will make up the different groups? Teacher-selected groups usually have the best mix, but this can only happen after the teacher gets to know his/her students well enough to know who works well together. Heterogeneous groupings are most successful in that all can learn through active participation. The duration of the groups' existence may have some bearing on deciding the membership of groups.

d How should the room be arranged? Practicing routines where students move into their groups quickly and quietly is an important aspect. Having students face-to-face is important. The teacher should still be able to move freely among the groups.

e What materials and/or rewards might be prepared in advance?

II Setting the Lesson

a Structure for Positive Interdependence: When students feel they need one another, they are more likely to work together—goal interdependence becomes important. Class interdependence can be promoted by setting class goals which all teams must achieve in order for class success.

b Explanation of the Academic Task: Clear explanations and sometimes the use of models can help the students. An explanation of the relevance of the activity is important. Checks for clear understanding can be done either before the groups form or after, but they are necessary for delimiting frustrations.

c Explanation of Criteria for Success: Groups should know how their level of success will be determined. Some examples include: Completing the design of a different experiment in 30 min; score of 80 or more will be "excellent"; give at least 5 descriptors of unsafe laboratory behaviors; give at least 4 positive comments to each of your teammates.

d Structure for Individual Accountability: The use of individual follow-up activities for tasks or social skills will provide for individual accountability.

e Specification of Desired Social Behaviors: Definition and explanations of the importance or values of social skills will promote student practice and achievement of the different skills.

III Monitoring/Intervening During Group Work

a Monitoring Students' Behaviors: Through monitoring students' behaviors, intervention can be used more appropriately. Students can be involved in the monitoring by being a "team observer," but only when the students have a very clear understanding of the behavior being monitored.

b Interventions During Group Work: Interventions to increase chances for success in completing the task or activity and for the teaching of collaborative skills should be used as necessary—they should not be interruptions. This means that the facilitating teacher should be moving among the groups as much as possible. During interventions, the problem should be turned back to the students as often as possible, taking care not to frustrate them.

IV Evaluating the Content and Process of Cooperative Learning Group Work

a Evaluating Achievement of Academic Objectives: Assessment of the achievement of content objectives should be completed by both the teacher and the students. The teacher can use a variety of common methods for determining understanding or the completion of a task. Students can go to their groups after an assignment or a test to review the aspects in which they experienced difficulties.

b Evaluating Achievement of Social Objectives: Two aspects become important when assessing the accomplishment of social objectives: how well things proceeded and where/how improvements might be attempted. Student involvement in this evaluation is a very basic aspect of successful cooperative learning programs.

Some Further Thoughts on Cooperative Learning

After learning that research supports the use of cooperative learning because of the various gains that seem to be associated with it in the areas of content achievement, growth of self-esteem, and growth in social and collaborative skills, more and more teachers seem to be attempting to use it. However, after becoming involved with it as with any teaching/learning approach or strategy, one invariably asks, how often and how extensively should I use this approach? We expect that if this question were asked of different educators, there would be a variety of answers along with appropriate rationales. Our view is that, as with any teaching/learning approach, there will be some (teachers and students) who gain more than others when they use cooperative learning. So, it is important, as you might do with any teaching/learning approach, to involve the students in appraising the approach and in determining what approaches they prefer to use. When students have a say in the choices of teaching/learning approaches to be used in their class, they are more likely to accept even those approaches that they do not feel enthusiastic about when they know that you, as the teacher, will also use those approaches that are preferred.

Evaluation in Biology

The goal of evaluation is to improve both student learning and program effectiveness. The science, technology, and society (STS) approach advocated by *Nelson Biology* encourages an integrated approach to science in which students:

- Develop an understanding for the manner in which scientific information is gathered, communicated, and tested (verified or falsified).
- Gain an appreciation for scientific and technological thinking and develop or refine problem-solving and communication skills within and beyond the context of the science classroom.
- Recognize the limits of scientific and technological problem-solving strategies within a social milieu. The scientific paradigm provides one way of knowing and is appropriate for answering certain questions within defined boundaries.

For an STS program to be implemented, assessment strategies must support the philosophy and rationale of the program. Clear expectations help provide the basis for performance standards and provide meaningful connections between what is expected to be learned and the level of skill development that takes place during the learning. Unlike traditional science programs, which stress content and process skill development, the STS model requires a wider range of assessment strategies that promote greater student interactions and emphasize communication. Student-generated checklists, performance-based activities, journal entries, debate presentations, project presentations, and portfolios provide some of the alternatives for student assessment. We believe that the wider spectrum for student assessment will provide a more authentic basis for evaluations to be used in making decisions about teaching methods, depth of content for instruction, classroom management, and grading procedures.

Formative Evaluation

Formative evaluation involves the ongoing evaluation of students' levels of understanding and their development of skills and attitudes. This evaluation is designed to provide students and teachers with diagnostic information that identifies learning and planning for learning. Formative evaluation provides a chart for student progress and achievement by recording: difficulties that have been overcome, learning strategies that have been employed, and the refinement and development of investigative skills, interpersonal skills, and communication skills. Formative evaluation is not designed for formal student assessment, but rather as a strategy for providing assistance for individual students and small groups of students on a regular basis. A few alternatives for formative evaluation are provided in the examples below.

Group Evaluation

Laboratory: Acid Rain and Aquatic Ecosystems (Chapter 3)

Members of laboratory group	Skills/processes observed	Follow-up
#1 John A Anna C Carol P Henry V	• Hypothesis stated • Field guide used to identify aquatic organisms • Graph and data table constructed • pH paper used to test acidity of aquatic ecosystem • Safety procedures observed	• Assistance given identifying variables • Assistance given developing procedure
#2 Nancy H Lee J Aaron R Sanjay T	• Variables identified • Field guide used to identify aquatic organisms • Graph and data table constructed • pH meter used to test acidity of aquatic ecosystem • Safety procedures observed	• Assistance provided for hypothesis • Assistance provided for developing conclusions

Comment Card

Student: *Mario M* Date: *Oct. 5/93*
Activity: *Laboratory: Acid Rain*

Comments:

- Identified problem clearly, hypothesis stated in a manner which is testable.
- Variables identified and control used during experiment.
- Uncomfortable with having to design an experiment rather than following predetermined procedure. Some assistance provided during the development of a procedure.
- Cooperates well with group members. Considers alternatives before deciding on a plan.

Summative Evaluation

This evaluation is designed to determine the extent to which instructional objectives have been achieved for any topic in biology. The use of formal information gained in tests, laboratory write-ups, homework assignments, preparation for and communication during debates, and journal entries may be used to determine the quality of student performance. This information is designed for student report cards and attempts to reflect student achievement in learning. The information is used to determine the granting of course credit and provide students with advice on selection of future courses.

Sample Unit Evaluation

Portfolios	**Total 35%**
• Two laboratory assignments selected by teacher	Suggested subtotal 10%
• Two laboratory assignments or case studies selected by student	10%
• Two review-question assignments selected by student	5%
• One critical-thinking question selected by student	5%
• Preparation of information including a bibliography for the debate	5%
Journal entries	**Total 10%**
• Include concept mapping • Outline student and group thinking during laboratory experiment design and problem-solving sessions • Outline student assessment of group cooperation • Include student comments on readings assigned	
Unit project	**Total 5%**
• Research topic • Case study	
Exams and quizzes	**Total 30%**
Science skill development	**Total 20%**
• Science inquiry • Problem-solving strategies • Decision-making	

Science Portfolios

Although portfolios have had a rich history in art classes and more recently in language-arts classes, they are somewhat new to science classrooms. A portfolio is a collection of a student's work, usually selected by both student and teacher. The work provides a record of a student's progress during the year and provides an authentic and personal profile of initial skills and their refinement during the course. The portfolio also provides students with a chronicle of their commitment to learning and successes through the program. Because students are asked to select various assignments that best reflect their learning, the portfolio encourages students to think about what constitutes excellent scientific work. It also encourages greater commitment on the part of students. Students are encouraged to believe that their next assignment will reflect their best work.

In traditional science classes, students hand in laboratory assignments to be marked by the teacher. In this system, students are encouraged to become involved in assessing their answers only after the teacher has provided his or her interpretation and given his or her judgment. A portfolio system provides the advantage that students can be given the opportunity to interpret their answers and conclusions during post-laboratory discussions. Because biology is largely understood through language, answers can be expressed in many different ways. Research by science educators who hold a constructivist viewpoint indicates that students construct meaning according to their personal references. Although these references are often logical within a narrow context, they can promote conceptions of science not consistent with modern scientific theories. Despite the ubiquity of alternate conceptions, students are rarely provided with opportunities to confront these conceptions within science classes. By comparing answers provided by classmates and their teacher, students are provided an opportunity to identify and confront their alternate conceptions. Important communication skills that require students to recognize subtle differences in expression can be strengthened within a classroom in which students are required to assume a more active role in evaluating their interpretations and conclusions. It should be noted that the teacher still has the opportunity to assess selected questions within the portfolio format; however, the teacher's primary role shifts from matching student answers to a predetermined key, to identifying alternate conceptions expressed by student answers. Less teacher time is spent judging whether or not the answers are correct; more is spent profiling student conceptions of scientific ideas in order to provide remedial assistance.

Assessment cards can be placed in the student portfolio to provide feedback on skill development and clearer direction on the teacher's expectations of the learner. The portfolio also provides students with a record of successful problem-solving strategies and a metacognitive map of social decision-making. For the student, the emphasis for learning shifts from a record of so many test marks to a profile of his/her skill development, theory reconstruction, and knowledge acquisition.

Journal

The journal is designed to provide students with an opportunity to begin thinking about the manner in which they construct knowledge and develop personal meaning. By increasing focus on metacognition, students become more aware of problem-solving strategies that they employ as individuals and members of groups during laboratory investigations, case studies, interactive debates, and the reading of biology. Their journals also provide them a record of how their thinking has changed over time and how they have developed and refined scientific, technological, and social skills. In addition, the journals provide a record of their evolving attitudes about learning and doing biology. Daily entries not only formalize student understanding and the interrelationships of biological concepts, but provide an avenue for students to organize knowledge in a manner that they find personally meaningful. Sample journal entries and a rationale for considering the journal entries are provided below. (Journal entries are taken from the various chapters of the *Teacher's Resource*.)

- Scientific theories are often described as tentative explanations of natural phenomena. Theories are socially constructed explanations to open-ended questions. Ask students to find examples of open-ended questions presented in the chapter. Why might scientists disagree about particular explanations?

Rationale: Journal entries can focus student attention on the nature of science. Many times students develop a distorted view of science as an accumulation of factual knowledge. The idea that theories are tentative, the idea that many pieces of the puzzle are missing, or the nature of scientific discourse are rarely introduced. This journal entry allows students an opportunity to reflect on how scientific knowledge, like other forms of knowledge, is socially constructed.

- Students may be asked to express concerns about ozone depletion, acid deposition, or global warming. Do they believe that these problems really exist? Do they believe that any of these environmental problems affect their health? How would they go about changing things?

Rationale: This journal entry encourages students to place the information in a personally meaningful context. It also provides the teacher with insight into why some things may be more important to some students than to others. The journal entry encourages students to begin thinking about personal applications of knowledge in terms of problem-solving.

- Construct a concept map from chapter 1, Equilibrium in the Biosphere. The following terms may be useful: biotic, abiotic, biosphere, population, community, atmosphere, photosynthesis, biogeochemical cycle, and cellular respiration.

Rationale: Concept maps are important ways of organizing knowledge. Students who find concept maps useful gain insight into the manner in which they organize knowledge. The concept map also provides teachers with an important diagnostic tool. Students who are unable to link ideas by concept maps often approach biology as a series of unrelated terms, events, or ideas. In general, the greater the number of interconnecting ideas, the more relevant the material is to the student.

- Students might be asked to write a dialogue between the text and themselves, as they refute the idea that energy is transformed during photosynthesis and cellular respiration. By acting as a "devil's advocate," they can challenge their own learning and push their understanding to a higher level.

Rationale: By taking up an approach that opposes knowledge provided by the text, students are encouraged to become active readers and critical thinkers. An argument that adopts an opposition stance requires a careful examination of the underlying assumptions of a theory. This approach also provides students with an opportunity to engage in a Popperian-falsification model of theory acceptance.

- Students can be encouraged to express the difficulties they experienced in formulating an understanding of this challenging chapter. What things did they employ to aid them in constructing their knowledge? For example, some students may indicate that they were guided by Figures 7.17, 7.18, and 7.25 before re-reading the text. Other students may have attempted a series of their own drawings or constructed concept maps. The journal entry can provide students with more information on how they learn.

Rationale: This journal entry encourages students to become more aware of the manner in which they learn. Assuming that each learner has individual skills and limitations, the time spent reflecting on learning should provide clues for future learning. Summary tables, flow chart diagrams, concept maps, and visual symbols provide alternate pathways for problem solving.

- Decision-making strategies may be recorded as each group prepares for the debate. Did the students change their minds during the preparation for the debate? Students may even be asked to record their initial positions about the social issue and to reflect upon these feelings after the debate has been completed. Did they change their minds after listening to opposing arguments?

Rationale: Social decision-making and group dynamics are given special focus by this journal entry. Students gain an appreciation for compromises made by groups and develop an understanding of why different groups express divergent opinions. The journal entry also provides a springboard for students to begin addressing the underlying assumptions that support divergent worldviews. Most divergent positions are supported by rational arguments that express divergent priorities and values.

- Ask students to comment on the snow-ecology laboratory investigation. What aspects of the investigation did they enjoy? What aspects of the investigation did they not enjoy?

Rationale: This type of journal entry encourages students to express positive attitudes developed or acquired during the activity.

- Despite winter being the dominant season in most of Canada, snow ecosystems are not well studied. Students might be asked to speculate about why these ecosystems are neglected. Would they ever consider pursuing a career that involved investigating snow ecosystems?

Rationale: This journal entry encourages students to reflect upon their learning and begin considering science as a career.

- Ask students to indicate how they think David, shown in Figure 11.1, would experience the world. How would his daily routines differ from their routines? What would it be like to be David, or David's mother?

Rationale: This type of journal entry attempts to heighten student awareness of and concern for others, and make the learning of biology more relevant to their lived experiences.

Science Skills Framework

Activities presented in the textbook and *Classroom Resource* encourage scientific process skill development. Because problem-solving skills are developed in concert with knowledge presented in the text, an effort has been made to integrate inductive and deductive thinking. Activities are organized according to three levels of skill development. Level 1, referred to as teacher-directed, provides comprehensive structures for investigations, with special concern for laboratory safety and attention to basic procedures, such as techniques for focusing the microscope. In addition, Level 1 activities stress communicating findings, either numerically, by using data tables, graphs, or statistical analyses, or in words, by using laboratory journals and summaries of conclusions, and by writing about possible implications. Level 2, referred to as teacher-guided, provides students with a problem and appropriate clues about procedures useful in solving the problem. Level 2 activities, however, stop short of providing detailed procedures. Level 3 activities, referred to as teacher-facilitated, range from open-ended laboratory designs based on prior procedures to collaborative student/teacher posed problems for investigation. In Level 3 investigations the teacher becomes an active, but not dominant, member of the research group.

Skill	Level 1: Teacher-directed	Level 2: Teacher-guided	Level 3: Teacher-facilitated
Initiation and planning of experiment	Teacher identifies problem as a simple statement.	Teacher provides an example that illustrates a problem. Students guided toward identification of the problem.	Students identify and clearly state problems from independent reading, prior investigation, or case study.
	Teacher supplies necessary background information.	Teacher helps direct students toward necessary background information.	Students use investigative skills to uncover pertinent background information. Teacher helps students focus on selecting background information.
	Laboratory directions and write-up are structured.	Students generate hypothesis and follow clues provided in the procedure. Variables are identified and a control is used.	Students propose procedure and reflect on it. Students may rely on past procedures. Teacher checks procedure and provides insight.
Collection and recording of data	Students carry out simple measurements. Data tables and charts are provided for students to organize data collected.	Students carry out simple measurements. Data tables and charts are constructed. Teacher guides students in data selection.	Students determine which measurements will be taken and what instruments will be used. Data tables, charts, and graphs are determined by students.
	Equipment is supplied and assembled by teacher.	Equipment is provided by teacher, but selected and assembled by student.	Equipment is provided by teacher or student, but selected and assembled by student.
	Safety regulations are made explicit. Teacher continuously monitors procedure.	Safety regulations are made explicit. Teacher continuously monitors procedure.	Students use past experience with laboratories to determine safe procedures that are discussed prior to beginning the laboratory. Teacher continuously monitors procedure.
Organization and communication of data	Teacher provides basic organizational schemes, such as data tables, format for graphs, and formulae used for quantitative analysis.	Teacher-guided organization, where students design their own data tables, charts, and graphs. Formulae used for data analysis supplied by teacher.	Students decide on the appropriateness of using data tables and select an appropriate type of graph to represent their data. Students use data-analysis formulae developed in previous laboratories to analyze quantitative data.
	Laboratory activities attempt to avoid discrepancies of data. Teacher will point out discrepancies when they occur.	Teacher-guided discussion about discrepancies in data collection or procedure. Teacher assumes responsibility for guiding students in discussions of scientific inaccuracies.	Students identify discrepancies in data during discussions with laboratory group and some consultation with teacher. Students assume responsibilities for initiating discussions about scientific inaccuracies.

Skill	Level 1: Teacher-directed	Level 2: Teacher-guided	Level 3: Teacher-facilitated
Analysis	Students identify simple cause-and-effect relationships.	Students identify cause-and-effect relationships in multivariable problems. They assess trends and relationships.	Students identify cause-and-effect relationships resulting from data collection or quantitative analysis.
	Teacher gives assistance to identify effect of sources of error.	With some guidance, students identify sources of error and suggest amendments to procedures and/or data manipulation.	Students independently identify sources of error, re-structure laboratory activity, and determine the reliability of the data.
	Teacher provides direction in identifying the assumptions related to measurement.	Students discuss assumptions relating to measurement and/or analysis with guidance from the teacher.	Students identify assumptions relating to measurement and/or analysis.
Synthesis	Students follow laboratory directions to record observations. Students answer questions in formulating conclusions in post-laboratory discussion or written work.	Teacher guidance is required to relate results to scientific theories and/or laws. Deductive thinking is stressed.	Students look for trends or relationships in developing scientific theories and/or laws. Inductive thinking is stressed.
	Textbook is the primary source for finding answers to laboratory questions.	Information must be contextualized before coming to conclusions. With teacher guidance, students evaluate assumptions and the effect of bias.	Information in the textbook serves as a springboard for further research. Proposed explanations and rationale for explanations is provided. Evaluate assumptions and the effect of bias.
Evaluation	Students explain the problem investigated, provide explanations, and draw conclusions related to the hypothesis.	Students explains the problem investigated, provide explanations, and draw conclusions related to the hypothesis.	Students explain the problem investigated, provide explanations, and draw conclusions related to the hypothesis. Students modify the theory and/or hypothesis on the basis of results.
	Teacher-directed discussion takes place of limitations of the experiment.	Teacher-initiated reconnaissance of experimental design and/or data collection takes place. Students discuss the limitations of the experimental design or data collected. With teacher guidance, students will make suggestions for improvements.	Student-initiated reconnaissance of experimental design and/or data collection. Students discuss the limitations of the experimental design or data collected; evaluate assumptions and the effects of bias; and evaluate experimental design and re-structure experiment.

The phases of problem solving may be extended to projects found in the *Classroom Resource*. Individual assessment of a student's process skill may be completed using the chart to the right.

Problem-solving Phases

Student name: Maria M.	**Activity: Effects of Humans on Snow Ecosystems**		
Skill	Level 1 Teacher-directed	Level 2 Teacher-guided	Level 3 Teacher-facilitated
Initiating and planning	√		
Collecting and organizing		√	
Organizing and communicating	√		
Analyzing	√		
Synthesizing		√	
Evaluating	√		

Safety in the Laboratory

The laboratory is a dynamic and exciting setting for learning biology; however, it is also a setting that can be potentially dangerous. The laboratory exercises in the *Nelson Biology* textbook have been designed to minimize any hazards. However, to prevent placing students in a harmful situation, it is important that all safety procedures be reviewed and followed.

Safety Equipment

Acquaint the students with the safety equipment found in the laboratory. Familiarize yourself with the use of the following safety items:

- Eyewash bottle or eye bath
- Chemical shower
- Fire extinguisher
- Fire blanket
- First-aid kit

Safety Symbols

The following symbols are used throughout the textbook to alert students to the use of specific safety equipment and procedures.

Safety goggles

Eye protection must be worn during any laboratory that involves flames, chemicals, or the possibility of broken glass or other small particles. Since safety goggles are shatter resistant and protect the side of the eyes, they must be worn even if you wear glasses.

Laboratory apron

A laboratory apron must be worn when handling any materials that could damage students' skin or clothing.

Wash your hands

Students should always wash their hands and forearms thoroughly with soap after any experiment.

General Safety Rules

1 Before students begin a laboratory, they should clear all unnecessary items from their work areas and remove them to a designated place. Ensure that their hair, clothing, or jewelry will not interfere with their work. Students must maintain a clean and tidy work area during the laboratory.

2 Students must read all the directions for an experiment carefully several times before they begin their laboratory work. Before students begin any step, make sure that they understand what to do. Review with students any safety procedures that are specific to the exercise. Have students follow the steps exactly as they are written or modified by you. Students should not experiment on their own unless you have approved their procedures and they have received permission to proceed.

3 Make sure that students report all injuries, accidents, and spills, no matter how minor, to you immediately.

4 Students should notify you of any piece of equipment that is broken or does not work. They should not attempt to fix any broken or defective equipment on their own.

5 Before students use any glassware, they must inspect it for chips or cracks. If it is chipped or cracked, they should not use it. If glassware breaks, students should notify you and dispose of it in the proper waste container using a safe procedure. Remind students that they should never force glass tubing into a rubber stopper; they must use gloves and a turning motion and a lubricant to help them. When students are heating a test tube, they must always point the open end away from themselves and others and move the tube back and forth through the flame. They must use a clamp or tongs when handling glassware that has been heated. Before students put glassware away, they should clean it thoroughly as instructed by you.

6 When students smell something, they should do it by a wafting of the fumes. They should not put their noses directly over the substance.

7 Students must not taste or touch any material unless they are told to do so by you. They should not eat or chew gum in the laboratory.

8 Students must exercise caution when dissecting an organism.

9 Students must disconnect electrical equipment by removing plugs from sockets and not pulling on the cord. Also remind students that it is hazardous to have water or wet hands near an electrical source.

10 Ensure that students are aware that in microbiology laboratories, it is standard practice to assume that all microorganisms are potentially hazardous.

11 Students must follow carefully all directions given regarding handling, sterilization, and disposal of cultures and associated materials.

12 Students should dispose of any unused or waste materials, specimens, chemicals, etc., as instructed by you.

13 When students have completed their laboratory exercise, they must clean up their work area, and return all equipment to its proper place and wash their hands thoroughly before leaving.

U N **1** I T

Life in the Biosphere

SUGGESTIONS FOR INTRODUCING UNIT 1

Unit 1 begins with the investigation of life at the level of the biosphere and progresses toward studying individual organisms at a species level. Chapters 1 to 3 introduce the exchange of matter and energy in the biosphere and individual ecosystems. Chapter 4 examines the manner in which organisms adapt for the role they play in ecosystems.

Students can be encouraged to select a particular animal for study and begin developing a profile of the animal, moving from specific adaptations to examining its ecological niche and finally evaluating its place in the biosphere. Information about the animal should be accumulated as students move through the unit and a final report may be presented to the class after completion of the unit. Students should learn about habitat selection, important vegetation, reproductive capacity, food chains, and the role of the animal in biogeochemical cycles. The case study approach allows students to move from the individual organism to more complex interactions of organisms and environment in a manner which reverses the strategy taken by the unit. The case study approach complements the approach presented in the unit by indicating how scientists use individual studies to construct theories and models. The case study also provides students with a different view of ecosystems and the biosphere. Each view has advantages but also limitations. The conclusions derived by many scientific studies often depend upon the initial perspective or viewpoint taken for interpretation.

KEY SCIENCE CONCEPTS

Key science concepts introduced in the Science 10 program are amplified and enriched within *Nelson Biology*. Beginning with investigations of the biosphere, themes of **matter** and **energy** are introduced in a more global perspective. Students have the opportunity to study the transformations of matter through the hydrological cycle and various biogeochemical cycles. Food webs and food pyramids provide an opportunity to follow energy transformations through various ecosystems. Water and snow ecosystems ensure that students have an opportunity to study relevant examples throughout the school year.

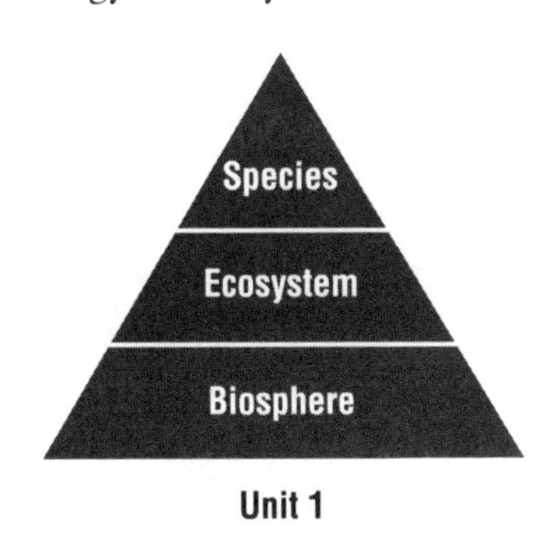

Unit 1

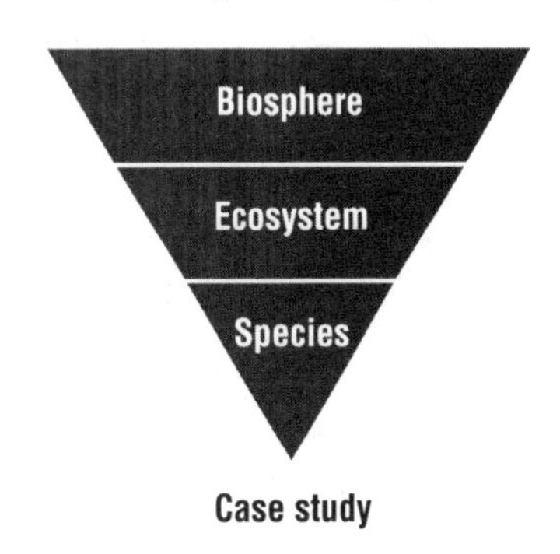

Case study

Although the initial direction of Unit 1, Life in the Biosphere, is in the development of scientific models, which could be classified as nature of science, the impact of humans on biogeochemical and hydrological cycles draws student learning toward the added dimensions of technology and society. The study of various ecosystems also provides a basis for the study of the **diversity** of life. Adaptation to winter presents a major focus for the students giving the textbook a distinctly Canadian identity. This unit provides a synthesis of the principles developed within Units 1 and 3 of the Alberta Biology 20 Program of Studies. ■

CHAPTER 1

Equilibrium in the Biosphere

INITIATING THE STUDY OF CHAPTER ONE

There are several significant concepts in chapter 1. The following suggestions address some of them.

1. During the first class, ask the students to keep a record of all the things that their families throw into the garbage over a one day period. Some of these items can be identified as recyclable, reusable, or reducible. Other categorizations might be biodegradable and non-biodegradable. The students can then brainstorm about whether their families can change their practices and how this might be done.
2. In groups, the students can consider such questions as: How can they and their family members change their behaviors in attempts to:
   - **a)** Decrease the greenhouse effect?
   - **b)** Decrease the destruction of the ozone layer?

ADDRESSING ALTERNATE CONCEPTIONS

- The chapter begins by introducing the Importance of Equilibrium in the Biosphere. The Gaia hypothesis, described by British scientist James Lovelock, provides an analogy of planet earth and the human body. Many times students find it difficult to conceptualize relationships on macro-environmental levels. The importance of an abiotic factor to the continued life of the biotic community is often overlooked. Students can be encouraged to extend the analogy by comparing the importance of equilibrium in the biosphere with that of the human body, or even with that at a cellular level. Regardless of the level of complexity, acquisition of nutrients, transformation of nutrients to other forms of matter, and the removal of wastes are essential life processes.
- During the development of the chapter, the earth is compared with a spaceship. Like a spaceship, planet earth can be considered a closed system with limited resources. During their study of chapter 1, students become aware that the supply of nutrients is limited and that the cyclic flow of matter is necessary to support life on earth. Water, carbon, nitrogen, and phosphorus cycles are discussed. Students may be asked to consider that carbon molecules which make up their body were once part of a complex starch molecule fixed by a tree, or part of the carbon dioxide exhaled by prehistoric dinosaurs. In the oxygen that they breathe

may be the same oxygen molecule used by Julius Caesar or Charles Darwin. The groundwork for developing the idea of conservation of mass can be presented during this chapter.

- Some students tend to underestimate the importance of decomposers in an ecosystem. Decomposers provide a link between the abiotic environment and the biotic community. Without decomposers, much of the organic matter needed for metabolism would remain tied up in dead organisms. Students learn that matter is cycled through the biosphere, but not energy. Some students tend to interchange terms like "matter" and "energy," indicating that decomposers return energy to the ecosystem.
- Although it is difficult to define, energy is understood through transformations. By introducing the concepts of cellular respiration and photosynthesis, students begin thinking about energy transformations and the manner in which waste energy leaves the biosphere. In chapter 2, Energy and Ecosystems, and chapter 7, Energy within the Cell, students will learn more about the importance of photosynthesis in sustaining the ecosystem. Energy, unlike matter, must be continuously added to the ecosystem.
- Although the biosphere is presented as a closed ecosystem, the laboratory Investigating the Albedo Effect, provides students with a dramatic example of how energy is added to the ecosystem. The albedo of snow cover is extremely high. In chapter 3 students will link how the presence of snow is a contributing factor to the low temperatures experienced during winter. Snow also delays warming in the spring, even though there is more solar radiation available per unit area. This will help students understand why the hottest and coldest days of the year do not coincide with the longest and shortest durations of sunlight. Some students will reason that June 21 and December 21 should be the warmest and coldest days of the year.

MAKING CONNECTIONS

- The expression "dynamic equilibrium," as used to describe order in the biosphere, can be linked with a physiological meaning of homeostasis as introduced in chapter 8, Organs and Organ Systems, and again in Unit 4, Coordination and Regulation in Humans. "Dynamic equilibrium" is used to describe any system in which changes are continuously occurring but the components have the ability to adjust to these changes without disturbing the entire system. Many principles of biology apply at the cellular level, at the organ-systems level, and at the level of the biosphere.
- Recycling of nutrients in biogeochemical cycles, as described by the water cycle, carbon cycle, nitrogen cycle, and phosphorus cycle, can be studied thematically. Acid deposition provides an STS context for studying the water cycle. The greenhouse effect provides an STS context for studying the carbon cycle, and the study of agricultural fertilizers provides a context for studying the phosphorus and nitrogen cycles.
- The biogeochemical cycles presented in chapter 1 provide a framework for understanding food chains and food webs and the relationship between producers and consumers, presented in chapter 2, Energy and Ecosystems. The biogeochemical cycles also provide a basis for understanding the importance of the principle of the conservation of matter, which is developed further in Unit 2, Exchange of Energy and Matter in Cells. The cycles of matter also stress the importance of decomposers.
- The environmental impact of the ozone layer may be linked with Blood and Immunity, as presented in chapter 11, and Breathing, as presented in chapter 12. Low levels of ozone in the upper atmosphere or high levels of surface ozone are deleterious. The relationship of ultraviolet radiation levels to skin cancer can be linked with the study of cell division, as presented in chapter 18, and oncogene regulation and cancer, as presented in chapter 23. Dr. Ram Mehta has conducted research, described in chapter 23, on establishing a link between UV radiation from the sun and cancer.
- The laboratory, Investigating the Albedo Effect, provides essential information for students who investigate snow ecosystems, as presented in chapter 3, Aquatic and Snow Ecosystems.

POSSIBLE JOURNAL ENTRIES

- Students might be asked to critically reflect on Lovelock's Gaia hypothesis. In what ways is a comparison of the earth to the human body limited?
- Scientific theories are often described as tentative explanations of natural phenomena. Theories are socially constructed explanations of open-ended questions. Ask students to find examples of open-ended questions presented in the chapter. Why might scientists disagree about particular explanations?
- The city of Los Angeles is very concerned about emissions from automobiles. Automobile emissions have environmental implications for acid deposition and the greenhouse effect, as well as a number of related health problems. By the year 2000, 10% of sales by automobile companies who wish to sell in Los Angeles must comprise cars with non-internal combustion engines. Should your community follow Los Angeles' lead?
- Students may be asked to express concerns about ozone depletion, acid deposition, or global warming. Do they believe that these problems really exist? Do they believe that any of these environmental problems affect their health? How would they go about changing things?
- Construct a concept map from chapter 1, Equilibrium in the Biosphere. The following terms may be useful: biotic, abiotic, biosphere, population, community, atmosphere, photosynthesis, biogeochemical cycle, and cellular respiration.
- Students may be asked to describe home recycling projects or to devise a plan for recycling household refuse.

USING THE VIDEODISC

Section Title	Page	Code	Frame (side)	Description
Importance of Equilibrium in the Biosphere	20 – 21		3302 (3)	A Drop of Fresh Water—movie sequence of organisms found within a mini-ecosystem
Biogeochemical Cycles	30 – 31		2767 (all)	Biogeochemical cycles—single frame
The Hydrologic Cycle	31 – 33		3292 and 3293 (all)	The water cycle—single frames
The Hydrologic Cycle	31 – 33		2698 (all)	Acids—single frame
The Hydrologic Cycle	31 – 33		3108 (all)	pH—single frame
The Carbon Cycle	34 – 37		2787 (all)	Carbon cycle—single frame
The Carbon Cycle	34 – 37		3114 (all)	Photosynthesis—single frame
The Carbon Cycle	34 – 37		2812 (all)	Chlorophyll—single frame
The Carbon Cycle	34 – 37		2813 (all)	Chloroplast
The Nitrogen Cycle	38 – 39		3066 (all)	Nitrogen cycle—still frame

IDEAS FOR INITIATING A DISCUSSION

Table 1.1: Organizational Levels of an Organism and the Biosphere

Students might be asked to compare the biosphere with other organizational structures. In addition to cells, students might consider factories, government bodies, or clubs.

Figure 1.5: Pictures of the antarctic showing levels of ozone from 1979 to 1990

Students might be asked to draw a picture of the planet, showing how ozone depletion might look in the year 2000, should the trend continue.

Research in Canada: Dr. Jill Oakes

Encourage students to bring any traditional clothing they have to school. Visits to museums or by guest speakers can also be considered.

Figure 1.19: Percentage change in carbon dioxide emissions from 1979 to 1989

The data show that Alberta and Newfoundland have increased carbon dioxide emissions. In Alberta much of the increase can be attributed to hydrocarbon electrical generating plants. What the data don't tell us is how the other provinces meet their energy demands. Ontario, for example, has nuclear power plants, which some individuals believe have even greater potential for environmental harm than increasing carbon dioxide levels.

REVIEW QUESTIONS

(page 25)

1 In what ways is the earth like a spaceship?

■ The earth is like a spaceship in that it is a closed system. The only materials available for use are the raw materials already present and the energy from the sun in the form of light.

2 What is meant by a closed system?

■ A closed system is an environment in which there is no outside source of raw material and no area to store waste material which does not come into contact with the system.

3 What are the abiotic and biotic components of the biosphere?

■ The abiotic components of the biosphere are the nonliving components (the chemical and physical factors). The biotic components of the biosphere are the biological components (living organisms).

4 In what way does a community differ from an ecosystem?

■ "Community" describes the populations of all the species found in a habitat. "Ecosystem" describes a community and its associated chemical and physical environment.

5 Name the levels of organization in the biosphere.

■ The biosphere is organized by organism, population, community, and ecosystem. (An organism is categorized by increasing levels of complexity. The organization starts with individual cells, then tissues, organs, organ systems, and finally an integrated body.)

6 Name the specific gases found in the atmosphere. What other materials are found there?

■ The specific gases found in the atmosphere are nitrogen (78%), oxygen (21%), argon (<1%), carbon dioxide (0.03%), and trace gases (e.g., ozone, methane, ± 0.01%). The atmosphere also contains water vapor, pollen, dust, spores, bacteria, and viruses.

7 State where the ozone layer is located in the atmosphere and explain why it is important.

■ The ozone layer is located in the stratosphere. It absorbs a large amount of the ultraviolet radiation emitted by the sun.

8 In what ways do the ionosphere and the magnetosphere protect living organisms?

■ The ionosphere absorbs X rays and gamma rays produced by solar radiation. The magnetosphere deflects potentially lethal charged particles produced by the sun and interstellar space.

9 What is meant by an ozone hole?

■ An ozone hole is a thinning of the ozone layer. This thinning is most prominent in the antarctic and arctic.

(page 37)

10 List the factors responsible for determining how much heat from solar radiation is retained by the earth.

■ The amount of solar radiation the earth absorbs is a function of the earth's albedo. Dust, snow, ash, deforestation, and water vapor all increase the earth's albedo and lower the amount of solar radiation absorbed.

11 What is the effect of the tilt of the earth's axis on the absorption of heat at the earth's surface?

■ Seasons occur.

12 How does water help in maintaining global temperatures?

■ Water holds vast amounts of heat. It can absorb heat during the day and release it at night. (This explains why temperatures in coastal areas or regions that are close to large bodies of water do not vary as much as temperatures in regions that are far from water. Similarly, this explains why the day and night temperatures in deserts (no water) vary so much.)

13 The albedo of a rocky surface is 0.6 (60%). What does this mean in terms of heat absorption?

■ The rock will reflect 60% of the solar radiation that strikes it. (The rock will therefore not absorb a great deal of heat.)

14 What is the greenhouse effect?

■ It describes a phenomenon where light rays pass through a medium, hit objects and are converted into longer wavelengths, and thus are unable to exit through the medium. This causes the temperature to increase. In the atmosphere, carbon dioxide acts as the medium that traps heat. (The increased carbon dioxide content of the atmosphere is expected to contribute to an overall global warming.)

15 Describe some of the possible effects of global warming.

■ The melting of the polar ice caps would flood most of the world's major cities. The prairie and tropical regions of the world would become deserts, and the arctic regions could become temperate.

(page 40)

16 What is a biogeochemical cycle? Why are these cycles important to living organisms?

■ A biogeochemical cycle is (describes) the process whereby nutrients are passed from the environment to an organism and from the organism back to the environment. These cycles are crucial to living organisms because they constitute the only way in which material can be reused after organisms die.

17 In terms of biogeochemical cycles, why is photosynthesis critical to the biosphere?

■ Photosynthesis is the process that traps energy from the sun, thus providing energy for biogeochemical cycles. (The biogeochemical cycles are not 100% efficient; energy is lost as material is passed from the environment to an organism and as material is passed from an organism to the environment.)

18 Describe the role of plants in the water cycle.

■ Plants absorb water from the soil by their roots and release it from their leaves through the process of transpiration. If plants are removed from an area, the surface runoff patterns and water-holding capacity of the soil may be disturbed, leading to a net loss of water from the environment.

19 What do nitrogen-fixing bacteria and lightning have in common?

■ Both nitrogen-fixing bacteria and lightning play a major role in nitrogen fixation—the conversion of free nitrogen in the atmosphere to nitrates—which all plants require to synthesize proteins.

20 Why is carbon critical to the biosphere?

■ Carbon is in the molecule that is used to store the energy trapped by photosynthesis. The energy is stored as glucose, a six-carbon molecule.

21 Where is most of the biosphere's carbon dioxide stored after it is released into the atmosphere?

■ Most of the carbon dioxide released into the atmosphere is stored in the oceans of the world.

22 Why are photosynthesis and cellular respiration often called the biotic phase of the carbon cycle?

■ These complementary reactions are carried out by living things. (Students may recognize that the transformation of N_2 to nitrates by lightning, characteristic of the nitrogen cycle, is an abiotic process. The transformation of CO_2 to carbonates in oceans is also abiotic.)

LABORATORY
INVESTIGATING THE ALBEDO EFFECT

(page 28)

Time Requirements

The time required to complete this laboratory would vary and depend on whether the students would be designing and completing a laboratory exercise to determine the effect of color and texture on the absorption of heat.

The laboratory investigation, as presented, would require approximately 45 min if the equipment and materials are made available for the students, and the students are working in teams of three to four people. This time period would allow the students to complete the investigation and some of the laboratory questions.

Notes to the Teacher

- Whenever students use electrical equipment, they should develop and practice a routine of checking the wires for any frayed sections. They should also check the plugs for any loose connections.
- Most physics laboratories have a supply of photocells and voltmeters (or galvanometers). If not the photocells are reasonably inexpensive and can be purchased at most electronic supply outlets.

Observation Questions

a) Why is it necessary to keep the photocell, wire, and voltmeter out of the direct line of light?

■ In order to prevent them from interfering with the reflected light.

b) Why does the voltmeter register an electrical charge?

■ Since the photocell is activated by light, a weak current is produced. Therefore, the galvanometer is indirectly measuring the amount of reflected light.

c) What variable must be controlled?

■ The distance between the light source and the reflecting material and the distance between the reflecting material and the photocell must be controlled.

d) Why are black and white card stock colors used in this experiment?

■ These cards would serve as controls.

e) List the colored surfaces in order from least to greatest reflected light.

■ The sequence would vary in relation to the colors used.

f) Use your data to make comments on the albedo effect of the materials used.

■ The answers will vary in relation to the materials used in this laboratory.

Laboratory Application Questions

1 Design an experiment to measure the effect of color or surface conditions on the absorption of heat. State your hypothesis and the predictions resulting from it.

■ (The intent of the question is to promote divergent thinking.) Students might design a great variety of experiments. But, every design should include a hypothesis that predicts the amount of heat absorption in relation to either color or surface conditions. In other words, the manipulated variable would be either the color or the surface texture, and the responding variable would be the amount of heat absorbed. All other variables would have to be controlled. It is likely that a chart might be made to record the data.

2 Perform the investigation (approval of instructor required) and draw conclusions based on the collected data.

■ With regard to color, the expected results would indicate that the darker the color, the greater would be the amount of heat absorbed. With regard to surface conditions or texture, it is likely that the conclusion would be that the more the surface is rough and thus less reflecting, the greater would be the amount of heat absorbed.

3 What variable were you unable to control in part 8 of the laboratory procedure? How could you redesign these steps to account for the variable?

■ The distances between the surfaces and the light or the surfaces and the photocell were not controlled. A measuring tape might have been included with the materials used in this experiment so that the distance of the light and photocell from the surfaces could always be adjusted and be the same for all surfaces.

4 In order to melt a pile of snow more quickly, a researcher sprayed dye-colored water on it. Which color was probably selected and why?

■ The answers will vary according to the data already collected.

LABORATORY

RECYCLING PAPER

(page 41)

Safety Precautions

The temperature of a boiling starch solution will be higher than that of boiling water—precautions should be taken in working with it.

Note: Check first aid kit to insure it contains appropriate ointments to treat minor burns.

Time Requirements

The time required for this laboratory would vary depending on whether the students would prepare their own 10% starch solution. Even if they do, the exercise can be completed to the point of drying the paper within a 50 min time period. While the students wait for their starch solution to heat, they can be encouraged to begin answering the questions and preparing for the other parts of the exercise.

Laboratory Application Questions

1 Why was the starch solution used?

■ Starch is used to bind the pieces of paper and hold the mulch together.

2 Why was the starch solution boiled?

■ Boiling breaks down the starch molecules into shorter molecules which can better be absorbed by the bits of paper, thereby producing a smoother mixture. (Starch is made up of a large number of six-carbon glucose sugar molecules which are bonded together in long chains with branches of glucose molecule chains.)

3 Propose a method for recycling paper on a larger scale.

■ Students might develop a great variety of suggestions. These might include better and easier ways of breaking down the paper when mixed with the starch solution. One example might be to use a food blender or mixer. Students might also suggest other ways for spreading the mixture on the screen such as a spraying device.

APPLYING THE CONCEPTS

(page 43)

1 Without a greenhouse effect, life could not exist. Discuss this statement in terms of the present concern with global warming.

■ The present concentration of greenhouse gases in our atmosphere (gases which can trap heat beneath them) maintains a temperature which allows life on our planet as we know it. However, if the concentration of these gases, such as carbon dioxide, increases in our atmosphere, some scientists believe that we will experience enough global warming that would produce irreversible effects and many organisms would not be able to survive the changes.

2 The albedo of the planet Venus is very high. At the same time, the atmosphere of the planet has an exceptionally high concentration of greenhouse gases. How might these two factors affect the surface temperature of Venus?

■ A high albedo means that a great amount of sunlight would be reflected from the surface of the planet. The high concentration of greenhouse gases means that the reflected heat would not escape and the atmospheric temperature of Venus would be very high.

3 List several tactics that would help reduce acid depositions. Predict areas of opposition and suggest reasons for the opposition.

Tactic to reduce acid depositions	Effect	Reasons for opposition
Scrubbers in smoke stacks	Acid-deposition causing gases are removed and prevented from being released into the atmosphere.	Increased expenses and decreased profits Increased taxes
Better grades of coal in electrical generating plants	Less acid-deposition causing gases would be produced.	Increased expenses and decreased profits
Better exhaust emission standards for cars	Less nitrogen oxides would be released.	Increased expense and decreased profits
Restriction of the use of gasoline-powered motorized vehicles	Less emissions that cause acid deposition would be released.	Inability of North Americans to break the habit of driving their vehicles at every opportunity

■ Students might well extend this table considerably. Certainly, many other tactics can be added.

4 Hypothesize how radiation from the Chernobyl nuclear power plant may have entered the water cycle.

■ During the accident, radioactive debris was sprayed into the atmosphere and it attached to anything that might have been present such as dust particles and other floating material. Eventually this matter came to the surface of the earth as dry deposition or in association with different forms of precipitation. In either case, it may eventually be washed away in the water and become part of the water cycle.

5 Explain how chloroflurocarbons (CFCs) can contribute to the depletion of the ozone layer.

■ When CFCs diffuse into the atmosphere, they are broken down by ultraviolet radiation and release a reactive chlorine molecule that is capable of reacting with and breaking down thousands of ozone molecules. (CFCs have commonly been used as aerosol propellants and as coolants in air conditioners, refrigerators, and freezers. They are also waste products during the manufacture of some foam plastics.)

6 Suggest a strategy for ridding your lawn of clover. Provide a scientific explanation for the strategy.

■ Clover cannot compete with the grass in a healthy lawn. However, when nitrogen levels drop, clover has an advantage with nitrogen-fixing bacteria providing the needed nitrates. By increasing the nitrogen levels (by using fertilizers), the clover loses its advantage.

7 A town that obtains its water supply from deep-well drilling discovers toxic wastes in the water. Blame is placed on a toxic waste dump over 600 km from the town. The waste has been dumped into an abandoned mine shaft.

a) How could you prove that the toxic wastes had come from the dump site?

■ (Students might research this topic and suggest several viable solutions.) One approach would be to use a water-soluble dye at the dump site and determine whether that dye also appears in the town's water at some later date. It is, of course, important that the water-soluble dye be nontoxic.

b) Assuming the dump site was the source of the pollution, explain how it could have contaminated the town's water supply.

■ In order for the toxic waste to reach the town's deep-well water, the drainage from the dump site must find its way into the groundwater and this groundwater must be common with the groundwater from which the town is getting its water.

CRITICAL-THINKING QUESTIONS

(page 43)

1 A plowed field is adjacent to the fairway of a golf course. During the winter, equal depths of snow cover the field and the fairway. Assuming that both fields are level and there is no disturbance to the snow pack, explain why the plowed field loses most of its snow before the fairway even begins to lose its cover.

- As the snow begins to melt in the spring, clumps of dirt will begin to show in the plowed field. The black soil will immediately begin to absorb more heat and melt the snow surrounding it.

2 To provide land for agriculture, the tropical rain forests of the Amazon are being destroyed by fire at an alarming rate. Describe any possible effects this clearing process may have.

- As large areas of rain forest are cleared, several things happen. The carbon dioxide that the rain forest used is no longer taken from the atmosphere. This increased carbon dioxide would increase the global greenhouse effect. The increased exposure of ground surface would similarly attract the sun's rays and increase surface temperatures. Rain would no longer be absorbed by the plants present in a rain forest, and the amount of soil erosion would increase. Evaporation that normally occurs from rain forest vegetation would be absent causing further climatic changes. The animal life that depends on a rain forest is extremely large and very diverse—all of it would be lost.
  Note: The possible responses to this question are extremely extensive and diverse. Students should realize that the effects that result when rain forests are lost are very far reaching.

3 A farmer's field became waterlogged during a very wet spring. This delayed planting for about five weeks. Planting was followed by excellent weather for the remainder of the growing season, yet the crop growth was poor. A local expert suggested that the reason for this poor growth lay in the nitrogen cycle. Study the nitrogen cycle carefully. Then write a report explaining to the farmer why the crop failed.

- While the soil was waterlogged, it is likely that it became compacted, producing anaerobic conditions for the denitrifying bacteria. Under such conditions, the bacterial breakdown of nitrates to free nitrogen increases and the amount of nitrates available for plants is decreased, resulting in crop failure.

4 Provide evidence to support Lovelock's Gaia hypothesis.

- (The intent is for students to speculate about the interdependence of living things in an ecosystem. The variety of acceptable answers is infinite.) By simply having pet dogs or cats, the bird population in an area is affected, which in turn affects the amount of insects in that area and the number of breeding insects. Populations of insects that prey on other insects are altered. With changes in insect populations come changes in plant populations. The demonstration of interdependence continues indefinitely.

CHAPTER 2

Energy and Ecosystems

INITIATING THE STUDY OF CHAPTER TWO

Brainstorming Session

1 Have students in small groups brainstorm on reasons that would explain why we do not gain the same weight as the weight of the food we eat. What is happening in the body when one gains or loses weight?

2 Have students in small groups brainstorm on ways by which humans have an effect on "nature." Encourage them to describe both positive and negative effects.

ADDRESSING ALTERNATE CONCEPTIONS

- Students may have difficulty identifying what an ecosystem actually is. By studying chapter 2, Energy and Ecosystems, students will come to realize that ecosystems are small regions within the biosphere. The boundaries of an ecosystem are defined by the researcher. The scale and complexity of an ecosystem vary, depending not only on the organisms found but on abiotic factors such as climate and local geology.
- Linking large themes, such as matter and energy, is often challenging for students. The concepts of matter and energy are so large and encompassing that students may develop a tendency to use them interchangeably or to collapse the meaning of one term into the other. By studying food pyramids, students develop a context for constructing knowledge about matter and energy. Food pyramids provide a perspective for studying the relationships between organisms in terms of populations, matter (nutrient acquisition), and energy (nutrient transformations).
- By studying food chains and food pyramids, students learn that usable energy is lost at each trophic level. This leads some students to conclude that energy is actually lost or destroyed. However, the actual amount of energy flowing through the biosphere must obey basic scientific principals of thermodynamics. Each time energy is transformed, some energy is lost from the system. In this case energy is lost from the ecosystem, but not the universe. Energy is not destroyed. The problem arises because students are not accustomed to thinking on such a large scale. An ecosystem is not a closed system. However, the universe can be described as a closed system.

MAKING CONNECTIONS

- By studying a variety of ecosystems and combining the various data, students gain a more complete picture of the biosphere as introduced in chapter 1, Equilibrium in the Biosphere. Two details become clearer: One, there is a cycling of matter between the biotic and abiotic components of an ecosystem. Two, energy drives the ecosystem.
- Whether students study a functioning organism or an ecosystem, they begin to recognize that both require energy to operate. They have already considered how solar energy drives many abiotic activities of the biosphere such as weather and biogeochemical cycles. Chapter 2 provides additional support for the idea that unlike matter, energy cannot be recycled. Students learn that if energy were not continually added to an ecosystem, the primary producers would not be able to support a food chain.
- The laws of thermodynamics, as presented within the context of an ecosystem, are visited again when students study chapter 7, Energy within the Cell. The second law of thermodynamics helps to explain why heat energy is classified as waste energy and why energy transformations yield less usable energy.
- The study of food webs provides valuable information for understanding diversity in different biomes. The most stable ecosystems, such as mature forests, have complex and well-developed food webs. Any reduction in numbers or even the complete removal of one of its organisms will have a minimal effect on the overall web. However, where natural conditions limit the number of organisms, the webs become simplified into structures more like food chains. This is particularly true in the high arctic. A limited number of organisms means that relationships are well defined. In these situations the loss of any one member will have a profound effect on all the remaining organisms. The study of snow ecology, as presented in chapter 3, Aquatic and Snow Ecosystems, and the study of chapter 4, Adaptation and Change, is supported by understanding why arctic ecosystems are so fragile.

POSSIBLE JOURNAL ENTRIES

- Students may be asked to comment on their understanding of the laws of thermodynamics. Why must energy be continually added to an ecosystem? Is energy destroyed within an ecosystem?

- Students may be asked to classify a number of ecosystems in their own schoolyard. How many microecosystems can they classify?
- Students may be asked to express concerns about artificial ecosystems. For example, do they have allergies? Do they feel that rugs in the school contribute to health problems? Do they believe that plants are really useful? How would they go about changing their artificial ecosystem?
- Construct a concept map from chapter 2, Energy and Ecosystems. The following terms may be useful: autotrophs, heterotrophs, producers, consumers, food chains, pyramids of energy, pyramid of biomass, pyramid of numbers, and biological amplification.

USING THE VIDEODISC

Section title	Page	Code	Frame (side)	Description
Autotrophic and Heterotrophic Nutrition	45		635 – 746 (all)	Various examples of autotrophs
Autotrophic and Heterotrophic Nutrition	45		747 – 1526 (all)	Various examples of heterotrophs
Trophic Levels and the Food Chain	47 – 48		2920 (all)	Food chains
Trophic Levels and the Food chain	49 – 50		2921 (all)	Food webs
Ecological Pyramids	56 – 57		3270 (all)	Trophic level

IDEAS FOR INITIATING A DISCUSSION

Table 2.1: Energy Used in Photosynthesis

Students might be asked to calculate energy efficiency at various stages of energy conversion.

$$\text{Efficiency} = \frac{\text{Usable energy} \times 100}{\text{Total energy}}$$

Figure 2.5: Examples of animals classified according to feeding relationships

Students might be asked to brainstorm to develop a list of animals which can be classified according to herbivore (primary consumer), carnivore (second, third, etc. order consumers), and omnivores.

Table 2.2: A Simple Grazing Food Chain

Students might be challenged to come up with their own example of a grazing food chain. Later, the food chain can be converted into a food web.

Research in Canada: Janet Edmonds

Students might be asked to speculate about the answers to the questions posed by Edmonds' research project.

REVIEW QUESTIONS

(page 50)

1 Why are producer organisms called autotrophs?

■ They are called autotrophs because they are capable of producing their own food. (The word autotroph literally means "self-feeder.")

2 How does a heterotroph differ from an autotroph?

■ A heterotroph cannot produce its own food, but must ingest other organisms in order to obtain organic molecules.

3 What is the role of the decomposer group in an ecosystem?

■ Decomposers break down complex organic molecules found in the wastes and bodies of organisms. (This serves to return raw materials to the environment, where they can be used again.)

4 Explain the term "trophic level."

■ Trophic level refers to the level an organism is away from the original solar energy that entered the system. (For example, autotrophs (producers) are always in the first trophic level. An organism that normally feeds on a heterotroph, which in turn normally feeds on an autotroph, would be in the third trophic level.)

5 Distinguish between a food chain and a food web.

■ A food chain is a step-by-step sequence of producers and consumers. (For example, a common food chain would be plant, deer, wolf, decomposing bacteria, plant.) A food web is a series of interlocking food chains. (A plant can be eaten by many different consumers as can a deer be attacked and eaten by different carnivores. A food web more accurately represents the flow of energy through the different trophic levels in the biosphere.)

6 How does an omnivore differ from a herbivore and a carnivore? Give an example of each.

■ An omnivore feeds on both plants and animals, while herbivores feed only on plants and carnivores feed only on animals. (Cows are herbivores, lions are carnivores, and humans are classified as omnivores.)

7 What type of food would be consumed by a secondary consumer? Explain your answer.

■ A secondary consumer feeds on a primary consumer, which is a herbivore. (Any carnivore or omnivore that feeds on herbivores is a secondary consumer.)

8 What is meant by the term "top carnivore"?

■ "Top carnivore" refers to the final carnivore in any food chain.

(page 62)

9 State the first law of thermodynamics.

■ Even though energy can be transformed from one form to another, it cannot be created or destroyed. (One of the most common examples of this is the way electrical energy is used in so many ways—for light, heat, sound, etc. But, the electrical energy was produced from other forms of energy.)

10 Describe the second law of thermodynamics in your own words.

■ During any energy change, some of the energy is converted into unusable forms, usually heat. (As our muscle cells use chemical energy to contract the muscles, some energy is lost as heat.)

11 Why is less than 20% of available chemical energy transferred from one organism to another during feeding?

■ Not all the food eaten is digested—some is eliminated in the feces. A portion of the energy from the food eaten is used in cellular respiration and other metabolic reactions where some energy is lost as heat.

12 Why do energy pyramids have their specific shape?

■ Energy flow in ecosystems obeys the laws of thermodynamics. As energy is transferred from one trophic level to the next, over 80% of the original energy is lost (converted into unusable forms).

13 What would be the best source of energy for an omnivore: the plant or animal tissue it feeds on? Explain.

■ Plant tissue would be the best source of energy. Being at the bottom of the food (energy) pyramid, there is only one transfer of energy and one point at which energy is lost.

LABORATORY

INVESTIGATING A MICROECOSYSTEM

(page 51)

Preparations

It is imperative that the teacher visit the site a few days in advance of the trip and be confident that all requirements for a successful field trip are available at the site. The teacher should also examine the area and be aware of the hazards associated with the site (poisonous plants, animals) and take the necessary precautions. Permission slips should be obtained from school boards and parents prior to the field trip. Consider taking field guides along to identify species.

Safety Precautions

- Insure that all students who are taking any medication carry it with them.
- Pack a complete first aid kit.

Time Requirements

The actual time required for the activities at the site is rather little. You may wish to prepare for other activities to be included during this trip if you have to travel any distance from your school. In such a situation, you would likely want to take at least a half a day.

Students might be requested to complete this laboratory data collection on their own, outside of school time, depending on your school's location.

Laboratory Application Questions

1 Compare the abiotic factors such as sunlight, moisture, wind, and temperature that influence life on the top and the bottom of the log. Use a chart to record your observations.

■ Students may use a chart similar to the following:

Abiotic factor	Top of log	Bottom of log
1 Sunlight		
2 Moisture		
3 Wind		
4 Temperature		

Note: The logs that will be studied during this laboratory will obviously have a great variety of stories to tell. The students should be encouraged to try to learn as much as they can about those stories from what they can see and, possibly, research. For example, they might want to learn something about the rate at which different fungi grow, or the food webs that exist among different invertebrates that they find associated with the log. They may find that there are some other data that they might want to record. For example, a log may be found at the edge of the lower branches of an evergreen, so that rain water might run off the branches onto the log. In such cases, they should be encouraged to extend the table above as they see fit.

2 Categorize the members of the biotic community as autotrophs, heterotrophs (carnivores or herbivores), or decomposers. Distinguish between those found on the top and the bottom of the log. Use a table to record your observations.

Biotic community	Top of log	Bottom of log
1 Autotrophs		
2 Heterotrophs		
a) Carnivores		
b) Herbivores		
3 Decomposers		

■ *Note:* As students become more involved with this activity, some may suggest that they might extend their study into a microscopic one. Certainly, many microorganisms would be associated with their logs and you will have to decide whether this is a worthwhile venture in relation to this investigation.

3 Using concept-mapping techniques, construct a food web showing the manner in which energy and matter cycle through an ecosystem. You may present a food web of organisms found either on the top or the bottom of the log.

■ *Note:* These concept-mapped food webs can become quite extensive as students research the different organisms they locate. You may want to delimit the number of organisms that are required for the web.

4 Select two logs at different stages of decomposition and combine the class data. Which log has the greater number of fungi? Explain your observations.

■ It is very likely that the log with the greatest amount of decomposition will have the greatest number of fungi. However, it is also possible that a few different kinds of fungi can cause a great amount of decomposition as well. Typically, fungal growth is closely associated with the availability of organic nutrients. And, as a log decomposes, a greater variety of organic foods are available to provide the appropriate nutrients for different fungi.

5 Provide a hypothesis that explains why bracket fungi are most often found on top of the log, while the yellow and white threads from molds, another type of fungus, are found under the log.

■ The molds found under the log require more moisture than bracket fungi. Other conditions such as light and wind may also affect their growth.

6 In what ways are the microecosystems under a log and inside a running shoe similar?

■ Both microecosystems have darkness and moisture in common. Certainly, the inside of a running shoe may be warmer if it is being worn. (Some differences also exist in that the running shoe may be allowed to dry out from time to time. It may also have different fungicides applied to it by a caring owner.)

LABORATORY

MEASURING ENERGY LOSS DURING METABOLIC ACTIVITIES

(page 54)

Safety Precautions

Use thermometers which do not contain mercury.

Time Requirements

The time required to set up this laboratory is about 30 min. In order to take temperature readings over three consecutive days, the laboratory would have to be set up on either a Monday or Tuesday. If the lab is set up on a Monday, answers to the questions can be discussed on the Friday of that week.

Observation Questions

a) What is the function of a control in an experiment?

■ A control has no manipulated variables and serves as a basis for comparison with other aspects of the investigation where some variables are manipulated.

b) What is meant by germination?

■ The resumption of growth of a spore or seed from a dormant state is called germination.

c) Why was it necessary to soak the seeds in water to initiate germination?

■ Germination involves numerous chemical reactions—most of these reactions require water.

d) Which is the manipulated variable? the responding variable? Explain.

■ The manipulated variables include the presence or absence of seeds and water. The responding variable is the temperature. Changes in temperature are expected to occur as a result of the germination of the seeds.

e) What would a temperature change in the control imply about the temperatures in the other two bottles?

■ If there was a temperature change in the control, a reason for this change would have to be determined. Further, changes in temperature in the other two bottles would not be as meaningful.

Laboratory Application Questions

1 Account for any temperature changes during the experiment.

■ With the availability of water, the seeds are able to metabolize stored food. This process involves numerous chemical reactions which release heat into the environment. (The heat produced by moist seeds can be so great that they may ignite.) "Dry" seeds may also be carrying on some cellular respiration so that with a very good seal and a sensitive thermometer, it may be that even the bottle with the "dry" seeds may show a slight increase in temperature.

2 How did your predicted results compare with the actual temperatures? Comment on any similarities and/or differences.

■ Answers will vary. The important aspect of this question is for the students to arrive at reasonable explanations for their results.

3 Why were dry seeds used in one thermos?

■ The dry seeds will serve as a control for the amount of respiration that is taking place in dry seeds—seeds that are essentially dormant. (This should be minimal and any heat released by these seeds might be so little that it would simply dissipate and not produce a measurable temperature change inside the thermos bottle.)

4 What do the results of the experiment indicate about the energy available to a potential consumer if it selects either the dry or the germinating seeds? Assume each seed has the same amount of bond energy at the start of the experiment.

- The amount of heat produced (the temperature change) is a measure of the amount of chemicals that have been broken down during the respiratory process. So, if a consumer chooses the moistened seeds, it would be choosing the seeds that have a decreased amount of chemical energy.

5 Since all the seeds are alive, explain the temperature difference between the dry and the moistened seeds.

- The moist seeds are carrying on cellular respiration at a much faster rate with a correspondingly greater release of energy in the form of heat.

6 In a similar experiment, germination did not occur in the moistened seeds. Nevertheless, there was a distinct rise in the temperature in the thermos. How could you account for this observation?

- There may not have been sufficient water for germination to occur. So, all that happened was the preliminary stages of germination where stored sugars were being broken down through the process of cellular respiration, with some release of heat, which caused the temperature change.

APPLYING THE CONCEPTS

(page 64)

1 In underground caves, where there is permanent darkness, a variety of organisms exist. In terms of energy flow explain how this can be possible.

- It is likely that energy is assimilated by chemosynthesizers, such as bacteria and blue-green algae, from different inorganic chemicals.

2 Based on what you have learned about energy pyramids, comment on the practice of cutting down rain forests to grow grain for cattle.

- Since 80% of the energy present in the grass grown (where the rain forests were) to feed the cattle is lost in the transfer from grass to animal flesh, we would be much better off to grow edible herbs and eat them rather than use them to feed animals.

3 Design complex food webs for a tundra ecosystem and a middle-latitude woodland ecosystem. Reference books can be used to determine the other members of the food web.

- The food webs that the students develop will depend on the resources that are available. This activity is one that you may consider completing in groups after each member of the class has made some effort in developing his/her food web.

a) Which ecosystem has the greatest biomass? Provide your reasons.

- The middle-latitude woodland ecosystem has the greatest biomass, because this is the more varied ecosystem with a greater variety of carnivores.

b) Which ecosystem has the greatest number of organisms? Provide your reasons.

- The middle-latitude woodland ecosystem has the greatest number of organisms.
  Note: The reasons that the students provide for this answer will vary. But, their answers should include some common entities such as: the growing season being longer; and that a greater variety of plant life satisfies a greater variety of herbivores, which in turn provides food for a greater variety of first, second, and third level consumers.

c) Which ecosystem has the greatest energy requirement? Provide your reasons.

- The middle-latitude woodland ecosystem has the greatest energy requirement. This is because the food chains would be more extended. There is a much greater loss of energy through the loss of heat from cellular respiration by the many organisms and through the loss of energy that occurs in moving from one trophic level to another.

d) Comparing the tundra and middle-latitude woodland ecosystems, indicate which is more susceptible to environmental pollution. Explain your answer.

- The tundra ecosystem is more susceptible to environmental pollution. The food webs are simple and the limited number of organisms means that a reduction or loss of one will have a profound effect on the remaining organisms.

4 By law, the cutting of forests must be followed by replanting. Why do some environmentalists object to monoculture replanting programs?

■ Monocultured plants would use a limited variety of nutrients and very rapidly deplete the soil of those nutrients, which would, in turn, result in poorer growth. Similarly, monocultures attract pests that are best adapted to live on that monoculture. When there is a variety of trees planted, different nutrients are used for the different trees. And, while some trees may attract certain pests, other trees may serve as deterrents to those pests.
Note: The students can research specific examples that support their different answers.

5 Assume that the plant material in a plant—deer—wolf food chain contains a toxic material. Why would the wolf's tissue contain a higher concentration of the toxin than the plant tissue?

■ Most materials that are toxic to living organisms are organic. As such they tend to dissolve in organic matter such as fat. Because the deer eats a large amount of plant matter in its normal diet, it accumulates the toxic material. And, since a wolf would eat more than one deer during its life, it would similarly accumulate the toxic matter from them in its fatty tissues, producing the tissues with the greatest concentration of the toxic material.

6 Provide examples of the two laws of thermodynamics in terms of some common everyday events.

■ Students will likely develop a great variety of examples. The combustion of gasoline in automobiles may be one example that they could develop.

7 Of the three basic energy pyramids, which best illustrates energy transfer in a food chain? Explain.

■ A pyramid of energy is the most accurate representation since this pyramid clearly illustrates the loss of energy at each trophic level.

8 Provide examples of how technological innovations have altered ecosystems.

■ Students will provide a very wide variety of examples that can be discussed in class in terms of which might have been the most devastating. One example might relate the increased use of motorized vehicles. As motorized vehicles became faster and increased in numbers, more, and wider roads were required. These roads that connect all communities and people with those communities have used up vast amounts of land of all kinds. Bodies of water have been altered. Hills have been levelled, and valleys filled.

9 Assuming an 80% loss of energy across each trophic level, state how much energy would remain at the fourth trophic level if photosynthesis makes available 100 000 kJ of potential energy. Show your reasoning. Construct a properly labelled pyramid to represent this situation. Could a fifth level organism be added to the chain? Explain.

■ 20 000 kJ remains at the second trophic level, 4000 kJ at the third, and 800 kJ at the fourth. The amount that would be available at the fifth would only be 160 kJ—an amount that would be insufficient to maintain the welfare of a carnivore.

CRITICAL-THINKING QUESTIONS

(page 65)

1 Assume that a ski resort is proposed in a valley near your favorite vacationing spot. What type of environmental assessment should be done before the ski resort is built? In providing an answer, pick an actual location you are familiar with and give specific examples of studies that you would like to see carried out.

■ Vegetation would be affected, which in turn would affect first-order consumers. Whenever vegetation is affected, entire food chains and food webs suffer the consequences. The answers to this question can vary considerably with the location that the students choose to describe.

2 Atmospheric warming may cause drought in some parts of the world. Illustrate the impact of drought by drawing an energy pyramid before and after the drought. Explain why the two pyramids are different.

■ **Before Drought**

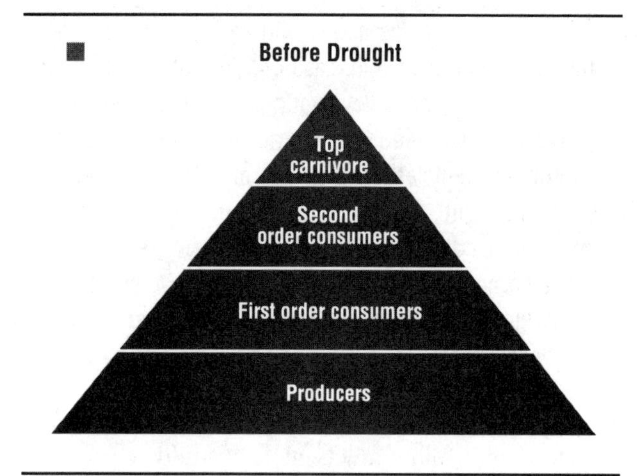

After Drought

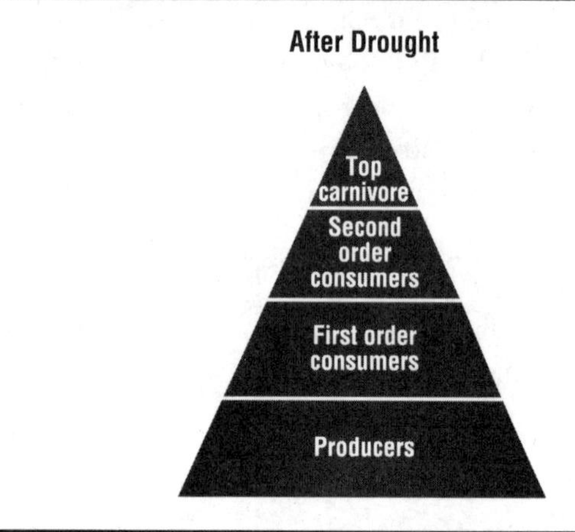

The number and variety of producers would be greatly decreased. As a result, there would be a responding decrease in the consumers that are dependent on the availability of producers. Similarly, the variety of consumers would be decreased.

3 Insect-eating plants such as the sundew are commonly found in bogs all across the country. Although often referred to as "carnivorous" plants, they are still considered to be members of the first trophic level. Is this the proper trophic level to assign to these plants? Research information on carnivorous plants, then state the trophic level you think is most appropriate. Present the reasoning behind your choice.

■ Green carnivorous plants would still be referred to as being producers in that they convert light energy into chemical energy, and as such, they belong to the first trophic level.

4 Some ecologists have stated that to maximize the food available for the earth's exploding human population, we must change our trophic level position. What is the probable reasoning behind this statement? Could any potential biological problems occur if this switch were actually made?

■ Depending on what we eat, our trophic level changes. When we eat plant matter, we are at the second trophic level. When we eat meat, we are probably eating a herbivore (second trophic level organism), and are representative of the third trophic level. In the process of moving up by one trophic level, more than 80% of the energy level is lost. As such, we would be better off in terms of energy use by eating plant matter.

The possible adverse effects of this are quite varied and the students should be encouraged to develop as many as they can. Some of these effects might include:

- If we become second trophic level herbivores, we would be in competition with other herbivores. Herbivores would become pests.
- The carnivores that fed on those herbivores (that we ate as second-order consumers) would no longer be our competitors. Their numbers would increase, and if their food became scarce, we might become quite appetizing to them!
- Health problems might develop among those human populations where the vegetable matter is carefully washed of any animal matter, because such vegetation may lack in certain vitamins essential to the health of humans.

5 A team of biologists notes that the population of white-tail deer is decreasing in a particular area. Hunting pressures have remained stable and the biomass of producers in the area has not changed. What other factors might the biologists consider to determine why the deer population is decreasing?

■ They might consider organisms at the next trophic level (natural predators), to see whether their numbers had increased. They might also study the health conditions of the deer population, because if they had become too densely populated, they may have contracted some disease that is beginning to spread through the population of deer.

CHAPTER 3

Aquatic and Snow Ecosystems

INITIATING THE STUDY OF CHAPTER THREE

Brainstorming Session

1 Have the students in small groups consider the following question: If one were able to have a precise count of each and every water molecule that exists on the planet earth at a particular moment, would that number be greater or smaller than the number of water molecules that existed on the planet earth at some time in the past (100 000 years ago) or at some time in the future (1000 years from now)?

2 See question 8 in Applying the Concepts. This question can serve to reveal the level of understanding that students hold about the interrelationships (interdependencies) among different organisms and how human intervention can affect such interrelationships.
Note: Discussion of the answers to such questions should be done in small groups before the class discusses them in a plenary session.

ADDRESSING ALTERNATE CONCEPTIONS

- Few students develop a personal reference for how lakes change, since most changes tend to be subtle and long term. Lake and pond succession, as presented in the text, provides a framework for recognizing these changes. Soil erosion, the addition of organic matter, and climatic change affect the rate of eutrophication. Water temperature, oxygen levels, and indigenous organisms provide indicators for change. In this section of the text, students have an opportunity to examine indirect scientific evidence of change. The collection and evaluation of data to determine change is also emphasized.
- In general students tend to think of winter as a time of little activity. A snow ecosystem, however, is anything but inactive. A number of insects and small rodents move through insulated, subnivean environments. Animals well adapted to gliding across the snow, such as the lynx, are provided with special advantages for catching their prey.
- In many cases scientific terminology is developed by an academic community. Most of the scientific terminology has a Latin origin and long history. In contrast, the terminology used for snow ecology comes from Inuktitut (the language of the Inuit), a descriptive language which employs approximately 20 different words for snow. The nomenclature used for snow ecology was not developed from a series of rules which follow academic protocol, but from a context of community experience and necessity. The example serves to demonstrate the pragmatic purpose of nomenclature and the manner in which it is socially constructed. This example also illustrates the important contributions to our understanding of snow ecosystems made by the Inuit people.

MAKING CONNECTIONS

- Lake and pond succession, described in chapter 3, provides a dynamic case study for investigating how matter cycles in ecosystems, as described in chapter 1, Equilibrium in the Biosphere. The role of decomposers, as introduced in chapter 2, Energy and Ecosystems, is also reinforced by the example of succession. Decomposers help explain why oxygen levels decrease after organic matter is added to a lake.

- By examining the physical factors of a lake ecosystem, such as spring and fall overturns, students are provided with a vivid example of how the abiotic environment affects the biotic community, as described in chapter 1. The concept is extended in chapter 25, Population and Communities, as students begin to examine the impact of environmental change on population size and density.
- The laboratory experiment, Acid Rain and Aquatic Ecosystems, is an open-ended investigation that asks students to develop their own procedures to investigate how pH affects the biotic community. This investigation provides an important connection between the water cycle, as described in chapter 1, Equilibrium in the Biosphere, and food chains, as described in chapter 2, Energy and Ecosystems. In addition, the laboratory provides a context for beginning a study on adaptation, introduced in chapter 4, Adaptation and Change.
- Adaptations for Winter, as described in chapter 3, examines the physiological adaptations of the mammalian circulatory system to prevent heat loss. Students explore counter-current exchange systems in which warm arterial blood transfers heat to the cooler venous blood. The adaptation not only provides an important example of adaptations, a topic covered in detail in chapter 4, Adaptation and Change, but also provides a background for many of the concepts developed in chapter 10, Circulation.
- By studying the impact of humans on subnivean ecosystems, students are provided with a context for developing many of the concepts presented in chapter 25, Population and Communities, which explores how humans affect various populations and communities.

POSSIBLE JOURNAL ENTRIES

- Students may be asked to express concerns about water quality in a local lake or pond. Do they believe that the ecosystem is changing? Has fishing changed? Are there more weeds? The students might be asked to hypothesize about why changes that they have identified are occurring. How would they go about testing their hypothesis?
- Despite winter being the dominant season in most of Canada, snow ecosystems are not well studied. Students might be asked to speculate about why these ecosystems are neglected. Would they ever consider pursuing a career that involved investigating snow ecosystems?

USING THE VIDEODISC

Section title	Page	Code	Frame (side)	Description
Importance of Water in Ecosystems	66 – 67		3302 (all)	A Drop of Fresh Water—movie sequence of organisms found with in a mini-ecosystem
Importance of Water in Ecosystems	66 – 67		2528 (all)	Water molecule—single frame
Aquatic Communities	68 – 69		3292 and 3293 (all)	The water cycle—single frames
Aquatic Communities	68 – 69		35747 (3)	Movie sequence of rotifer, with drawal of the wheel organ
Aquatic Communities	68 – 69		36899 (3)	Movie sequence of Daphnia swimming
Characteristics of Standing Water	69		18824 (4)	Movie sequence of mosquito life cycle

Section title	Page	Code	Frame (side)	Description
Characteristics of Standing Water	69		26902 (4)	Fish swimming behavior—movie sequence
Characteristics of Standing Water	69		27650 (4)	Frog, underwater swimming behavior
Lake and Pond Succession	69 – 70		2880 (all)	Ecological succession—single frame
Adaptation for Winter	78		45808 (4)	Ptarmigan, animal camouflage—movie sequence

IDEAS FOR INITIATING A DISCUSSION

Figure 3.1: Photo of an arctic aquatic ecosystem

Ask students to consider the importance of ice flows in bringing animals to isolated islands. Polar bears stranded on ice flows have been carried from the high arctic to the shores of Newfoundland.

Figure 3.5: Decomposers in a lake's ecosystem

Challenge students to determine changes in lake water quality through a school year. Coliform kits can be purchased from most biological supply houses at reasonable cost. Why might fecal coliform counts be highest in spring?

Figures 3.7 and 3.9: Diagram showing graph of solubility of oxygen in water and photo of industries

Ask students to predict environmental problems brought about by using lake water to cool industrial machinery. They may be asked to identify industrial plants in their area which routinely use fresh water for cooling.

Frontiers of Technology: Echo Sounding

Ask students to consider the impact of using echo location devices for fishing. Will this technological instrument further cause a depletion of fish stocks?

Figure 3.31: Snowmobile trails and subnivean mammal movement

Brainstorm other ways in which humans affect snow ecosystems.

REVIEW QUESTIONS

(page 71)

1 Why is a water environment less subject to change than a terrestrial ecosystem?

■ The temperature in water ecosystems is much more stable than in terrestrial ecosystems.

2 Distinguish between freshwater and marine communities.

■ Freshwater communities are found in bodies of water that do not contain significant amounts of dissolved salts. Marine communities are found in oceans and other bodies of water with high concentrations of salts. Marine communities are far larger than freshwater communities.

3 What is the major difference between a lake and a pond?

■ Sunlight penetrates to the bottom of a pond, while a lake has a profundal zone where no sunlight is found.

4 Where would you expect to find brackish water?

■ Brackish water (a mixture of fresh and salt water) is found at the mouth of a river. This is the area where the fresh water from a stream or river mixes with the salt water of the ocean.

5 What is meant by temporary standing water?

■ This is an area where surface water is only temporarily present. It may be formed by heavy rains, spring runoff, or other types of flooding.

6 Compare an oligotrophic lake with a eutrophic lake.

■ An oligotrophic lake is usually deep and cold, with limited amounts of nutrients. A eutrophic lake is usually relatively shallow and warm with large amounts of nutrients.

7 What are some major causes of eutrophication?

■ Increased water temperature, increased plant growth, runoff from fertilized fields, runoff from livestock enclosures, and algal blooms are some major causes of eutrophication.

(page 75)

8 Distinguish between the littoral, limnetic, and profundal zones of a lake.

■ The littoral zone of a lake or pond is the area where the water is shallow enough to allow enough light to permit the growth of rooted plants. The limnetic zone is where the lake is so deep that no rooted plants exist. Plankton are common to this zone. The profundal zone is so deep that sunlight is absent and no photosynthesis occurs.

9 What is meant by the term "plankton"?

■ Plankton is the term which describes both photosynthetic and non-photosynthetic, microscopic organisms.

10 What is the effect of the fall overturn on a lake?

■ Fall overturn is the time when the thermal layers in a body of water break down, and highly oxygenated water is distributed to all depths.

11 What is unique about arctic lakes?

■ Arctic lakes may experience only one overturn in the year. If the summer is cold, there may not be a fall overturn because the lake water did not warm.

12 What is the relationship between dissolved oxygen and the temperature of a lake?

■ The lower the temperature, the greater is the amount of dissolved oxygen.

13 Why does a thermocline form in the summer?

■ The thermocline is formed in the summer because of the difference in density between water above 4°C and water below 4°C. The warmer water is less dense than the cooler water. As water warms at the surface of a lake it remains above the cooler water and does not mix with the water that is below 4°C.

14 Define algal bloom. What are its effects on a lake?

■ An algal bloom is the rapid growth of plants, particularly algae, caused by an increase in the amount of nutrients available. The rapid growth is followed by the rapid decomposition of the plants, which leads to a reduction in the oxygen levels of a lake. This process increases the rate of eutrophication.

15 Outline the steps of pond succession through to a mature forest.

■ The amount of detail that the students provide in this answer could vary considerably depending on the forest that is to be the mature climax community. The major stages include the pioneer stage with a bare bottom, submerged vegetation, temporary pond and meadow, and the climax stage, with black spruce and tamarack.

16 Explain how waste runoff from a cattle ranch can cause eutrophication.

■ The waste runoff from a cattle ranch would likely contain high levels of nitrogen and phosphates. These chemicals would serve as nutrients for plants and algae which increase the rate of eutrophication of the lake.

(page 91)

17 Provide examples of anatomical, physiological, and behavioral adaptations to a winter ecosystem.

■ A wide variety of answers are possible. A sample answer is provided. That mammals such as the lynx, have large, wide feet covered with fur is an example of anatomical adaptation. Physiological adaptation can be demonstrated by the counter-current blood flow, which prevents heat loss from the extremities. An example of behavioral adaptation is migration to southern climates during the winter months.

18 Define: chionophobes, chioneuphores, and chionophiles.

Classification	Description
Chionophobes	Animals that avoid snow and winter conditions.They do not inhabit snowy regions.
Chioneuphores	Animals that are able to withstand winter conditions.
Chionophiles	Animals that have adapted to snow. The range of the animal is limited to regions of long, cold winters.

19 Explain why snow acts as an insulator.

■ Six-pronged snowflakes trap air between the projections. The air acts as a poor conductor of heat.

20 Why does a pukak layer form?

■ Geothermal energy from the earth's crust and the anaerobic fermentation from plants cause sublimation in the snow-ground interface, and a pukak layer forms.

LABORATORY

ACID RAIN AND AQUATIC ECOSYSTEMS

(page 76)

Preparations

If this is the first time that your students are involved in designing an experiment to test a variable, you may have to provide a few guidelines. Student creativity can be the greatest when the guidelines are minimal. However, frustration can also increase when direction is absent. Through knowing your individual students, you might take time to put them into the most compatible groupings where they can learn from one another. Not only can they learn to design an experiment and complete it, but they can learn to persevere and accept a fairly difficult, first-time assignment as a challenge rather than a frustration!

Remind your students to control all but the responding variable(s), to measure or count all ingredients, and to maintain careful records of all measurements and observations.

Minimally, you will have to review the requirements for an aquatic ecosystem and, possibly, have the necessary components that the student groups can use in making their ecosystems. (If an aquatic ecosystem is handy to the school or to some of your students' homes, you might request their assistance in collecting these components.) You will also have to monitor student construction of the acid rain generator. Finally, you might remind students to carefully follow the guidelines and suggestions offered in the Procedure for the lab.

You may wish to establish the duration of this investigation or leave it as an open-ended experiment so that those students who might wish to attempt to neutralize their aquatic ecosystem and make further observations can be encouraged to do so.

Safety Precautions

- Some safety precautions are suggested in the lab outline.
- If your students have not had any experience in connecting glass tubing to rubber tubing or inserting glass tubing through stoppers, you will have to either demonstrate proper procedures or give them detailed directions for the safe construction of the acid rain generator.
- The handling of corrosive liquids should be reviewed and students should be reminded to report any spills.
- Make sure your fully supplied first aid kit is readily available.

Time Requirements

Different approaches can be taken for this laboratory exercise, and each approach might have different time requirements. One approach would be as follows:

Day 1—approximately 30 min to discuss the lab and preparations for it. Form groups. The assignment is to, individually, attempt to develop a design which would be discussed with the group members during the next class.

Day 2—group work for approximately 45 min to design the experiment and plan details of the procedure. During this session, the students would also determine details regarding the data that they would collect. (Assignment might be for the group members to collect the ingredients for the aquatic ecosystem.)

Day 3—approximately 45 min to set up the aquatic ecosystem, generate the acid rain, and begin the treatment of the ecosystem.

Day 4—acid rain treatments(?), and data collection (study of the responding variables).

Laboratory Application Questions

1 List the conclusions that you have drawn from the laboratory data.

■ Answers will vary.

2 Provide reasons for each of your conclusions.

■ Answers will vary.

3 Identify modifications or changes to the procedure that you would incorporate should you wish to carry out the investigation again.

■ This laboratory will have a variety of hypotheses and designs. It will also have a tremendous variety of pond water organisms involved. As such, the possible answers to these laboratory questions are very extensive.

LABORATORY
ABIOTIC CHARACTERISTICS OF SNOW

(page 93)

Preparations

Even though this investigation is particularly interesting in its revelations of abiotic characteristics of snow, its completion is totally dependent on the weather.

In advance of the investigation, the teacher should identify a location where the snow is at least 30 cm in depth. Open and sheltered locations can serve as good comparisons.

Each group should be responsible for carting its own equipment and supplies. Group members can also check on one another to see if they have enough warm clothing (see Safety Precautions below).

Safety Precautions

As indicated in the Procedure, students should be cautioned about their clothing and be prepared for below freezing temperatures. Proper footwear and headgear are particularly important. (We can lose a very large quantity of heat from our heads.) Because the students would be exerting themselves physically, they should be encouraged to wear layered clothing so that they can remove some layers prior to physical exertion and then replace these layers when finished. Such an approach would minimize perspiration and subsequent rapid body cooling. The teacher can also encourage students to bring extra sweaters and large shirts for those who might be lacking them (invariably some people forget to bring extra clothing). Clothing is particularly important if the excursion takes you some distance away from any shelters.

Time Requirements

This investigation can be completed in approximately one hour. The total time for the laboratory would depend on whether a large amount of time is required to get to the site to be used. If a field trip is planned, other activities might be included.

Observation Questions

a) Describe the snow in terms of appearance, hardness, and the size of the snow crystals at different depths.

■ The answers to this question will vary depending on the time of the year, the temperature conditions at the time of snowfall, and the extent of each snowfall during the accumulation of the snow. Students may have different answers even if they are in areas that are fairly close together. Wind conditions vary and microclimates may also exist.

b) Note whether a pukak layer is visible.

■ Pukak layers are most commonly present during a winter which started with a relatively early and heavy snowfall that covered the ground completely and remained for the duration of the winter. Students should be able to observe it quite easily.

c) Why is it necessary to wait two minutes before removing the thermometers?

■ Sufficient time must be given for the thermometers to cool or warm to the temperature of the environment.

d) Why must the temperatures be read immediately after the thermometers are removed from the snow?

■ Upon removal from the snow, the thermometer readings will begin to change to the atmospheric temperature. Readings should be taken before this happens.

e) Is there any evidence of snow strata? If so, record the thickness of each layer and its position in the snow pack. A diagram may be helpful.

■ Snow strata will vary from year to year with the frequency and amount of snowfall on each occasion.

f) Referring to the diagrams of snow crystals provided earlier in the chapter, determine the age of the snow.

■ Students will have some difficulty with this part of the lab activity if they want to be particularly precise. Their goal should be to simply realize that the snow crystals can help them determine snow of varying age.

Laboratory Application Questions

1 Is there any indication that snow has any effect on the temperature within and beneath it? Explain, using data from your investigation.

■ It is expected that the students should be able to observe the insulating powers of snow during this investigation. The data should support this property of snow.

2 Discuss the results you obtained from site 1 and site 2. Are any trends observable? Use your graph and tables of data to support your conclusions.

■ If the sites are both in a large field with similar terrain, it is rather unlikely that the data between the two sites would vary greatly. If the sites are found in very different terrain, the data could vary by a relatively large amount.

3 Would ground level conditions be any different if there were no snow cover. Explain.

■ If there were no snow cover, the temperature were below freezing, and there were cloud cover, it is likely that the ground temperature would be similar to the ambient temperature since there would be no insulating snow present. If there were no clouds and the sun were shining, it is likely that the ground would be warmer than the air temperature. Without the insulating presence of snow, the ground temperature is more directly related to the atmospheric conditions.

4 Would the effects have been any different if the air temperature had been above 0°C? How could you confirm your opinion?

■ The ground temperature under the snow would stay relatively constant. This could be confirmed on a day during early spring when the daytime temperature rises above freezing.

5 Survival experts often suggest that if you are stranded in a snowstorm you should dig a hole in a snowdrift and crawl in. Comment on this idea, using data from your observations.

■ Observational data confirm that snow has insulating properties. If a person would crawl into a hole in the snow, the insulating property of the snow along with the person's body heat would warm the hole to higher than the ambient temperature.

APPLYING THE CONCEPTS

(page 95)

1 Describe how a water-treatment plant slows down eutrophication.

■ Waste-water-treatment plants remove organic matter from the waste water before it is returned to the surface water. If the organic matter were allowed to remain in the water, it would provide food for the organisms that are involved in eutrophication.

2 Following a late spring, the ice cover on a lake lasts an extra four weeks past normal breakup. Describe the possible effects on the lake ecosystem.

■ The dissolved oxygen may be depleted and fish which are particularly susceptible to the decreased amount of dissolved oxygen may begin to die off. With these deaths, the amount of organic matter in the lake will increase, and this will greatly accelerate the process of eutrophication.

3 Explain why certain species of animals are used as markers for determining the oxygen levels of lakes. (For example, many leeches are found along the bottoms of lakes with low levels of oxygen; stonefly larvae are indicators of higher levels of oxygen.)

■ The metabolic processes of organisms vary. Some require more oxygen than others. As such, different organisms indicate the amount of dissolved oxygen that is present and are referred to as "markers."

4 How does an increased number of decomposers affect the life of a lake or pond?

■ The life of a lake will be shortened. An increase in the number of decomposers occurs as a result of an increased amount of organic matter and relates to a depletion in the amount of dissolved oxygen.

5 Explain why eastern lakes are more seriously affected by acid rain than their western counterparts.

■ Eastern Canadian lakes are deeper and colder and usually are found in deep depressions in granite rock. They have few chemicals which might serve to neutralize acidic chemicals.

6 After a wind, many of the projections break from snow crystals. Explain how winds affect the insulating value of snow.

■ Snow which is more loosely packed will have a greater insulating value. Winds, by breaking off the projections from snow crystals, will serve to decrease the insulating value of the snow by allowing the snow to become more tightly packed.

7 Does your paper carrier walk across your lawn? Explain why the grass often appears dead along the paper carrier's path even in early spring. Why does the grass next to the path appear much healthier?

■ Because the snow along the path becomes tightly packed, the insulation of the snow is decreased, so that the grass along the path is much more affected by winter temperatures than the grass that is next to the path and insulated by the less packed snow.

8 A conservationist feeds grain to a herd of starving deer during a particularly difficult winter. Discuss the implications of such practices.

■ This question can serve quite nicely to indicate how no one thing occurs without far-reaching consequences. For example, as the deer become healthier and fatter through the availability of grain, they can better survive the winter and better escape from their predators who are having an equally difficult winter. The predators can then turn to other available prey such as domesticated livestock. A larger number of healthy deer survive the winter, produce more healthy offspring, and more rapidly increase their population. As their population increases, more predators are attracted to the area. Similarly, because the deer population is greater, there is less food for those smaller animals which compete with the deer. The entire food web becomes affected!
Students can be encouraged to discuss this question at some length. Their discussions will serve to help them conceptualize the complexity of the interrelations that exist in different ecosystems.

9 Considering the low levels of solar radiation available for heating in the subarctic winter, would a change of coat color from dark to white really be an advantage to an animal such as the snowshoe (varying) hare? Explain.

■ Yes, particularly because there is little solar radiation. Snowshoe hares can find protection by burrowing into snowdrifts. The main danger that faces them lies in being seen and hunted by predators. Their white coat provides them with excellent camouflage.

10 Indicate how activities such as cross-country skiing and snowshoeing can affect snow ecosystems.

■ Cross-country skiing and snowshoeing will create paths where the snow becomes packed down. These paths or tracks will inhibit the movement of subnivean creatures such as mice, voles, shrews, and lemmings. When confronted with such barriers to their movement, the organisms might come to the surface of the snow. In so doing, they are vulnerable to their different predators. They may be affected, but their predators, of course, benefit. However, such benefit is quite short term, in that if the subnivean creatures survive the winter, they produce offspring which in turn provide food for more predators!

CRITICAL-THINKING QUESTIONS

(page 95)

1 Discuss the various roles that detritus can play in aquatic ecosystems. Why have some cottage towns insisted that their residents equip cottages with sewage-holding tanks?

■ As bacteria proceed to decompose detritus, the dissolved oxygen is depleted. The lower levels of oxygen may cause the death of those organisms requiring a more oxygen-rich environment—this further increases the amount of detritus. The rate of eutrophication increases greatly.

Human sewage from outhouses can seep into the lake through the groundwater. When this happens, such seepage acts like a fertilizer and increases the rate of plant growth in the lake. When the plants die at the end of their growing season, the plant matter adds to the amount of detritus and accelerates the eutrophication process. It's interesting that the very reason why the cottage owners chose their vacation place (the lake), is most seriously affected by the owners' presence.

2 If humans completely stopped interfering with lake ecosystems, would eutrophication be stopped? Explain.

■ No! Eutrophication is a natural aging process. Lakes normally receive nutrients such as phosphorus and nitrogen by drainage from the land that surrounds them. Such nutrients promote the growth of plants. Plants die at the end of their growing season and become detritus at the bottoms of lakes. This detritus is further broken down by bacteria that use oxygen in the process. As the amount of dissolved oxygen in the water decreases, the animal life also changes with those organisms that require less oxygen flourishing. Organisms that require a greater amount of oxygen die, adding to the amount of detritus. The process continues and accelerates as the lake becomes shallower and the water warms during the growing season. Eventually the shores of the lake begin to recede.

3 Thermal pollution has significant effects on Canadian lakes and ponds and the communities associated with them.

a) Why are Canadian standing waters particularly affected?

■ Most Canadian lakes are rather shallow, so the sun can have a significant heating effect. Also, because Canada's summer days are long, the heating effect of the sun is even greater and the cooling effect of the shortened nights during the summer is similarly decreased. The extended length of the days has a further effect in that the period of photosynthetic activities is also extended. The very rapid growth in such Canadian lakes can result in an acceleration in the process of eutrophication.

b) Comment on specific effects of thermal pollution.

■ Specific effects of thermal pollution include:

- Accelerated metabolic activities of lake inhabitants
- Accelerated growth rates among lake inhabitants
- Accelerated photosynthesis
- Accelerated production of detritus
- Accelerated activity by decomposers
- Accelerated eutrophication

Students should be encouraged to expand on such effects.

4 An early autumn storm dumps 30 cm of heavy, wet snow on a section of the boreal forest prior to leaf fall. What effects, both positive and negative, might this storm have on this section of the forest?

■ ***Note:*** For this question, the students should be encouraged to think more holistically. What might appear to be a negative effect from certain perspectives can be seen as a positive effect from others.

Because this snowfall is early, it is likely that the ground is not yet frozen. The moisture from the snow will have the opportunity to soak into the ground and increase the amount of soil moisture available for next spring's plant growth.

Such an immediate cover will insure that the ground will not freeze to great depths and allow for early growth of vegetation in the spring. If the snow remains for the winter, it is likely that the wet snow will freeze against the ground surface leaving little or no space for subnivean organisms.

Because the leaves will eventually fall on top of snow, they will serve to accelerate melting, by absorbing the heat from the sun in the spring, and further promote the growth of an early food supply in the spring. The leaves that fall on top of the snow will not be available as food for subnivean organisms. Further, such leaves will not have the same opportunity to begin the process of decomposition—they will simply not be available to the decomposers. These leaves will accelerate melting in the spring by absorbing the heat from the sun's rays which may result with a greater runoff of snow water rather than its absorption by the soil.

The heavy snow may break branches and promote the infection of trees. The infecting organisms will be benefitted. The breaking of some weaker branches may serve to "prune" some trees so that they would be benefitted by growing more branches. Weaker trees may be so devastated that they would die. This would leave more room for the healthier trees, and the fallen trees would provide food for decomposers and eventually enrich the soil.

Note: The above dialectic can be expanded when students are encouraged to look at each event from as many perspectives as they can envision.

CHAPTER 4

Adaptation and Change

INITIATING THE STUDY OF CHAPTER FOUR

This chapter focuses on "evolution." Different approaches can be taken to determine student views on the subject.

1. It would be worthwhile to know how students interpret the term evolution. An approach that can be taken is to ask the students to write down their definitions of evolution prior to a discussion on the term. When students are given the time to prepare a written statement, it is more likely that they will become involved in any associated discussion. You can have these definitions handed in to you so that you can review them to determine the variety of views and have them available to assist you in your lesson planning for the chapter. (You may wish to have them handed in prior to the discussion.)
2. Students can be given the assignment of preparing a brief statement or sketch of what changes might take place in humans (what humans might look like) over the next several thousand years as a result of different changes that are taking place in our environment. The assignment should also require the students to explain how the changes would come about. This part of the assignment would identify the beliefs that students hold about how evolution occurs. After completing the assignment, students could attempt to arrive at "the most successful human of the future" through group work, with each group required to present its answer to the rest of the class.

ADDRESSING ALTERNATE CONCEPTIONS

- Expect some students to reject the theory of evolution by natural selection. Some students will reject the theory on the basis of alternate scientific views or on the basis of religious views. The rejection of the popular scientific view of natural selection by some students provides an opportunity to discuss the tentative nature of scientific theories. As information is presented, which cannot be explained by the theory, the theory is most often modified or replaced by another theory altogether. Students should recognize that other scientific theories have been replaced. Students who accept a religious view which differs from the scientific view expressed in the text, can be assured that scientific investigations and religious inquiry are aimed at different questions. Each way of knowing uses a different approach, but different religions will also disagree in the same manner that different scientists will disagree. The confrontation is not between science and religion, but between different belief systems. The scientific theory which advocates evolution by natural selection cannot be considered irrefutable; it is merely the theory which most scientists have decided to accept. Natural selection, according to current thinking, is best supported by indirect scientific evidence.

- Some students will develop a Lamarckian view of evolution—that organisms adapt because of need. Although the text provides the Darwinian view along with the Lamarckian view, evidence which supports the Darwinian view is incomplete. Students may gather further evidence when they study genetics, the basis for variation. Although the case study on the peppered moth, which appears in chapter 4, provides an opportunity to confront the Lamarckian notion of need, the misconception may be re-addressed later when students study the Hardy-Weinberg law in chapter 24, Population Genetics.

MAKING CONNECTIONS

- Many examples that support the study of adaptation may be taken from chapter 3, Aquatic and Snow Ecosystems.
- The case study, Resistance to DDT, provides students with an opportunity to link information that they gained about pesticides and food chains, as presented in chapter 2, Energy and Ecosystems. The concept also provides a bridge to Unit 6, Heredity, as students learn how genetic variation provides a basis for resistance. The insects that survive carry genetic information which provides them with an adaptive advantage.
- Physiological and structural adaptations presented in the section Types of Adaptation provide an important link between evolution and human physiology and anatomy presented in Unit 3, Exchange of Matter and Energy in Humans, and Unit 4, Coordination and Regulation in Humans.

POSSIBLE JOURNAL ENTRIES

- Students may be encouraged to read mythological explanations for the origin of life. How do mythological explanations differ from scientific explanations?
- Students may be asked to view Figure 4.7 and comment on how embryological evidence supports the theory of evolution. Why are the early stages of development so similar? Do humans pass through an evolutionary ancestry as we develop?
- Ask students to comment on the theory that Pangaea was once a supercontinent. Why might someone remain unconvinced?

USING THE VIDEODISC

Section title	Page	Code	Frame (side)	Description
Importance of Adaptation	96 – 97	[barcode]	2705 (all)	Adaptation—single frame
Evidence for Evolution	97	[barcode]	2904 (all)	Evolution—single frame
Evidence for Evolution	97	[barcode]	2588 (all)	Homologous forelimb of whale, bird, and crocodile
Direct Evidence: Fossils	97 – 100	[barcode]	2922 (all)	Fossils

Section title	Page	Code	Frame (side)	Description
Direct Evidence: Fossils	97 – 100		2250 (all) 2251 (all) 2252 (all)	Fossil bone
Indirect Evidence: Living Organisms	101		2589 (all)	Homologous forelimb of human, bat, and horse
Indirect Evidence: Living Organisms	100 – 101		1739 – 1758 (all)	Starfish development
Indirect Evidence: Living Organisms	100 – 101		1759 – 1785 (all)	Sea urchin development
Indirect Evidence: Living Organisms	100 – 101		1786 – 1828 (all)	Chick embryology
Indirect Evidence: Living Organisms	101		2593 (all)	Evolutionary relationships
Indirect Evidence: Living Organisms	101		2641 (all)	Evolution of sex, isogamy
Indirect Evidence: Living Organisms	101		2642 (all)	Evolution of sex, anisogamy
Indirect Evidence : Living Organisms	101		2643 (all)	Evolution of sex, oogamy
Biogeography	102 – 104		2844 (all)	Convergent evolution
Biogeography	102 – 104		2717 (all)	Allopatric speciation
Biogeography	102 – 104		3235 (all)	Sympatric speciation
Theory of Natural Selection	108		3051 (all)	Natural selection
Types of Adaptation	110 – 111		45271 (4)	Movie sequence showing spider disappearing on a stem—camouflage
Types of Adaptation	110 – 111		45522 (4)	Movie sequence showing leaf fish—camouflage
Types of Adaptation	110 – 111		45624 (4)	Movie sequence showing how a frog blends in with its background
Types of Adaptation	111		41000 (4)	Movie sequence showing fiddler crabs using a structural adaptation—the large claw
Types of Adaptation	112		44112 (4)	Movie sequence of sage grouse showing territorial and courtship behavior—behavioral adaptation

Section title	Page	Code	Frame (side)	Description
Types of Adaptation	111 – 112		38467 (4)	Movie sequence showing honey bee waggle dance—physiological adaptation (pheromones)
Types of Adaptation	111 – 112		40155 (4)	Newts mating—pheromones used during courtship and spermatophore transfer

IDEAS FOR INITIATING A DISCUSSION

Figure 4.2: Zebras at a water hole

Students might be asked to consider how the striping of the zebras shown in the picture provides an adaptive advance. (Consider the stripes in a forest area.)

Figure 4.4: Geological time scale

Students might be asked to develop a time chart, beginning with the Cambrian era, using a 5 m string. Each 1 m can be calibrated to represent 100 million years. Mark the appearance of the fishes during the late Ordovician era, the land invertebrates during the Devonian era, the amphibians and reptiles during the Ordovician era, the mammals during the Triassic era, the birds during the Jurassic period, the land flowering plants during the Cretaceous era, and the marine mammals during the Quaternary period.

Figure 4.10: Drifting apart of Pangaea

Have students trace the map of the world, and fit the contents in together, as they existed in Pangaea, the supercontinent.

Figure 4.11: Marsupial and placental mammal

Challenge students to come up with pictures of animals that demonstrate convergent evolution.

REVIEW QUESTIONS

(page 104)

1 What does adaptation mean in biological terms?

■ Adaptation is the modification of an organism's structure, function, or mode of life to enable it to survive in a particular habitat or environment.

2 Why is evolution such an important scientific concept?

■ Evolution provides a plausible explanation of how a population of organisms changes over time—thereby explaining the diversity and unity of life on this planet.

3 What is a fossil? How are fossils formed?

■ A fossil is the remains or traces of an organism that lived in the past. Fossils may be formed in a number of ways. (See pages 97-98.) The answer should reflect an understanding of the process by which most fossils are formed in sedimentary rock.

4 How do homologous and analogous structures differ? Give examples.

■ In evolution, homologous structures have similar origin and structure, but may perform different functions (this implies a common ancestor). An example might be the flipper of a dolphin and the forelimb of a dog. Analogous structures function similarly in two different organisms, but have different embryological origins. An example might be the wing of a butterfly and the wing of a bird.

5 Why is the fossil record considered to be incomplete?

■ The fossil record contains many gaps. Many of the transitional forms or "missing links" between one major type of organism and another have not been found. Reasons for these gaps are many and varied.

6 Why is biochemical evidence of evolution considered to be indirect?

■ Biochemical evidence is gathered largely from living organisms, and must be extrapolated to previous forms. It is not, therefore, direct evidence in the same sense as are fossils.

7 What does the name "Pangaea" refer to?

■ "Pangaea" refers to the single supercontinent that existed up to approximately 225 million years ago.

8 Why are there greater numbers of species of organisms on landbridge islands than on oceanic islands?

■ Land bridge islands were once directly connected to continents, and have recently separated. They share a wealth of species from the past. Oceanic islands have always been isolated from continental land masses, and immigration of new forms has been much more difficult.

(page 112)

9 How does the modern view of evolution differ from the earlier beliefs?

■ The modern view of evolution is based on scientific evidence which combines both genetic information and the theory of natural selection. It implies a change over time. Earlier beliefs were based on opinion and nonscientific evidence. This earlier view held that living things had been "fixed" since the beginning, and were unchangeable.

10 What adaptation in the Galápagos finches made the greatest impression on Darwin?

■ The variety (adaptation) of beak types displayed by the islands' finch species made the greatest impression on Darwin.

11 What is the significance of the fourth step in Darwin's theory of natural selection?

■ Survival of the fittest means that individuals with traits best suited for an environment are better able to compete, survive, and reproduce. Adaptation is the basis through which natural selection can occur, with those individuals which are best adapted to particular environments becoming the most successful reproducers. The direction of evolution is determined by the adaptation.

12 What contribution did Lamarck make to our understanding of the mechanism of evolution?

■ Lamarck offered an explanation for the mechanism of evolution—that species change over time, and that the environment is a factor in that change. He showed that evolution is adaptive and that the diversity of life is the result of adaptation. (See pages 105-106.)

13 How did Buffon's theory influence Darwin's thinking about evolution?

■ Buffon's theory indicated that the creation of a species did not occur in a single place at a single time and that a species was not created in a perfect state. Later he wondered if certain species might develop from a common ancestor. Darwin accepted Buffon's ideas and went on to provide an explanation for the manner in which species change over time.

14 What are the three general types of adaptation?

■ Structural, physiological, and biochemical are the three types of adaptation.

15 What adaptation did the peppered moth make to the pollution around Manchester in the mid 1800s?

■ The moths adapted a color (pigmentation) change from light-colored to predominantly dark-colored (melanic) forms in response to the environmental changes caused by the industrial fumes of the mid 1800s.

CASE STUDY

ADAPTATION AND THE PEPPERED MOTH

Observation Questions
(pages 109–110)

a) What might have caused the appearance of the first melanic form? Explain.

■ The melanic form probably first appeared because of a mutation. However, prior to the industrial revolution, this trait had no selective advantage for the moth population. In fact, they would have been easy prey for insect-eating birds which saw them against the light-gray speckled lichens that grew on tree trunks.

b) If the environment caused the selection pressure for change, what was the actual selecting agent in this case?

■ The selecting agents were predatory (insect-eating) birds and the environment.

c) What might have happened to the moths that were not recaptured?

■ Moths not recaptured had likely been eaten by predators before recapture was possible.

d) How can you account for the differences in the recapture numbers for polluted and nonpolluted sites?

■ The contrast between the moth's pigmentation or color and the color of the tree bark in both the polluted and unpolluted areas leads to different selection pressures.

e) What generalization do the results suggest about environmental selection for the two forms of moth?

■ Environmental selection favors melanic forms in polluted areas and light-colored forms in unpolluted areas.

f) Explain why the melanic form is more abundant today than in the early part of the 19th century.

■ Increased pollution since the mid 1800s has created an adaptive advantage for the melanic forms over a wider area.

Case-Study Application Questions (page 110)

1 Even a population which is 98% melanic retains the factor for light color in some of its members. What would happen if the environmental conditions were again reversed?

■ Light-colored moths would have an adaptive advantage and the population would shift in favor of the light-colored forms. Consequently, there would be a sharp decline in melanic forms.

2 Explain the following statements as they apply to this case study:

- Evolution and adaptation need not always involve long periods of time.
- While the change was quick it was actually quite small.
- Evolution and adaptation usually occur by means of small changes.

■
- Over a short time period (approximately 50 years), the population shifted from predominantly light-colored forms to predominantly melanic forms.
- The only apparent trait involved was one of color or pigmentation.
- (Answers may vary.) Although the diversity of life on earth is usually associated with major differences in structure and function, the case study demonstrates that evolution and adaptation may occur through small modifications such as a change in color or pigmentation.

CASE STUDY
RESISTANCE TO DDT

Observation Questions (pages 114–115)

a) Where did the resistant trait come from in the original population?

■ The resistant trait arose from a mutation.

b) What happens to the nonresistant trait over the three stages of the population? Why?

■ The frequency of the nonresistant trait declines as a result of selection pressure against it. It is killed by DDT.

c) What happens to the resistant trait over the course of evolution in this hypothetical situation? Why?

■ The frequency of the resistant trait increases due to its selective advantage. It is able to survive or withstand DDT treatments.

d) Explain the difference in the selection pressure (the trait natural selection is favoring) in stages 1 and 3.

■ In stage 1 (prior to a change in the environment), selection pressure is neutral, but the nonresistant forms have an advantage as they are present in much higher numbers. In stage 3, following repeated DDT sprayings, natural selection favors the resistant forms, as illustrated by their increased frequency in the population.

e) Is this a case of "evolution in action"? Explain.

■ Yes, it might be a case of "evolution in action." The population has shifted from a low frequency of resistant forms in stage 1 to a high frequency of resistant individuals in stage 3. These observed changes reflect changes in the population gene pool.

f) What explanation can be given for the early success of the malarial spray program and its eventual failure in later years?

■ The first populations exposed to spray programs contained low frequencies of resistant individuals and were very susceptible to the spray. Later populations had greater resistance due to the increased frequency of resistant forms.

g) What combination of factors most likely contributed to a strong selective force for developing resistant strains in the mosquito population?

■ (Answers may vary; several options are offered in the case study.) Examples include repeated sprayings, multiple- and cross-resistance, etc.

h) Because the anopheles mosquito has developed resistance to DDT, what might be the long-term effectiveness of other known synthetic chemicals in controlling malaria? Explain.

■ Under the selective pressures exerted by DDT, mosquitoes might develop cross- and multiple-resistance to other known (or newly developed) synthetic chemicals. The case study outlines how this phenomenon has already occurred in similar circumstances.

Case-Study Application Questions (page 115)

1 The resistance of mosquitoes to DDT is a good example of natural selection. Explain how this case study supports Darwin's observation that

a) hereditary variations exist among species

b) natural selection acts on organisms with these variations.

■ **a)** If there was no hereditary variation in the mosquitoes, then they would either all be resistant to DDT (clearly not the case) or none would be resistant (also cannot be true due to the observed results of spray programs).

b) The shift in frequency of resistant vs. nonresistant strains over time is clearly linked to the "selection pressure" exerted by the spray program. The gene pool is being modified over time—a signal that natural selection is occurring.

2 Discuss the following statement: "Natural selection as the cause of evolution has been neither proved nor disproved."

■ "Cause" is very difficult to prove when dealing with a process that occurs over the very long time scales required for evolution. Natural selection can be shown to result in changes in the population gene pool, but that demonstrates only a mechanism by which evolution can occur, not a cause of evolution. Natural selection is just one piece of evidence supporting the theory, but it is consistent with the theory, and therefore cannot be used as evidence to refute it. (Answers might also indicate that "theories," by definition, cannot be proven or disproven. A proven theory is a law.)

APPLYING THE CONCEPTS

(page 119)

1 In what way is the geological time scale both a biological calendar and a framework that describes major geological events?

■ The geological time scale provides a framework for the development of earth's geological features, as well as a time frame for the corresponding evolution of life on earth. It allows us to place the development of life in an orderly sequence from which we can infer the development of more modern forms from older organisms.

2 Contrast the views on evolution held by modern scientists with those of earlier proponents.

■ Many answers are possible here, but some possibilities include:

Early proponents

- based on non-scientific opinion and hearsay
- considered organisms to be "fixed" (unchanging)
- felt that there had not been sufficient time for evolution to have occurred
- rejected theories of evolution (Darwin's and others)

Modern view

- views based on scientific evidence
- organisms considered to change over time
- evolution presumed to have been going on for millions of years
- most scientists recognize natural selection and genetics as the basis for evolution

3 How is the theory of continental drift useful in the study of evolution?

■ Continental drift explains the changes in the position and orientation of the continental land masses over the past 540 million years. These changes profoundly affected the evolution of life forms. For example, when land masses separated, adaptation and formation of new species occurred separately, producing a diversity of organisms on land and along shores. When land masses came together (collided, e.g., India colliding with Asia and creating the Himalayas), habitats were lost and the overall diversity of organisms declined. Shifting of continents also moved land masses into different climatic regions, with different selection pressures on populations.

4 How did Darwin use the idea of Lamarck in formulating his theory of evolution?

■ Lamarck had recognized the impact of environmental factors on the course of evolution. He recognized that the mechanism for evolution was natural selection through adaptation. This became the basis for Darwin's theory of evolution.

5 What does the peppered moth case illustrate with respect to the rate of evolutionary change?

■ The traditional and widely accepted model for the rate of evolutionary change holds that change occurs slowly or gradually within populations of organisms. The peppered moth case (and others mentioned in the text) demonstrated that certain changes (e.g., coloration in moths) can take place rapidly.

6 Using examples, demonstrate from an evolutionary viewpoint that the survival of species is more critical than the survival of specific individuals.

■
- Some populations (e.g., mosquitoes exposed to insecticides) show wide fluctuations in number, indicating the loss (or gain) of tremendous numbers of individuals, but the species continues (although it may evolve in the process).
- The loss of individuals may reduce the gene pool and modify the population, but the species continues.
- The life span of individuals is trivial in comparison to the "life span" of species. Hence, specific individuals contribute very little to the survival of the species. On the other hand, if the species does not survive, there will of course be no more individuals.

7 How could supercomputers and computer simulation play a useful role in predicting how plants and animals might adapt to a changing environment? Give examples.

■ Adaptation to environment is a very complex process and very difficult to study in the natural environment. It is impossible, for example, to devise an experiment to show the effect on a caribou population of another ice age. Adaptation also takes a very long time, and may not be suitable for direct study (e.g., even an adaptation taking hundreds of years would be difficult to study directly). Carefully constructed computer models mimic the response of organisms in the "real world," and allow scientists to run simulations that show, in minutes, the effects of environmental changes that might take years of real time. They can also vary different aspects of the environment and look at their effects on populations. Supercomputers allow such models to contain far more complexity and to be more realistic. Examples will be various.

8 Give an example of how environmental impact caused by human activity can alter the natural course of evolution.

■ Answers will vary, and might include some of the following:
- Roads can be a barrier to some species, and their construction can subdivide populations. Each smaller population may experience slightly different selection pressures, and it may contain a slightly different gene pool than the original, continuous, population.
- Hydroelectric dams cause extensive flooding and disruption of natural populations, creating new sets of selection pressures which can affect the direction and rate of evolution. Some major projects have even been stopped because they would cause the extinction of a specific organism found only in the area which would be disrupted by the dam.
- The building of navigable waterways connecting previously unconnected (or poorly connected) bodies of water allows the spread of species beyond previous limits, and alters community composition and selection pressures (e.g., the lamprey entering the Great Lakes).
- Pumping of waste water from ships' bilges has been implicated in the introduction of the zebra mussel to the Great Lakes, introducing a new and fast-growing grazer to the community, with implications both for existing species and for man's use of the waterways.

9 Explain why organisms with a high reproductive capacity can adapt more readily than many organisms that have a lower reproductive capacity.

■ High reproductive capacity is normally linked with high egg number and short development time (e.g., many species of fish and insects). Where there are many young, there will be many different combinations of parental genes, increasing the chance that there will be some combinations that are better able to withstand a particular selection pressure. Such populations can also respond more rapidly to sudden pressures (compare the response of humans, with a nine-month gestation period and approximately 20 years between generations, with the response of flies, which may have only days between successive generations).

CRITICAL-THINKING QUESTIONS

(page 119)

1 The wombat, a marsupial mammal found in Australia, and the rabbit, a placental mammal that was brought into Australia by explorers, have many of the same requirements. A biologist studying both animals suggests that rabbits will eventually take over the ranges occupied by wombats because they reproduce at a faster rate. What evidence would you want to gather before accepting this conclusion?

■ How close are their requirements? To what extent do their niches overlap? Do they exploit the same habitat in the same way? Are there selection pressures that will limit the maximum population size of rabbits, despite their higher rate of reproduction (e.g., differential predation and disease)?

2 Many of Darwin's ideas about evolution came from his observations on the Galápagos Islands. Today, scientists recognize that islands are literally "laboratories of evolution." Discuss the validity of this viewpoint.

■ Answers should recognize that, in the Galápagos, similar populations of organisms invaded a series of islands on which there were different selection pressures. It is almost the type of situation a researcher might set up if she were interested in evolution in natural situations, and if she had hundreds (perhaps thousands) of years to observe the results. Answers should include the idea of isolation of one island population from populations on other islands and/or the mainland.

3 Scientists suggest that theories provide tentative explanations of the natural world and should never be interpreted as absolute fact. Provide an example which indicates how biological explanations of the natural world would change if a Lamarckian explanation of evolution were endorsed by the scientific community rather than the Darwinian explanation for evolution by natural selection.

■ If a Lamarckian explanation of evolution were to be endorsed, then evolutionary changes in an organism would be interpreted as meeting the needs of the individual organism.

4 Adaptations are so numerous that they embrace practically every structure or occurrence in biology. Even when organs have fallen into disuse they often become adapted to a second purpose. Investigate the present function of the pelvic girdle of snakes and the three hinge bones of reptile jaws.

■ Answers will vary, but can be evaluated in relation to the literature used to research them.

UNIT 2

Exchange of Energy and Matter in Cells

SUGGESTIONS FOR INTRODUCING UNIT 2

During the study of Unit 2 (chapters 5, 6, and 7), it is anticipated that students will steadily increase their understanding of the complexities and dynamics of living cells. They will begin to realize that there are a great many different types of cells, with each type specialized to perform specific structural or diverse functional roles. The following activities are suggested as means by which you can promote student interest in the study of cell structure and function, with the added dimension of determining the conceptions that your students hold of cell structure and function prior to, during, and at the end of the unit of study. This activity will start with the students attempting to design the most perfectly functioning cell and continue during the unit with the students continually assessing their designs and redesigning them as they see most appropriate. Their involvements can be described in four parts as follows.

1 Prior to the unit:
On an individual basis, students are requested to begin their design of a "perfect cell." This activity can be done in class or as an out-of-class assignment. Following the individual efforts, which students would hand in for assessment and possible feedback, the students can be grouped so that they can attempt a cooperative design of the "perfect cell." Such efforts might be presented to the rest of the class and/or handed in for your review and possible feedback.

2 After chapter 5: Development of the Cell Theory
Immediately following the completion of chapter 5 and prior to the beginning of chapter 6, the original individual "perfect cell" designs would be returned to the students, and they should then be requested to continue the design, revising their original efforts in response to their learning from chapter 5. The groups can be reformed to continue with their design of the "perfect cell," with or without class presentations of group efforts. If the groups are reformed, their efforts can be handed in for review and feedback.

3 After chapter 6: Chemistry of Life
Continue the activity as described in 2, above.

4 After chapter 7: Energy within the Cell
Continue the activity as described in 2, and 3, above. After the groups meet to complete their final redesigning of their "perfect cells," each group should have a chance to briefly describe or show their drawing of the cell, explaining different structures and functions. At this point, discussion of the possibilities and probabilities of the designed cells should occur. Answers to questions such as the following should be discussed for each of the group-designed cells.

a) What are the strengths of such a cell?
b) What are the weaknesses of such a cell?
c) Is such a cell possible?

KEY SCIENCE CONCEPTS

Unit 2, Exchange of Energy and Matter in Cells, enables students to study the nature of **energy** and **matter** flow at the cellular level. This unit builds students' knowledge gained from Science 10: Matter and Energy in Living Systems, and the previous unit of the textbook, Life in the Biosphere. By studying the flow of matter and energy through subcellular systems, students are provided with an opportunity to relate many of the macro concepts introduced at a "macro level," the biosphere, to a "micro level," the cell. This unit provides a synthesis of the principles developed within Unit 2 of the Alberta Biology 20 Program of Studies. ■

CHAPTER 5

Development of the Cell Theory

INITIATING THE STUDY OF CHAPTER FIVE

Students often attempt to compare the size of a cell with other things studied in science classes. Some students will equate cell size with other things which are not visible to the human eye; however, cells are much larger than viruses, large protein molecules, or the tiny hydrogen molecule. Refer students to Figure 5.8 which provides a reference for comparison. Students may be asked to compare the relative size of a cell with a hydrogen molecule. The DNA molecule with a diameter of 2 nm is about 50 000 times smaller than a typical plant cell. Using a dot placed on the chalkboard to represent a hydrogen molecule, they may be asked to estimate the size of a cell. Would the cell be as large as a closet? the classroom? a wing of classrooms? the entire school?

ADDRESSING ALTERNATE CONCEPTIONS

- Occasionally, students may express beliefs about life that approximate those of spontaneous generation. These beliefs are often related to the microbial world, a world which provides students with few empirical clues. John Needham's experiment provides an avenue for examining beliefs about spontaneous generation. Students can be encouraged to repeat Spallanzani's experiment, using beef broth extracts.
- Students often express the idea that greater magnification is required to view objects in more detail, while ignoring the importance of resolving power. The laboratory activity, An Introduction to the Microscope, provides students with an opportunity to investigate resolving power. Figure 5.3 also illustrates the importance of maintaining resolving power with increased magnification.
- Some students will confuse the terms cell wall and cell membrane. The laboratory activity, Plant and Animal Cells, asks students to draw and label plant and animal cells. Students are also asked to compare plant and animal cells during the laboratory. Plant cells contain both a cell wall (nonliving) and a cell membrane, while animal cells only have a cell membrane.

MAKING CONNECTIONS

- The presentation of nuclear and cytoplasmic functions provides the underpinnings for learning physiology and genetics. For example, nuclear structure is addressed in Unit 5, Continuity of Life, and Unit 6, Heredity. The function of mitochondria is reinforced and expanded upon in chapter 7, Energy within the Cell, and chapter 12, Breathing, while ribosome function is expanded upon in chapter 23, Protein Synthesis.

- An understanding of cell membrane structure provides the framework for developing an understanding of the antibody formation, blood typing, and tissue rejection, as presented in chapter 11, Blood and Immunity.
- An introduction to plant cell structure provides a basis for developing greater understanding about photosynthesis, as presented in chapter 7, Energy within the Cell.
- The principles of passive and active transport of materials across cell membranes, as presented in chapter 5, provide a basis for understanding kidney function, as presented in chapter 13, Kidneys and Excretion, and understanding how nerves conduct impulses, as presented in chapter 15, The Nervous System and Homeostasis.

POSSIBLE JOURNAL ENTRIES

- Students can be encouraged to develop a concept map which includes all key terms presented in the chapter.
- Students can be encouraged to outline difficulties and triumphs experienced during the microscope laboratory. What types of things were they able to see?
- Group thinking and decision-making strategies may be recorded as students prepare for the debate. Did the students change their minds during the preparation for the debate? Students may even be asked to record their initial positions about the social issue and to reflect upon these feelings after the debate has been completed. Did they change their minds after listening to opposing arguments?

USING THE VIDEODISC

Section title	Page	Code	Frame (side)	Description
Importance of the Cell Theory	122 – 123		1598 – 1612 (all)	Living cells
A Window on the Invisible World	124 – 125		1534 – 1554 (all)	Microscopes
A Window on the Invisible World	124 – 125		1555 – 1568 (all)	Various objects under the microscope
Laboratory: An Introduction to the Microscope	127 – 128		14497 (1)	Movie sequence showing how the microscope works
Overview of Cell Structure	131 – 133		33499 (1)	Movie sequence of electron micrographs showing cell structure
Overview of Cell Structure	131 – 132		38600 (1)	Movie sequence showing flagella movement
The Second Technological Revolution	132		2519 (all)	Electromagnetic spectrum
The Second Technological Revolution	132		2231 – 2254 (all)	Microscopes and staining techniques

Section title	Page	Code	Frame (side)	Description
The Second Technological Revolution	133		2255 – 2285 (all)	Scanning electron microscope
The Second Technological Revolution	133		2286 – 2300 (all)	Transmission electron microscope
The Second Technological Revolution	133		20013 (1)	Movie sequence showing the preparation of tissue for the electron microscope
The Living Cell Membrane	142 – 143		2797 (all)	The cell membrane
Passive Transport	143 – 144		2871 (all)	Diffusion
Passive Transport	144 – 146		3086 (all)	Osmosis
Passive Transport	144 – 146		3302 (2)	Movie sequence of cytoplasmic streaming
Passive Transport	144 – 146		5406 (2)	Movie sequence of osmosis—in plant cells of the leaf
Passive Transport	144 – 146		6646 (2)	Movie sequence of osmosis—controls the shape and development of cells
Active Transport	146 – 148		39272 (1)	Movie sequence showing contractile vacuole of paramecium
Active Transport	148		3118 (all)	Pinocytosis

IDEAS FOR INITIATING A DISCUSSION

Figure 5.2: Early microscopes

During the 17th century most scientists displayed a microscope in the study as a symbol of research. What symbols do scientists use today?

Figure 5.3: Magnification with resolution

Students can speculate about why resolution is more important than magnification.

Figures 5.12 and 5.13: Transmission and scanning electron microscopes

Students can brainstorm about the advantages of transmission and scanning electron microscopes.

Figure 5.15: Animal cell

Students can be encouraged to write a job application for a cell of the body.

REVIEW QUESTIONS

(page 131)

1 What is a scientific hypothesis?

- It is a possible solution to a problem or a proposed explanation of an observed phenomenon.

2 Why do scientists test hypotheses by experimentation?

- Experiments provide empirical evidence. Because experiments are reproducible, scientists gain insight into prediction of natural events. Scientists test hypotheses by experimentation to determine if they are valid. A hypothesis that is logical, for example, the theory of abiogenesis, does not necessarily prove to be correct when tested by experimentation.

As students progress in the course, a more complete answer would address the notion that even though controlled experiments where only one variable is tested at a time are ideal, there are situations where such experimentation is not possible for various reasons (e.g., study of evolution, or to study of the effects of toxic chemicals on humans). Nevertheless, various approaches, including the use of mathematical models and computer simulations, can be included as examples of appropriate experimentation.

3 What observations led to the theory of abiogenesis?

■ The theory of abiogenesis was presented by Aristotle, who observed that flies were always found on rotting meat. The logical assumption for him was that the flies were produced from the meat. He extended his hypothesis to argue that nonliving things could be spontaneously transformed into living things.

4 How did Francesco Redi's experiment challenge the theory of abiogenesis?

■ Francesco Redi showed by experimentation that rotting meat could not turn into flies by itself. The flies had to be produced by other flies which laid eggs on the meat. When flies were prevented from coming into contact with exposed meat, the meat did not become infested with maggots and flies. His experimental results directly contradicted the theory of abiogenesis, which had been accepted for several thousand years.

5 Using Redi's experiment, differentiate between experimental variables and controls.

■ The "control" has all variables controlled, while the "experimental" has one variable that is being tested. In Redi's experiment, the control did not allow flies to come into contact with the meat, while the experimental set-up did.

6 Who was one of the first contributors to the development of the microscope?

■ Van Leeuwenhoek is often credited with inventing the first microscope, but in reality, the microscope appears to have been an ongoing development occurring simultaneously in different parts of the world. Van Leeuwenhoek was definitely very much involved with its development.

7 How did the work of John Needham cause a resurgence of the theory of abiogenesis?

■ Needham used sterile broth to show that, within a few days, the broth was teaming with microorganisms. He proposed that even though multicellular organisms might not develop by abiogenesis, microorganisms could. Needham's conclusions were accepted for many years, despite the many flaws associated with his experimental methods.

8 Explain how Pasteur refuted the theory of abiogenesis.

■ Proponents of the theory of abiogenesis maintained that air had to have access to organic material in order for new life forms to spring from it. Pasteur designed a flask which was open to the air, but did not allow bacteria and other microbes to enter after the flask and its contents were sterilized. As expected, no organisms grew in the flask. To prove that the flask was capable of sustaining life, Pasteur tipped it over so that its contents ran up the long neck and allowed microbes to enter the broth. Microbes then started growing in the flask, proving that only life begets life.

9 Why were Robert Hooke's discoveries important to the development of the cell theory?

■ Hooke was the first person to observe and describe cells.

10 What are the two components of the cell theory?

■ All living things are composed of one or more cells. (1) The cell is the basic living unit of organization, and (2) all cells come from pre-existing cells. Cells do not come from nonliving things.

(page 139)

11 How do prokaryotic cells differ from eukaryotic cells?

■ Prokaryotic cells are more primitive. They lack a nuclear membrane which is present in eukaryotic cells. They also lack mitochondria and lysosomes which may be present in eukaryotic cells.

12 Identify the two structures of protoplasm and give a generalized function of each.

- The two structures of protoplasm are nucleus and cytoplasm. The nucleus contains the genetic information for the cell, while the cytoplasm contains all the organelles responsible for the ongoing functions of the cell.

13 What are chromosomes and genes?

- Chromosomes are threadlike structures in the cytoplasm which contain the cell's genetic material. Each chromosome contains a number of different characteristics or genes. Genes are units of instruction which determine the specific traits of an individual.

14 Discuss the contributions of the Canadian scientist James Hillier to the study of cell biology.

- Hillier was a codesigner of the first useful transmission electron microscope, which could magnify objects to 7000 times.

15 List the cytoplasmic organelles found in animal cells and state the function of each.

Animal cell cytoplasmec organelles

Mitochondria	Centers for oxygen-consuming cellular respiration which converts chemical energy into forms that are used for the normal functioning of the cell and organism.
Ribosomes	Proteins are synthesized in the ribosomes.
Endoplasmic Reticulum	There are two types of endoplasmic reticulum, rough and smooth. Rough endoplasmic reticulum has many ribosomes associated with it, while smooth endoplasmic reticulum has none. The endoplasmic reticulum is a pathway along which cellular material can be carried throughout the cell.
Golgi Apparatus	Storage organelle for protein molecules.
Lysosomes	Saclike structures which store digestive enzymes used for breaking down large food particles and for destroying harmful substances within the cell.
Microtubules	Tiny threadlike fibers that transport materials throughout the cytoplasm.
Microfilaments	Tube-like structures that provide support and help cells move.

16 What are chloroplasts?

- Chloroplasts are plastids which contain the green pigment chlorophyll. They are responsible for the process of photosynthesis: the combination of carbon dioxide, water, and light energy to produce sugar and oxygen.

17 Why would you expect to find more amyloplasts in a potato tuber (root) than in a potato leaf?

- Amyoplasts store starch. Potato tubers are where the product of photosynthesis (starch) is stored in potato plants.

(page 148)

18 How have cell fractioning techniques and the use of radioisotopes helped advance the knowledge of cell biology?

- Cell fractionation allows for different cell parts to be separated so that study of the chemical reactions associated with each part can occur.
  The use of radioisotopes allows scientists to follow chemicals in different chemical reactions as they occur in the cell organelles.

19 In what ways does a cell membrane differ from a cell wall?

- The cell membrane acts as a boundary between the cell and its external environment. The cell membrane regulates the movement of molecular traffic into and out of a cell. Cell membranes are made of lipids and proteins.
  Unlike cell membranes, cell walls are nonliving. Composed of cellulose, cell walls protect and support plant cells. Gases, water, and some minerals can pass through small pores in the cell wall. Cell walls are also found in prokaryotic cells. Animal cells do not have a cell wall.

20 Describe the structure of a cell membrane.

- A cell membrane is made up of two layers of lipids, with the water-soluble heads of these molecules facing the outer surface and inner surface of the cell. Embedded in the lipid bilayer are a variety of proteins. The proteins control the permeability of the cell membrane, act as receptor sites for hormones, and function to transport material in and out of the cell.

21 List three factors that alter diffusion rates.

- The concentration difference between the inside and outside of the cell (the greater the concentration difference, the greater the rate of diffusion); temperature (as temperature increases, the rate of diffusion increases); and pressure (the greater the pressure difference, the greater the rate of diffusion).

22 Define isotonic, hypertonic, and hypotonic solutions.

- Isotonic solutions are solutions in which the concentration of solute molecules in the solution is equal to that inside the cell.
  Hypotonic solutions are solutions in which the concentration of solute molecules is less than that inside the cell.
  Hypertonic solutions are solutions in which the concentration of solute molecules is greater than that inside the cell.

23 How does facilitated diffusion differ from normal diffusion and osmosis?

- Diffusion, osmosis, and facilitated diffusion are forms of passive transport in which molecules move from an area of higher concentration to an area of lower concentration. Diffusion involves the passage of any molecules from a region of high concentration to a region of lower concentration. Osmosis refers to the diffusion of water molecules, through a selectively permeable membrane.
  In facilitated diffusion, protein carrier molecules in the cell membrane transport molecules with the diffusion gradient at a much faster rate than is found with normal diffusion.

24 Why does grass wilt if it is over-fertilized?

- The concentration of solutes outside the grass leaves is greater than that inside the leaf cells and water is drawn from the leaves, causing the grass to wilt.

25 How does passive transport differ from active transport?

- Passive transport occurs when solute molecules move *with* the diffusion gradient without the expenditure of energy. Active transport occurs when solute molecules move *against* the diffusion gradient. This process requires energy.

LABORATORY
AN INTRODUCTION TO THE MICROSCOPE

(pages 127–128)

Safety Precautions

Even though appropriate safety precautions are given in the laboratory outline, a moment should be taken to review them at the beginning of the laboratory exercise. Of particular importance is the precaution that students should take when using a microscope with a mirror—*they should not use direct sunlight as their light source.*

Time Requirements

This is probably one of the most important laboratories if your students will be using the microscope for other investigations after this laboratory. It should not be rushed! Students should have ample time to not only be able to develop a skill, they should have time to gain confidence in their ability. This takes practice. If you allow two laboratory periods for this laboratory and the students realize the value of this activity, you will probably save time over the long run!

Observation Questions

a) Describe its appearance with the naked eye.

- Students will likely simply claim to see a portion of a picture.

b) Describe the appearance of the picture under low-power magnification.

- Under low magnification, the students will be able to see dots of black, gray, and white which make up the picture.

c) Are both threads in focus?

- Under low power, the two threads will be quite visible but one will be in better focus than the other.

d) Measure and record the distance (in mm) between the bottom of the objective lens and the top of the cover slip.

- The distance will vary in relation with the strength of the low power objective and eyepiece lenses. The lower the power, the greater will be the distance. If you have microscopes that have different power eyepieces, it may be of interest to compare the distances.

e) Measure and record the distance between the cover slip and the objective lens.

- This distance, under medium power, will be less than the distance measured under low power.

f) Measure and record the distance between the cover slip and the objective lens.

- This distance, under high power, will be less than the distance measured under medium power.

g) As you move from the low- to higher-power objectives, describe the change in light intensity.

■ There will appear to be less light as you move from low- to higher-power objectives.

h) Which objective is the best for showing the detail of the threads?

■ The high power objective is best for showing the detail of the threads.

i) Under which objective is the bottom thread clearest when the top thread is in focus? (You may wish to re-examine the threads with each of the magnifications.)

■ The low power objective shows the bottom thread clearest.

j) Compare the length of thread seen under each objective.

■ The length of thread seen will be shortest under high power and longest under low power.

Laboratory Application Questions

1 Explain why microscopes are stored with the low-power objective lens in position.

■ When the low-power objective is in place, the weight of the ocular tube can rest at the bottom of its adjustment gears, and there would be no danger of the objective lens being scratched by stage clips or anything else that might be put on the stage because the low-power objective lens is the shortest.

2 *Astigmatism* is a common disorder in which the lens of the eye has an asymmetrical shape. Most people have symmetrical lenses—the top half is identical in shape to the bottom half. Explain why individuals who have an astigmatism may experience difficulties distinguishing fine detail with the naked eye.

■ Without any correction for the astigmatism, parts of that person's field of vision will not be in sharp focus.

3 Explain why resolving power decreases as the thickness of the objective lenses increases.

■ The thicker the lens, the more light is bent. The bending of light from various angles increases the probability that multiple focal points will occur. Multiple focal points cause the image to become blurry, which decreases resolving power.

4 Why should the coarse-adjustment focus not be used with a high-power objective lens?

■ When a high-power objective lens is being used, the lens must be very close to the object to be focused. Because the coarse adjustment moves the lens over a large distance with even a small turn of the adjustment knob, the lens could quite easily break the slide or cover slip and become scratched.

5 The microscope invented by van Leeuwenhoek consisted of a single lens. What advantages do compound microscopes have over single-lens microscopes?

■ A much greater magnification is possible with compound microscopes.

LABORATORY
PLANT AND ANIMAL CELLS

(pages 140–141)

Time Requirements

Approximately 50 min are required. This laboratory can be done over a two-day time frame.

Notes to the Teacher

- We recommend demonstrating the technique prior to the laboratory.
- Special care should be given prior to using the scalpel.
- Prepared slides should be used for epithelial tissue. Do not permit students to use their own cells.

Observation Questions

a) Draw and describe what you see.

■ Students should be able to see cell walls and cytoplasm. The nucleus is visible in some cells. Some students will have a tendency to draw structures only seen by the electron microscope. Mitochondria, ribosomes, and endoplasmic reticulum are not visible. Depending on the cell, however, a vacuole may be seen.

b) Which light intensity reveals the greatest detail?

■ High light intensity reveals the greatest detail.

c) Why was iodine used?

■ Iodine makes cell parts more visible.

d) Draw a four-cell grouping and label as many cell structures as you can see.

■ Make a diagram showing cell arrangement. Plant cells remain attached.

e) Estimate the diameter of one cell.

■ The diameter is approximately 50 μm to 100 μm.

f) How does the arrangement of plant and animal cells differ?

■ Animal cells are not touching; plant cells are. Animal cells appear to have irregular shapes.

g) Draw three different cells and label those cell structures that are visible.

■ Diagrams will vary.

h) Estimate the diameter of the cells.

■ Generally animal cells are smaller: 10 μm to 30 μm.

Laboratory Application Questions

1 What function is served by the cells of the epithelial tissue of plants and of animals?

■ The cells of the epithelial tissue serve as a protective covering; they prevent injury to internal cells.

2 In what ways do the onion cells differ from those of the animal epithelium?

■ Onion cells have a cell wall and animal cells do not. Onion cells also have chloroplasts and leukoplasts, although none are usually seen. The onion cells have regular rectangular shapes, while the animal cells tend to be irregularly shaped.

3 Explain why the cells of the onion bulb do not appear to have any chloroplasts.

■ The part of the onion examined, the bulb, is found underground. Chloroplasts only function when they are exposed to sunlight.

4 Why are cells of the onion and animal epithelium classified as eukaryotic cells?

■ Both contain a true nucleus defined by a nuclear membrane.

APPLYING THE CONCEPTS

(page 150)

1 Biology, like other sciences, progresses by observation. Unfortunately, the observations of nature are often flawed by interpretation. Provide two examples of faulty conclusions that supported the theory of abiogenesis and explain the source of the error.

■ Early scientists observed that dried out water beds had fish and other water creatures shortly after there was rain. They believed that the creatures must have fallen with the rain.
Aristotle believed that flies came from rotting meat, because he had always observed flies on rotting meat.

2 Explain why John Needham came to an incorrect conclusion about abiogenesis. What was his error?

■ Needham prepared different nutrient broths and sterilized them by boiling. After a few days, he would observe that the broths were teeming with microorganisms. He concluded that these microorganisms were spontaneously created by the broth. His error resulted from the fact that the broth was accessible to contaminated (spore-containing) air.

3 Those who supported the theory of abiogenesis used the critical factor of fresh air to refute Spallanzani's experiments. Explain why Spallanzani was unable to overcome the challenge from those who supported the theory of abiogenesis. How did Pasteur finally disprove the theory of abiogenesis?

■ If Spallanzani left his flasks open, they too would become clouded by the growth of microorganisms because his broth, like that of Needham, would be accessible to contaminating spores.
Pasteur was able to use flasks that allowed the circulation of fresh air, but trapped the contaminating spores before they could reach the broth and begin to grow.

4 A student suggests that flour sealed in a jar has been transformed into flour beetles after it has been sitting for six weeks. Provide a probable explanation for the flour beetles and then design an experiment to test your hypothesis.

■ Students may produce different answers.
An acceptable answer would be to claim that some beetles must have gained access to the flour and laid eggs which hatched and produced further beetles.
A test of the hypothesis would be to have an apparatus which would physically prevent the beetles from having access to the flour.

5 A cell is viewed under low-power magnification. When the revolving nosepiece is turned to high-power magnification, the object appears to disappear, despite many attempts to refocus the slide. Provide a possible explanation for the disappearance of the cell.

■ When the nosepiece is turned to high power, only the central portion of the first field of view is magnified. If the cell is outside of this new, smaller field of view, it could not be seen.

6 By comparing a bee's body mass to its wing span, a physicist once calculated that bees should not be able to fly. Cell biologists have found that the muscles that control the wing of the bee have an incredible number of mitochondria. Indicate why this finding may help explain why bees can fly.

■ Because of the large number of mitochondria present, these muscles have access to great amounts of energy released through cellular respiration.

7 Explain why stomach cells have a large number of ribosomes and Golgi apparatus?

■ Stomach cells would need to have and store digestive enzymes, which are protein molecules. Ribosomes are where proteins are synthesized and the Golgi apparatus store these proteins.

8 Hormones are the body's chemical messengers. Protein hormones, such as insulin, must attach themselves to a receptor site on the cell membrane; fat-soluble steroid hormones, such as sex hormones, pass directly into the cell. Explain why steroid hormones pass directly into the cytoplasm of a cell.

■ Because steroids dissolve in droplets of fat, they, along with fat, can be engulfed into cells through the process of pinocytosis.

9 Identify some limitations of the light microscope and explain why the transmission and scanning electron microscopes have had major impacts on the study of cell biology.

■ Light microscopes are limited in the magnification that they can reach. Very small viruses cannot be studied with light microscopes. The transmission electron microscope is capable of 2 000 000 x magnification. What makes the scanning electron microscope so valuable is that it produces three-dimensional images.

10 A marathon runner collapses after running on a hot day. Although the runner consumed water along the route, analysis shows that many of the runner's red blood cells have burst. Why did the red blood cells burst? (Hint: On hot days many runners consume drinks that contain sugar, salt, and water.)

■ The runner must have drunk pure water so that his or her blood fluid became hypotonic to the blood cells. So, water was absorbed by the blood cells to the point where they began to burst.

CRITICAL-THINKING QUESTIONS

(page 151)

1 The statement "We are not alone" is often used in science fiction movies. Explain how this statement applies to the cells within your body.

■ All cells in a human body are part of a tissue. Different cells or tissues have different functions. Thus, the different cells in the body are dependent on each other.

2 Early scientists often proposed hypotheses but did not test them by experimentation. Instead, they used logic to translate the hypotheses into an explanation or theory. Using the theory of abiogenesis, explain why this method of inquiry is susceptible to errors.

■ When hypotheses were not tested by cautious observations or through experimentation, scientists sometimes moved further from the truth rather than toward it. For example, when water accumulated in what was a dried out depression in the land as a result of rain and/or flooding, and fish and other water creatures became evident, early scientists believed that the fish and other creatures must have fallen with the rain! Aristotle hypothesized that such creatures came from the mud. He also proposed that because flies were always present with decaying meat, they must come from that meat. Such were the beginnings of the theory of abiogenesis—the spontaneous generation of life from nonliving matter.

3 Why was Francesco Redi's experiment considered to be a significant turning point for the way in which scientific experiments were performed?

■ Redi attempted to "control" variables so that he could observe the effects of other "experimental" variables. He used empirical evidence to validate his conclusions.

4 Many people argue that technology follows science. In many cases, this may be true; however, technology can assume a leading role. Using the development of the lens as an example, explain how technology changed the manner in which humans perceived themselves and their environment.

■ Early lens makers fooled with different materials and realized that lenses had properties which could magnify objects or bring distant objects into better view. They didn't start out with such goals in mind. However, as a result of their discoveries, fields of science such as microbiology and astronomy were very significantly advanced. The microcopes showed that we are composed of many cells working together, while the telescopes showed that we are but one planet in our solar system, which is part of a larger universe.

5 Some unrelated diseases may have an interesting link: the mitochondria. Disorders of the liver (i.e., viral hepatitis, obstructive jaundice, and cirrhosis) and some muscle disorders (i.e., muscular dystrophy) are characterized by abnormally shaped mitochondria. Large amounts of unprocessed chemicals appear to accumulate in the mitochondria. Indicate why this link may be important. Does it provide any clues to the cause of the diseases? How might scientists use this information to develop a cure?

■ It may be that the unprocessed chemicals that accumulate in these abnormally shaped mitochondria are metabolic intermediates which are not further broken down because of a lack of appropriate enzymes. The diseases may be due to these accumulated chemicals. If this were so, an appropriate approach for scientists would be to attempt to develop ways by which appropriate enzymes might be provided to the cells with the abnormally shaped mitochondria and to determine first, if the chemicals stop accumulating, and, second, if the effect of the disease is lessened.

CHAPTER 6

Chemistry of Life

INITIATING THE STUDY OF CHAPTER SIX

1 Have the students survey a number of diet books and place a statement about the advice taken from each of the diet books on a bulletin board. (Sample: avoid saturated fats; diets should be high in carbohydrates.) Evaluate each of the statements following the completion of the chapter.

2 Have students survey cholesterol-free foods from the local supermarket or corner grocery store. Display cholesterol-free or cholesterol-reduced labels on the bulletin board. Ask students to comment on the use of cholesterol at the end of the chapter. Not all cholesterol is bad. One beer company, eager to gain approval from health-conscious individuals, once advertised a cholesterol-free beer; the fact that other beers do not contain cholesterol escaped them.

ADDRESSING ALTERNATE CONCEPTIONS

- Having rejected spontaneous generation, some students will construct a view of organic matter that closely parallels the view once held by a group of scientists referred to as the vitalists. The vitalists believed that nonliving things were distinctly different from living things. Some even believed that life had a vital force that was neither chemical nor physical. These alternate conceptions can be addressed by way of discussion. Students should become aware that the same principles of chemistry apply in both the physical world and the living world. Understanding life in part comes from an understanding of how chemical reactions are regulated within cells.
- Students often think of molecules and atoms as solid spheres; however, most of the volume of an atom is taken up by the space in which electrons orbit. Figure 6.2 compares the size of the nucleus with a bee in SkyDome.
- Students often think of cholesterol in negative terms, but cholesterol has many positive aspects. Cholesterol is found in cell membranes and acts as the raw material for the synthesis of certain hormones. The difference between males and females would be somewhat obscured if it were not for cholesterol. Sex hormones are made from cholesterol. High-density lipoprotein (cholesterol) or HDL, is often called "good cholesterol." The HDLs carry cholesterol back to the liver, which begins breaking it down. The HDLs lower blood cholesterol.

MAKING CONNECTIONS

- The organic compounds introduced within the chapter, namely, carbohydrates, lipids, proteins, and nucleic acids, provide important background information for understanding how cells use energy, as addressed in chapter 7, Energy within the Cell. This informtion is also pivotal in developing an understanding of Unit 3, Exchange of Matter and Energy in Humans, with special reference to chapter 9, Digestion.
- The section on anabolic steroids provides a cause-and-effect case study which provides a bridge between chapter 6 and chapter 14, The Endocrine System and Homeostasis. Anabolic steroids are tissue-building messengers. Most performance-enhancing steroids are either synthetic or natural versions of the male sex hormone, testosterone. Androgens from the blood attach themselves to a binding site on muscle cell membrane and cause the muscle cells to grow and develop. The fact that males generally have more muscle tissue than females can be explained by higher levels of androgens. However, women can increase muscle development by injecting the male sex hormone. Females grow bigger and stronger, but as they do, many experience side effects associated with maleness. The development of facial hair under the influence of testosterone is one common effect. A lowering of the voice and retardation of growth by younger females and males has also been documented. Males do not escape without side effects. Males who rely on large doses of testosterone cannot utilize all of the testosterone in the body. Consequently, the excess testosterone is gradually converted into estrogen, the female sex hormone. Although every male contains some estrogen, few contain the same levels as women. Breast enlargement and a shrinking of the testes have been documented. Liver and kidney dysfunctions are also associated with steroid use. Current studies have shown that prolonged use of anabolic steroids decreases HDL-cholesterol (high-density lipoprotein) in the blood. The HDL-cholesterol is often referred to as good cholesterol because it decreases the harmful LDL-cholesterol which is responsible for the accumulation of plaque on the inside of blood vessels. The low levels of HDLs make steroid users prime candidates for heart attack.

POSSIBLE JOURNAL ENTRIES

- Students can be encouraged to develop a concept map which includes all key terms presented in the chapter.
- Students can be encouraged to express their ideas about diets.
- Group thinking and decision-making strategies may be recorded as each group prepares for the debate. Did the students change their minds during the preparation for the debate? Students may even be asked to record their initial positions about the social issue and to reflect upon these feelings after the debate has been completed. Did they change their minds after listening to opposing arguments? Should athletes who take steroids be banned from competition for life?

USING THE VIDEODISC

Section title	Page	Code	Frame (side)	Description
Organization of Matter	153 – 155		2749 (all)	Atom
Organization of Matter	153 – 155		2750 (all)	Atomic number
Organization of Matter	153 – 155		2751 (all)	Atomic weight
Organization of Matter	153 – 155		2525 (all)	Hydrogen atom
Organization of Matter	153 – 155		2527 (all)	Oxygen atom
Chemical Bonding	155 – 156		2526 (all)	Hydrogen molecule
Chemical Compounds	157		2528 (all)	Water molecule
Acids and Bases	157 – 159		2518 (all)	pH of various solutions
Acids and Bases	158		3108 (all)	pH scale
Carbohydrates	160 – 162		2532 (all)	Glucose
Carbohydrates	160 – 162		2533 (all)	Fructose
Carbohydrates	160 – 162		2786 (all)	Various carbohydrates
Carbohydrates	160 – 162		2942 (all)	Glycogen
Carbohydrates	160 – 162		3222 (all)	Starch
Carbohydrates	160 – 162		3232 (all)	Sucrose

Section title	Page	Code	Frame (side)	Description
Lipids	164 – 66		2997 (all)	Lipids
Proteins	168		2721 (all)	Amino acids
Proteins	169 – 170		3154 (all)	Proteins
Proteins	168 – 170		2535 (all)	Amino acid alanine
Proteins	168 – 170		2536 (all)	Amino acid valine
Denaturation and Coagulation	170		2863 (all)	Denaturation
Anabolic Steroids	172		3225 (all)	Steroids

IDEAS FOR INITIATING A DISCUSSION

Figure 6.2: Photograph of the SkyDome

Students might be asked to compare the size of the nucleus of an atom to a bee in SkyDome. Ask students to provide other analogies.

Figures 6.13 and 6.14: Structure of glycogen and cellulose

Students can be asked to compare the chemical structures of glycogen and cellulose. Humans cannot digest cellulose.

REVIEW QUESTIONS

(page 159)

1 Define the following terms: atom, element, molecule, compound, and ion.

■ Atoms are the smallest particles of matter and are composed of smaller subatomic particles, which include neutrons, protons, and electrons.
Elements are pure substances that cannot be broken down into simpler substances. There are 109 different elements.
Molecules are units of matter that can be composed of one or more different atoms.
Compounds consist of molecules. They contain two or more kinds of atoms.
Ions are either atoms or molecules that have either more or less than the number of electrons required to be neutral. Ions have either a negative charge (excess electrons) or a positive charge (lack of electrons).

2 Element X has 6 protons, 8 neutrons, and 6 electrons. Predict the atomic mass of element X.

■ The atomic number is 6 and the atomic mass is 14.

3 How does carbon-14 ($^{7}_{14}C$) differ from carbon-12 ($^{7}_{12}C$)?

■ Carbon 12 has two fewer neutrons than carbon 14.

4 Differentiate between an ionic bond and a covalent bond.

■ Ionic bonds are formed when electrons are transferred between two atoms and covalent bonds are formed when electrons are shared between two or more atoms.

5 What are polar molecules?

■ Polar molecules are molecules that have positive and negative ends.

6 How does hydrogen bonding help explain the different boiling and freezing points of water and carbon dioxide?

■ Hydrogen bonding is responsible for forces holding water molecules close to one another and resisting the separation of those molecules that occurs during boiling.

7 How do acids differ from bases?

■ Acids are substances that release hydrogen ions in solution.
Bases are substances that release hydroxide ions in solution.

8 How is the buffering of an acid different from its neutralization?

■ In neutralization, hydrogen ions are neutralized by hydroxide ions, and the pH is maintained at the neutral point of 7.0. Buffers absorb excess hydrogen or hydroxide ions without a change in pH. A buffered solution can be maintained at a pH other than 7.0.

(page 172)

9 Differentiate between organic and inorganic molecules. Provide an example of each.

■ Organic compounds have carbon while inorganic compounds do not. Water or sulfur dioxide are examples of inorganic molecules. Sugars are examples of organic molecules.

10 Name three classes of carbohydrates.

■ Monosaccharides, disaccharides, and polysaccharides are three classes of carbohydrates.

11 What are the two structural components of triglycerides?

■ Glycerol and fatty acids are two structural components of triglycerides.

12 In what ways do plant and animal fats differ?

■ Plant fat molecules have more double bonds between carbon atoms and are less stable than animal fat molecules.

13 Differentiate between monoglycerides, diglycerides, and triglycerides.

■ Mono-, di-, and triglycerides have one, two, and three fatty acids per molecule, respectively.

14 What is cholesterol, and why has it become so important to health-conscious consumers?

■ Cholesterol is an example of a lipoprotein—a molecule that has characteristics common to fats and proteins. Accumulation of certain kinds of cholesterol and other fats in blood vessels causes atherosclerosis—a situation which restricts the flow of blood.

15 State the function of HDLs and LDLs.

■ Both high-density lipoproteins (HDLs) and low-density lipoproteins (LDLs) are fatty compounds which store large amounts of energy. However, the LDLs can clog arteries and restrict blood flow. HDLs can attract LDLs and move them to the liver where they are broken down.

16 In what ways do proteins and nucleic acids differ from carbohydrates and lipids?

■ Besides carbon, hydrogen, and oxygen, which are the main elements found in carbohydrates and fats, proteins and nucleic acids also contain nitrogen.
Proteins consist of amino acids which are held together by peptide bonds.
There are 20 different amino acids. When you consider that proteins can differ from one another by the length of the amino acid chains, their three-dimensional relation with one another, and their sequence, the number of different proteins that is possible is infinite.
There is similarly a great variety of possibilities in the structuring of nucleic acids.
There are carbohydrates and fats that are common to many living things; the same is not true for proteins and nucleic acids.

17 Why do so many different proteins exist?

■ Proteins are very large molecules which consist of chains of amino acids. There are 20 different amino acids. The length of the chain, its three-dimensional shape, and the sequence of amino acids are variables responsible for producing the many different protein molecules.

18 Using examples of carbohydrates, lipids, and proteins, explain the process of dehydrolysis synthesis.

■ Dehydrolysis synthesis is the process by which larger molecules are formed by the removal of water from two smaller molecules.
With carbohydrates, a water molecule is extracted from the two monosaccharide sugar molecules to form the disaccharide.

With lipids, a water molecule is extracted when each fatty acid combines with a glycerol molecule. When a triglyceride is formed, there would be three molecules of water produced.
With proteins, a water molecule is extracted when each amino acid is added to the chain of amino acids to produce the protein molecule.

19 Define denaturation and coagulation.

■ Denaturation is the temporary altering of a protein's shape. It is caused by a change in temperature, pH, or radiation. A protein that has been denatured may return to its original form if the environmental conditions are returned to normal. Coagulation is a permanent change in protein shape caused by extreme changes in environmental factors. A protein that has been coagulated cannot return to its original form. A cooked egg is a good example of coagulated protein.

20 What are nucleotides?

■ Nucleotides are the functional units of nucleic acids. They are composed of a five-carbon sugar (either ribose in RNA or deoxyribose in DNA), a phosphate group, and a nitrogen base.

LABORATORY
IDENTIFICATION OF CARBOHYDRATES

(page 163)

Time Requirements

The basic laboratory could be completed in approximately one hour. If you should choose to extend the investigation into determining whether different carbohydrates are present in different foods, you would need more time, unless you organize the lab exercise so that different lab groups are working with different substances. If this latter approach is taken, it would be worthwhile to have more than one group testing each substance so that results can be compared and tests can be repeated if necessary.

Notes to the Teacher

It is important to have clean glassware for this laboratory. The confidence in one's results when doing chemical analyses depends in part on being certain that you do not have any contaminating chemicals present in the vessels in which the reactions occur. Students should practice sound laboratory procedures by being sure that they wash and rinse all glassware carefully if they should be using the same apparatus for different tests.

Solutions X, Y, and Z might include different fruit juices, milk, and other liquids that the students may wish to test for the presence of different carbohydrates. You might consider to also use different concentrations of X, Y, and Z (e.g., 20% X or 50% Y). These solutions should be tested in advance so that you know what to expect when students test them under laboratory conditions. The starch test can also be carried out on solids such as pieces of potato, or other vegetables and fruit. So, this laboratory can be extended into an investigation of foods for the different carbohydrates that they might contain.

Prior to the lab date, all first-aid equipment and supplies relevant to the procedures involved in this exercise should be checked. For example, electrical wiring and switches should be checked, and eye wash station(s) should be checked.

Safety Precautions

The laboratory outline offers some safety precautions. These should be emphasized to the students before they start the investigation. This laboratory provides a good opportunity to review laboratory behaviors and procedures that address safety and the avoidance of accidents that might occur in relation to the procedures used. These could include electrical wiring accidents, burns, and splashes of known and unknown liquids onto skin and clothing or into the eyes. It similarly provides the opportunity to review actions to be taken when different accidents occur.

Observation Questions

a) Why should the graduated cylinder be cleaned and rinsed after the measurement of each solution?

■ If the graduated cylinder is contaminated by the presence of different solutions, the results of any test may be affected.

b) Record any color changes in a chart like the one below.

Solution	Initial color	Final color	% sugar
Water	Blue	Blue	0
Glucose	Blue	Red brown	+2.0
Fructose	Blue	Red brown	+2.0
Maltose	Blue	Red brown	+2.0
Sucrose	Blue	Blue	Neg.
Starch	Blue	Blue	Neg.

Note: Sucrose is a dissacharide that consists of glucose and fructose combined in such a way that it cannot reduce Cu^{2+} ions. Even though starch is a polymer of glucose, a reducing sugar, the active portion of each glucose molecule in the polymer is involved in the bonds which hold them together making them unavailable for reacting with Benedict's solution. However, if the starch is boiled for a long enough period, it may be digested to the extent that enough glucose molecules are released to produce a positive Benedict's test.

c) Record the color of the solution.

■ The color is a diluted iodine solution color.

d) Record the color of the solutions. Which solutions indicate a positive test?

■ The starch solution will turn a blue-black color, indicating the presence of starch.
The sugar solutions will take on the iodine solution color, indicating a negative result and the absence of starch.

e) Record your data.

■ Data in response to this question will vary with the unknowns used. You may have your class test a variety of substances with not all students testing all of the substances. If such is the situation, it is quite important that there are at least two groups testing each of the unknowns so that if there is a difference in results, the students can repeat their tests.

Laboratory Application Questions

1 Which test tube served as a control in the test for reducing sugars and starches?

■ The "negative control" for the Benedict's test was the tube that contained water. The "negative control" for the starch test occurred when the iodine solution was mixed with water. The "positive control" for the Benedict's test was the tube that contained glucose. The "positive control" for the starch test occurred when starch solution was mixed with the iodine solution.

2 What laboratory data suggest that not all sugars are reducing sugars?

■ The Benedict's test with sucrose indicates that all sugars are not reducing sugars.

3 A student decides to sabotage the laboratory results of his classmates and places a sugar cube into solution Z. Explain the effect of dissolving a sugar cube in the solution.

■ Table sugar cubes consist of sucrose. Since sucrose is not a reducing sugar, this action will not have any effect on the results of the Benedict's test. It will only sweeten solution Z.

4 A drop of iodine accidently falls on a piece of paper. Predict the color change, if any, and provide an explanation for your prediction.

■ Since most paper has some starch which serves to hold the fibers together, it is likely that the paper will turn black, indicating the presence of starch.

LABORATORY
IDENTIFICATION OF LIPIDS AND PROTEINS

(page 171)

Safety Precautions

Some safety precautions are presented in the laboratory outline. Students should be reminded of appropriate safety precautions.

Time Requirements

This laboratory as outlined should be completed in approximately one hour. However, if you want your students to test other foodstuffs for the presence of either fats or proteins or both, you would necessarily have to extend the time unless you have different lab groups doing different tests. If the latter is the case, at least two groups should test each test substance so that any discrepancies in results could be reviewed and retested if necessary.

Notes to the Teacher

Make sure that all glassware is clean or have your students carefully wash and rinse all glassware before they begin this laboratory.

Solutions X, Y, and Z should be prepared in advance and tested so that you know what to expect when the students do the tests. These solutions might include dilutions of milk or milk products, beef broth, dilutions of yogurt, etc. It is worthwhile to have solutions that will test negative. As with the previous laboratory, you might consider having students become involved in bringing different foods to test for the presence of either fat or protein or both.

Prior to the lab date, all first-aid equipment and supplies relevant to the procedures involved in this exercise should be checked.

Observation Questions

a) Record the color of the mixtures.

■ Sudan IV is a powder which has an affinity for, and is soluble in, lipids. As such, test tube C, which does not contain any lipids, will contain some Sudan IV which remains as bits of undissolved powder. Test tube T, which contains some lipid, will be red in color because the Sudan IV has dissolved in it.

b) Record whether or not the papers appear translucent.

■ The "C" paper will appear normal except for a water stain, while the "T" paper will have an area of translucency where the oil was dropped.

c) Record any color changes.

■ Test tube C will remain blue in color, indicating that there is no protein present (this is the negative control). Test tube T will rapidly go from blue to violet and possibly purple, indicating the presence of protein.
The following chart can be used as a type of quantitation and indicates that the biuret test can demonstrate different concentrations of protein.

Color	Symbol
Blue	-
Pink	+
Violet	++
Purple	+++

d) Record your results.

Solution	Sudan IV	Translucence	Biuret
5% glucose (dextrose)	No change	No change	No change
5% gelatin	No change	No change	Positive
Liquid soap	No change	Positive	No change
Egg albumin	No change	No change	Positive
Liquid detergent	No change	Positive	No change
Simulated blood serum	Reddish	Positive	Positive
Solution X			
Solution Y			
Solution Z			

Note: Results of analysis of solutions X, Y, and Z would vary according to the solutions that are prepared for this investigation. It may be of interest to prepare solutions that contain more than one of the test substances.

Laboratory Application Questions

1 Why were controls used for the experiments?

■ Controls were used so that if an analysis were completed and the results were different from that for the negative control, a clear indication of a positive result would be evident.

2 A student heats a test tube containing a large amount of protein and notices a color change in the test tube. Explain why heating causes a color change.

■ Heating a protein causes the protein to become different. The protein is either denatured or coagulated. Such changes may produce a color change, because the chemical structure of the protein is altered.

3 Explain the advantage of using two separate tests for fats.

■ One test can be a confirmation of the other. Also, because fats could be present in such small amounts, it may be that either one of the tests could be negative. A positive result for either test would provide cause for a re-examination of the sample.

4 Would you expect to find starches and sugars in the blood plasma? Explain your answer.

■ No, starches are digested by humans to individual units of glucose. Glucose is carried by the blood.

APPLYING THE CONCEPTS

(page 174)

1 Explain why lithium is more likely to react with fluorine than with beryllium or sodium. Use the information provided in the chart below.

Element	Symbol	Atomic number	Electron number	Energy level 1	2	3
Helium	He	2	2	2	0	0
Lithium	Li	3	3	2	1	0
Beryllium	Be	4	4	2	2	0
Fluorine	F	9	9	2	7	0
Sodium	Na	11	11	2	8	1

■ Lithium has only one electron in its outer energy level, and its nucleus has a weak attraction for that electron. Since fluorine needs only one more electron to complete its outer energy level, the electron from lithium could do this. So by sharing the electrons in their outermost energy levels, both fluorine and lithium would become more stable.
Beryllium and sodium both have a weak attraction for the electrons in their outer energy levels, like lithium, and like lithium they would be more likely to combine with atoms which have a high attraction for electrons like fluorine.

2 Use proteins to provide examples of catabolism and anabolism.

■ Catabolism refers to reactions in which complex chemical structures are broken down into simpler molecules. An example of catabolism would be: the breaking down through the process of digestion to release the amino acids present in the proteins that have been eaten.
Anabolism refers to chemical reactions in which simple chemical substances are joined to form complex chemical structures. An example of anabolism would be: amino acids resulting from the catabolism of proteins eaten are used to build our own body proteins through dehydrolysis synthesis.

3 Explain why marathon runners consume large quantities of carbohydrates a few days prior to a big race.

■ An excess of carbohydrates in our diet will produce an excess of glucose in the blood. This excess of glucose is transported to the liver where it is converted to glycogen and stored until it is needed. Marathon runners can use this stored energy as they run their race.

4 Indicate some of the symptoms of individuals who are deficient in carbohydrates, proteins, and lipids respectively.

■ Carbohydrates serve as a ready source of energy. When one's diet is lacking in carbohydrates, that person will feel lethargic and lacking in energy.
Proteins in one's diet provide a source of amino acids for rebuilding cells and producing hormones. When proteins are lacking in one's diet, this lack of amino acids may be seen through that individual's slow repair of tissues that have been damaged through cuts and scratches, dull, damaged hair or hormone deficiencies.
A lipid deficiency may be seen as a vitamin deficiency since lipids are responsible for carrying certain vitamins.
Note: Students may add to these examples through identifying different functions that carbohydrates, proteins, and lipids serve and then indicating that a lack of a necessary food in the diet will result in the corresponding function being affected.

5 Why is cellulose, or fiber, considered to be an important part of the diet?

■ Fiber, or cellulose, holds water and therefore helps in the elimination of wastes. It is also believed that fiber, or cellulose, can assist in the removal of cholesterol for some people.

6 Why are cows able to digest plant matter more effectively than humans?

■ Microbes assist cows in the digestion of cellulose. And, cows can regurgitate their food and get a second opportunity at digesting other macromolecules after they receive the assistance of the microbes.

7 Margarine is processed by attaching hydrogen atoms to unsaturated double bonds of plant oils. The oil becomes solid or semi-solid. Have you ever noticed that some margarines are stored in plastic tubs, while others are stored in wax paper? Compare the two types of margarine. Which would you recommend?

■ The margarine that has fewer saturated single carbon to carbon bonds (or more unsaturated double bonds) is less solid and must be stored in plastic tubs. The margarine that has more saturated carbon bonds is more solid and can be stored in wax paper; it is considered to be less healthy since it has been linked with atherosclerosis.

8 Why are phospholipids well suited for cell membranes?

■ Phospholipids are soluble in fats (nonpolar solvents) and water (polar solvents). This means that polar and nonpolar molecules can penetrate cell membranes.

9 Explain why many physicians suggest that the ratio of HDL-cholesterol to LDL-cholesterol is more significant than the cholesterol level.

■ There are two kinds of cholesterols or lipoproteins. Low density lipoproteins or LDLs are considered to be "bad cholesterol." About 70% of cholesterol intake is LDLs. High levels of LDLs have been associated with the clogging of arteries. LDL particles bind to receptor sites on cell membranes and are removed from the blood (principally by the liver). However, as the levels of LDLs increase and exceed the number of receptor sites, excess LDL-cholesterol begins to form deposits on the walls of arteries. The accumulation of cholesterol and other lipids on the artery walls is known as plaque. Unfortunately, plaque restricts blood flow to the essential organs and can lead to heart attack or stroke.The second type of cholesterol, high density lipoproteins or HDLs, is often called "good cholesterol." The HDLs carry cholesterol back to the liver, which begins breaking it down. The HDLs lower blood cholesterol. Most researchers now believe that the balance between LDL and HDL is critical in assessing the risks of cardiovascular disease. A desirable level of HDL is 35 mg/100 mL of blood or higher. If there are sufficient amounts of HDLs in the blood stream, the "bad" or LDLs will not accumulate in the blood stream to cause the problems with which they are associated.

10 List some of the side effects of prolonged anabolic steroid use.

■ Females who have used anabolic steroids may experience the development of facial hair. A lowering of the voice and retardation of growth by younger females and males has also been documented. Excess testosterone in males is gradually converted into estrogen, the female sex hormone. Although all males contain some estrogen, few contain the same level as women. Breast enlargement and a shrinking of the testes have been documented. Liver and kidney dysfunctions are also associated with steroid use. Many individuals who appeared to suffer no ill side effects found that prolonged use caused an accumulation of plaque in arteries. Current studies have shown that prolonged use of anabolic steroids decrease HDL-cholesterol (high density lipoprotein). The low levels of HDLs make steroid users prime candidates for heart attack.

CRITICAL-THINKING QUESTIONS

(page 175)

1 A student believes that the sugar inside a diet chocolate bar is actually sucrose because a test with Benedict's solution yields negative results. How would you go about testing whether or not the sugar present is a nonreducing sugar?

■ Add an enzyme that digests sucrose into its two component parts. The glucose and fructose will provide a positive test for reducing sugars.

2 Three different digestive fluids are placed in test tubes. The fluid placed in test tube #1 was extracted from the mouth. The fluids in test tubes #2 and #3 were extracted from the stomach. Five millilitres of olive oil are placed in each of the test tubes, along with a pH indicator. The initial color of the solutions is red, indicating the presence of a slightly basic solution. The solution in test tube #3 turns clear after 10 min, but all of the other test tubes remain red. State the conclusions that you would draw from the experiment and support each of the conclusions with the data provided. (Hint: Consider which substance is digested. What are the structural components?) See diagram to the right.

- Digestion of olive oil, a fat, would yield fatty acids and glycerol. The fatty acids would change the pH of a mixture to be more acidic than prior to any digestion.
  Digestion did not occur in test tube #1—lipases are not present in saliva.
  With regard to the results for test tubes #2 and #3, students will likely suggest various conclusions and support for the conclusions. The following offers some views that may be suggested.
  Since test tubes #2 and #3 contained the same mixtures, but digestion only occurred in test tube #3, different conclusions become acceptable. First, because of the conflicting results, one should repeat the experiment under more controlled conditions. For example, it may be that some of the glassware used for the investigation was contaminated by some substance which caused the conflicting results. Second, the source of the stomach extracts would have to be reviewed and different questions answered. For example, were the stomach extracts from the same person? If so, were they extracted at the same time? Was the tube taking stomach extract from the same portion of the stomach? Answers to such questions would provide further information from which to draw conclusions.

3 High cholesterol levels and high risk of atherosclerosis are in part related to diet and in part determined by genetics. The LDL-cholesterol receptor sites located on cell membranes are controlled by genetics. Explain how the number of receptor sites may cause a predisposition to atherosclerosis.

- If the number of receptor sites is great, even a small amount of LDLs would be problematic in that they would soon occupy the receptor sites and any excesses would begin to accumulate in, and clog, arteries, resulting in atherosclerosis.

4 Why do some athletes feel compelled to take anabolic steroids?

- The athletes who feel compelled to take anabolic steroids believe that such drugs enhance their performance. The pressures to win may have encouraged them to seek any advantage.

5 See the diagrams to the right. Three experimental designs are presented to determine the best pH for the hydrolysis of starch to smaller glucose units. Specified quantities of starch and 2 mL of an enzyme capable of initiating starch breakdown are added to each of the test tubes. Which experimental design would you choose? Justify your answer.

- The best design would be 3. Since the quantity of starch solution is constant, this design would allow the experimenter to determine whether an acidic or basic pH is most appropriate.

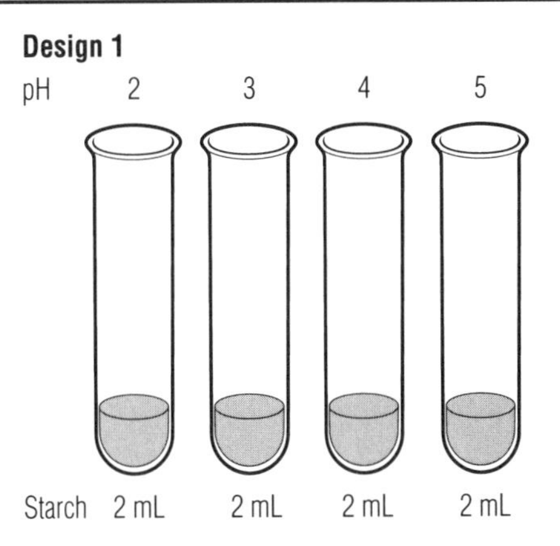

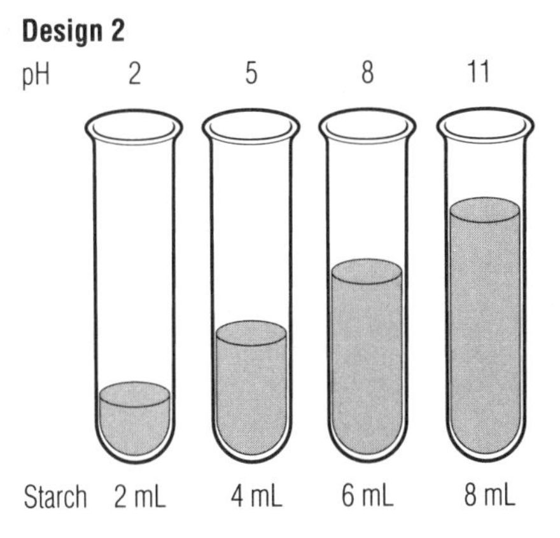

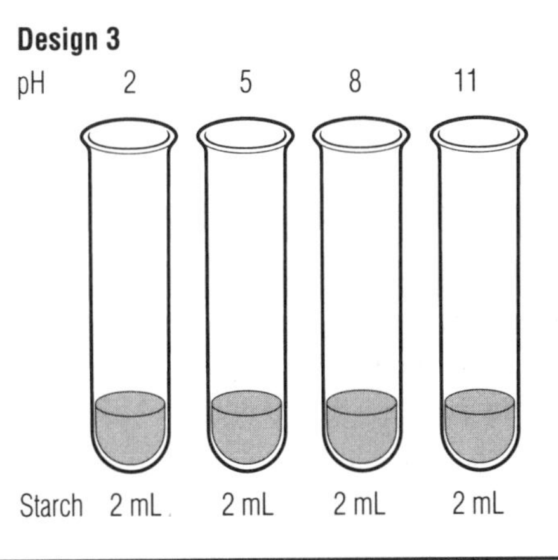

CHAPTER 7

Energy within the Cell

INITIATING THE STUDY OF CHAPTER SEVEN

Students can be asked to describe energy in a journal. Their ideas about energy can be checked once they have completed the chapter. The following questions may help them express some of the alternate conceptions they hold about energy. Do the mitochondria produce energy? If plants are capable of making their own food, does this mean that they can make energy? If animals cannot make their own food, does this mean that animals cannot make their own energy? Can a food chain exist without sunlight energy? (Some caves rely on chemosynthetic organisms which act as producers.)

ADDRESSING ALTERNATE CONCEPTIONS

- Energy can be defined as the ability to do work; however, the definition provided in a biology class differs from that given in a physics class. For living things, the definition of work must be considered in a broad sense. Cell reproduction, the synthesis of cytoplasmic organelles, the repair of damaged cell membranes, movement, and active transport are but a few of the things that must be considered as work.
- Enzymes do not change the direction of a chemical reaction, but provide alternate pathways. The reactants and products remain the same.
- Some students will conclude that plants carry out photosynthesis and animals perform cellular respiration. It is important for students to understand that plants also carry out cellular respiration.
- Many textbooks refer to "light" and "dark" reactions of photosynthesis. Although we continue to use the term "light reaction" to describe hydrolysis of the water molecule, we purposely avoided the label "dark reaction." Carbon fixation occurs during the daylight hours and hours of reduced light. Although the reaction does not require sunlight, it can proceed during the daytime. (Even the label "light-independent"reaction can be confusing, since carbon fixation will not proceed unless water molecules are hydrolyzed.)
- Students and teachers often fall into the trap of speaking about producing energy when describing exergonic reactions, such as cellular respiration. Although energy changes forms, it is never made. "Energy transformation"is a preferred description to either "producing energy"or "releasing energy"(for example, the mitochondria do physically hold and release energy). The same amount of energy was present at the beginning of the universe and will continue to be present until the end of time.

MAKING CONNECTIONS

- The laws of thermodynamics, as presented in chapter 7, are also explained in chapter 2, Energy and Ecosystems. By reinforcing these principles at both a cellular and an ecosystems level, students gain an appreciation for the fundamental laws of science.
- An understanding of enzyme function is essential for investigating how digestive enzymes cause hydrolysis, as explained in chapter 9, Digestion. The function of cofactors and vitamins, as introduced in chapter 7, also provides an important connection between metabolic functions and the products of digestion.

- The subsection Regulation of Enzyme Activity provides an essential foundation for investigating how hormones are regulated, as presented in chapter 14, The Endocrine System and Homeostasis, and for investigating gene regulation and oncogenes, as presented in chapter 23, Protein Synthesis.
- The presentation of photosynthesis provides a detailed description of a biochemical process at the cellular level. With a more detailed comprehension of photosynthesis, students gain a better appreciation for the intricacies of energy storage and transformation reactions in the biosphere. An understanding of the carbon cycle, as introduced in chapter 1, Equilibrium in the Biosphere, is enhanced by investigating the principles of photosynthesis. In turn, this section provides further understanding for contextualizing knowledge about energy in the biosphere, Unit 1, and the exchange of matter and energy in humans, Unit 3. The concept of cellular respiration complements the study of photosynthesis, enabling students to follow energy through a series of transformations.

POSSIBLE JOURNAL ENTRIES

- Students might be asked to write a dialogue between the text and themselves, as they refute the idea that energy is transformed during photosynthesis and cellular respiration. By acting as a "devil's advocate," they can challenge their own learning and push their understanding to a higher level.
- Students can be encouraged to develop a concept map which includes all key terms presented in the chapter.
- Students can be encouraged to express the difficulties they experienced in formulating an understanding of this challenging chapter. What things did they employ to aid them in constructing their knowledge? For example, some students may indicate that they were guided by Figures 7.17, 7.18, and 7.25 before re-reading the text. Other students may have attempted a series of their own drawings or constructed concept maps. The journal entry can provide students with more information on how they learn.
- Decision-making strategies may be recorded as each group prepares for the debate. Did the students change their minds during the preparation for the debate? Students may even be asked to record their initial positions about the social issue and to reflect upon these feelings after the debate has been completed. Did they change their minds after listening to opposing arguments? Why are people so concerned with the destruction of the rain forests?

USING THE VIDEODISC

Section title	Page	Code	Frame (side)	Description
Importance of Cell Energy	176 – 177		23455 (4)	Movie sequence of marine flatworm crawling and swimming
Importance of Cell Energy	176 – 177		27846 (4)	Movie sequence of Komodo dragon lizard feeding and walking
Importance of Cell Energy	176 – 177		25668 (3)	Movie sequence of geotropism in time-lapse photography
Metabolic Reactions	178 – 179		2702 (all)	Activation energy
Enzymes and Metabolic Reactions	179 – 184		2897 (all)	Enzymes
Enzymes and Metabolic Reactions	180		2703 (all)	Active site

Section title	Page	Code	Frame (side)	Description
Enzymes and Metabolic Reactions	184		2718 (all)	Allosteric activity
Enzymes and Metabolic Reactions	182		28381 (1)	Movie sequence presenting a graphic computer model of pepsin active site and the imposition of a repressor molecule
Research in Canada	185		29807 (1)	Movie sequence presenting computer graphic of the rapidly changing shape of a biological model
Energy Storage and Transformation	187		2707 (all)	ATP molecule
Energy Storage and Transformation	188 – 189		2886 (all)	Electron transport
Energy Storage and Transformation	188 – 189		3092 (all)	Oxidation
Photosynthesis and Energy	189 – 195		1855 (all)	Spirogyra showing chloroplasts
Photosynthesis and Energy	189 – 195		2812 (all)	Chlorophyll
Photosynthesis and Energy	189 – 195		2813 (all)	Chloroplast
Photosynthesis and Energy	189 – 195		3114 (all)	Photosynthesis
Photosynthesis and Energy	189 – 195		3117 (all)	Phytochrome
Photosynthesis and Energy	189 – 195		3227 (all)	Stoma
Photosynthesis and Energy	189 – 195		2789 (all)	Carotenoid
Photosynthesis and Energy	189 – 195		2780 (all)	C-4 pathway
Photosynthesis and Energy	189 – 195		2782 (all)	Calvin-Benson cycle
Photosynthesis and Energy	189 – 195		2809 (all)	Chemiosmosis
Cellular Respiration in Plants and Animals	197		3173 (all)	Respiration
Cellular Respiration in Plants and Animals	197 – 198		3112 (all)	Phosphorylation
Cytochrome Enzyme System	200 – 201		2858 (all)	Cytochrome enzymes
Cellular Respiration in Plants and Animals	197		3032 (all)	Mitochondrion

Section title	Page	Code	Frame (side)	Description
Cellular Respiration in Plants and Animals	197 – 202		3173 (all)	Cellular respiration in plants and animals
Anaerobic Respiration	199 – 200		2943 (all)	Glycolysis
Krebs Cycle	201 – 202		2988 (all)	Krebs cycle

IDEAS FOR INITIATING A DISCUSSION

Figure 7.3: Second law of thermodynamics

Students might be asked to speculate about the efficiency of their body. How much waste energy is released in the form of heat? Only about 15% of the energy released by oxidation of foods can be harnessed.

Figures 7.5 and 7.6: Exergonic and endergonic reactions

Ask students to compare exergonic and endergonic reactions and provide examples of each.

Figure 7.10: Graph of reaction rate vs. temperature for an enzyme reaction

Students may be asked to speculate why fevers can be problematic. (The high temperatures alter the shape of the enzyme (proteins denature).) Because the enzyme no longer combines with the substrate, reaction rates drop dramatically.

Figure 7.25: Mechanical analogy for photosynthesis

Students can explain the mechanical analogy for photosynthesis. Proton gradients are difficult to interpret without mechanical models because students have no experimental learning to link with biochemical pathways.

Figure 7.33: Conversion of glucose to CO_2 and H_2O

Students may be asked to explain why glucose is a high-energy compound and carbon dioxide is a low-energy compound. Many students will indicate that energy is released once chemical bonds are broken, but this is not true. Energy is required to break bonds (activation energy). Energy is released when lower-energy compounds are formed.

REVIEW QUESTIONS

(page 179)

1 Provide two examples of energy transformation within your body.

■ After you eat a large meal, excess food energy will be transformed into glycogen or fat by your body cells. Similarly, if you do not eat for a period of time, your body will begin transforming fat stores into energy that you need to maintain your body temperature, move, think, and eat.

2 Differentiate between heterotrophs and autotrophs.

■ Autotrophs are organisms that are capable of turning sunlight and inorganic material into organic energy-storage compounds such as glucose. On the other hand, heterotrophs are unable to make their own food, and therefore must ingest other organisms.

3 In what ways are photosynthesis and respiration complementary processes?

■ Photosynthesis is the production of carbohydrates from sunlight and inorganic materials. Respiration is the breaking down of the carbohydrates to yield energy which is used for cellular metabolism.

4 State the first two laws of thermodynamics.

■ The first law of thermodynamics states that energy can neither be created nor destroyed. The second law of thermodynamics states that no reaction is 100% efficient; all reactions produce heat, which is useless for doing work.

5 Define entropy.

■ Entropy is the measure of non-usable energy in a system. The higher the level of entropy, the more disorganized a system is.

6 Provide an example of a reaction that requires activation energy.

■ Many reactions require activation energy. A common example is the combustion of petroleum, as in a car's engine. A spark is required to begin combustion, after which the reaction continues until the reactants, oxygen and petroleum, are exhausted.

(page 184)

7 Explain the importance of enzymes in metabolic reactions.

■ Because cells cannot exist at very high temperatures, reactions must occur at relatively low temperatures. However, chemical reactions do not proceed fast enough at low temperatures to make life possible. This is why enzymes are crucial. Enzymes increase the rates of chemical reactions to levels at which life is possible, without the need for an increase in temperature.

8 How do enzymes increase the rate of reactions?

■ Enzymes increase the rate of chemical reactions by bringing the reactants together in the proper configuration for the reaction to proceed. This reduces the activation energy necessary for the reaction to proceed.

9 List and explain four factors that affect the rate of chemical reactions.

■ Four factors that affect the rate of a chemical reaction are:

Temperature—as temperature increases, the rate of the chemical reaction increases. In humans, the peak temperature is about 37°C. Above this temperature the enzymes begin to denature and the rate of the reaction is reduced because the active sites are altered.

pH—all reactions have an optimal pH. As the pH varies from the optimum, excess H+ or OH- ions interfere with the enzyme shape, reducing the rate of the reaction.

Substrate concentration—as substrate concentration increases, the rate of the reaction increases until all the enzyme molecules are occupied.

Inhibitor molecules—inhibitor molecules compete with substrate molecules for the active site of enzymes, reducing the rate of the reactions.

10 How do cofactors and coenzymes work?

■ Cofactors and coenzymes are molecules that help enzymes combine with substrate molecules. They alter the active site of the enzyme so the enzyme can bind with the substrate. Cofactors are inorganic and coenzymes are organic molecules.

11 What are competitive inhibitors?

■ Competitive inhibitors are molecules, other than the normal substrate, which have a shape that is capable of occupying the active site of an enzyme. Thus, they take the place of the substrate molecule, tying up the enzyme. If enough enzymes are "choked" with inhibitors, the rate of the reaction will be reduced.

12 What is allosteric activity?

■ Allosteric activity is the regulation of enzyme activity by the binding of a molecule to the regulatory site of an enzyme.

13 How are metabolic pathways regulated by the accumulation of the final products of the reaction?

■ As the final products of a metabolic reaction accumulate, the reaction begins to slow down. This is because the final product binds with the regulatory site of the enzyme, altering the active site. The enzyme is then no longer capable of binding with the substrate, and the reaction rate is reduced. This type of inhibition is known as feedback inhibition.

(page 195)

14 What is ATP?

■ ATP is the short name for the molecule adenosine triphosphate. ATP is composed of a ribose sugar, adenine, and three phosphate molecules. ATP is the usable form of chemical energy for all organisms; its phosphate bonds contain large amounts of stored energy.

15 Explain why oxidation reactions are often associated with the creation of ATP.

■ Oxidation-reduction reactions normally involve the transfer of electrons from one unstable molecule to another unstable molecule. In the process, both molecules tend to have increased stability, and therefore less energy. The energy that is thus released is available for work and is used to form ATP.

16 Differentiate between oxidation and reduction.

■ Oxidation involves the removal or loss of an electron, while reduction involves the gain of an electron.

17 Explain why electron transport systems are important in living things.

- Large amounts of energy can be released by chemical reactions in the body. However, this energy could potentially damage the cell if released in one burst. Electron transport systems provide for the slow transfer of energy from an energy-releasing reaction to an energy-absorbing reaction.

18 Explain the meaning of the word photosynthesis.

- Photosynthesis is derived from two words, "photo," meaning light, and "synthesis," meaning to make. Thus, it means to make something, in this case glucose, using light as the energy source.

19 Summarize the events of the light-dependent reaction.

- In the light-dependent reaction, light energy is converted to chemical energy, which is stored as ATP. This conversion is accomplished when light is used to split water into hydrogen and oxygen. The oxygen is released, and the hydrogen combines with the coenzyme NADP+. In an oxidation-reduction reaction, NADP+ becomes NADPH, releasing energy, which is used to form ATP.

20 In what part of the chloroplast does the light-dependent reaction take place?

- The light-dependent reaction takes place in the thylakoid membranes of the chloroplast.

21 What function does chlorophyll serve in the light-dependent reaction?

- Chlorophyll serves to trap light energy for use in the light-dependent reaction.

22 What is the Calvin-Benson cycle? Where does it occur?

- The Calvin-Benson cycle involves the fixing of CO_2 from the air into six-carbon chains using ATP and NADPH from the light-dependent reaction. The Calvin-Benson cycle occurs in the stroma of the chloroplasts.

(page 202)

23 Contrast ATP production in anaerobic and aerobic respiration.

- Anaerobic respiration occurs in the absence of oxygen, while aerobic respiration requires oxygen to proceed. In anaerobic respiration, glucose is broken down to lactic acid, yielding two ATP molecules per molecule of glucose. In aerobic respiration, glucose is completely oxidized in the presence of oxygen to yield carbon dioxide, water, and 36 ATP molecules.

24 Why is the phosphorylation of glucose necessary?

- Glucose is phosphorylated by ATP in order to make it into a higher-energy molecule. This extra energy provides the activation energy necessary for the oxidation of glucose to begin.

25 How does the cytochrome enzyme system provide the energy for the synthesis of ATP?

- The cytochrome enzyme system is a series of progressively stronger electron acceptors located in the mitochondria. Hydrogen and an electron removed from glucose are passed along this series of enzymes, giving up energy at every step. This energy is used to form ATP.

26 How does the Krebs cycle provide additional ATP?

- The Krebs cycle takes acetyl coA, and further breaks it down into carbon dioxide and water. The extracted hydrogen gives up energy, which is used to make ATP.

27 Under what conditions do plant cells undergo cellular respiration?

- Plant cells undergo cellular respiration when they require energy to drive metabolic processes.

LABORATORY

ENZYMES AND H_2O_2

(page 186)

Safety Precautions

- Cautions are identified in the lab outline. Even though the hydrogen peroxide solution used in this lab is fairly dilute, it should be handled carefully and *any* accidents reported.
- Eye wash facilities and equipment should be checked.
- Insure that a fully equipped first-aid kit is available to be used in case of splashes or small cuts.

Time Requirements

One hour should provide ample time for experimentation, clean-up, and possibly allow some time for discussion of some results.

Notes to the Teacher

The chicken liver used in this experiment should be fresh. It can be purchased at the local supermarket. In advance of the lab day, you might talk with the manager of the meat department at the market and determine when might be the best day for you to get the fresh chicken liver. If kept refrigerated, you could keep it for a few days. An extension of this laboratory would be to use liver of varying ages with different lab groups to determine the decomposition rate of the enzyme in the liver.

As with all experiments involving chemical reactions, you should insure that the glassware used is well cleaned.

Observation Questions

a) Record your results in a table similar to the following.

Test-tube number	Reaction rate	Product of reaction
#1	0	Some oxygen
#2	1	Oxygen
#3	1-2-3	Oxygen

Note: Reaction rate for tube #3 will vary with the age of the liver and temperature conditions.

b) Record your results.

Test-tube number	Reaction rate
#4	1
#5	3

c) Record your results.

Test-tube number	Reaction rate
#3	1
#6	4

Note: The reaction rate in test tube #6 is so very much more intense because the enzyme has been released from many of the liver cells.

Laboratory Application Questions

1 In part I, which test tube served as a control?

■ Test tube #1 served as the control because it provided an indication of how the hydrogen peroxide would behave before any organic matter was added to it.

2 Account for the different reaction rates between the liver and potato.

■ Plant tissue is much less reactive. Its requirement for oxygen is extremely small in comparison with that of warm-blooded birds such as chickens.

3 Account for the different reaction rates in test tubes #4 and #5.

■ The reaction rate will be faster in test tube #5 because of the second piece of liver and the addition of fresh hydrogen peroxide.

4 Why did the crushed liver in test tube #6 react differently from the uncrushed liver in test tube #3?

■ When the tissue is crushed, the enzyme is released from the cells and is able to more rapidly come into contact with its substrate, the hydrogen peroxide.

5 Predict what would happen if the liver in test tube #3 were boiled before adding the H_2O_2. Give the reasons for your prediction.

■ No reaction beyond that which occurs in the control would occur, because boiling irreversibly changes the character of proteins (enzymes are proteins), and boiling would irreversibly alter the enzyme's active site so that it would not react with the hydrogen peroxide.

LABORATORY

PHOTOSYNTHESIS

(page 196)

Safety Precautions

The lab outline provides some cautionary notes for the students regarding the handling of chemicals.

Prior to this lab, students might be instructed as to the safe handling of small fires involving flammable chemicals such as ethanol.

Insure that a fully equipped first-aid kit is available to be used in case of splashes or small burns.

Time Requirements

With some advanced preparations, this laboratory can be readily completed in one hour.

To save some lab time, you might consider heating some water immediately prior to the lab so that students can start with water that is already heated to some degree. The students might be encouraged to study the procedure in advance to determine the most efficient way of proceeding. They should be able to see that they can start heating their water for part II prior to doing part I.

Notes to the Teacher

Place one or more coleus plants in direct light for at least 72 h and one or more coleus plants in darkness for the same period of time prior to this experiment. The plants in the light will be able to produce and accumulate starch in their leaves. Those that were left in the dark should consume most, if not all, of the starch in their leaves.

If you have a number of plants you may also wish to extend the experiment into an investigation into the effect of the amount of light exposure on the accumulation of starch in the leaves by using plants which have been exposed to direct light for different periods of time and/or different intensities of light.

Observation Questions

a) Record the color of each test tube.

■ The test tube containing water and iodine will be the color of diluted iodine solution. The test tube containing starch solution and iodine will be a blue-black color indicating a positive test for the presence of starch.

b) Sketch the leaf, and label areas of red and green pigment.

■ Sketches will vary.

c) Describe the appearance of the leaf.

■ At this point, practically all of the pigment is removed.

d) Describe the appearance of the leaf.

■ Some portions of the leaf will be a dark blue-black color, indicating the presence of starch.

e) Sketch the leaf, and label the areas that contain starch.

■ Sketches will vary.

f) Describe the appearance of the leaf.

■ Because the leaf is from a plant which was in darkness for 72 h and not carrying on any photosynthesis, most, if not all, of the stored starch in the leaf was consumed for required energy. So, it is quite possible that there will be no areas of the leaf indicating the presence of starch. However, if any starch remains, those areas with starch will appear blue-black in color.

g) Sketch the leaf, labelling the areas that contain starch.

■ Sketches will vary.

Laboratory Application Questions

1 Why was iodine used in the experiment?

■ The iodine solution combines with starch in such a way as to change from its normal amber color to a blue-black color. Such a color change serves to qualitatively indicate the presence of starch.

2 From the data that you have collected, support the fact that light is required for photosynthesis.

■ When the leaf from the plant that was kept in the dark for 72 h was tested for starch, it had a very small amount of starch in comparison with the leaf from the plant that was kept in the light.

3 Explain why coleus was used rather than a plant that contains only green pigments.

■ Coleus plants produce carbohydrates in the areas that are green.

4 Predict how the results would have been affected if the plant exposed to light had been covered by a transparent green bag. State the reasons for your prediction.

■ The amount of starch produced would be greatly reduced because the green bag would absorb the same wavelengths of light that chlorophyll would absorb in the process of photosynthesis. This reduced light energy for the chlorophyll would greatly reduce sugar, and in consequence, starch production.

5 Design an experiment to determine which visible colors provide the optimal energy for photosynthesis.

■ Light from any source can be used. Coleus plants would be used and have different-colored filters to restrict them from receiving specific colors of light. The plants would receive filtered light for at least 72 h and leaves from the different plants would be tested for the presence of starch, as was done in this experiment.

6 Why was a Bunsen burner not used as a heat source for the ethanol?

■ If a Bunsen burner were used, it is quite possible that during the heating of the alcohol, the alcohol vapor could ignite and result in a fire in the beaker.

APPLYING THE CONCEPTS

(pages 205–206)

1 Explain how enzymes work in the lock-and-key model. How has the "induced-fit" model changed the way in which biochemists describe enzyme activities?

■ According to the lock-and-key model, each enzyme has a specially shaped active site that provides a place for specific substrate molecules. There was a temporary joining of the enzyme with the substrate molecules forming the enzyme-substrate complex which consequently resulted in the substrate molecule being broken down. The "induced-fit" model, which replaced the lock-and-key model, suggests that the actual shape of the active site is altered slightly when the substrate molecules are trapped, making the fit between enzyme and substrate even tighter during the formation of the enzyme-substrate complex.

2 Using the information that you have gained about enzymes, explain why high fevers can be dangerous.

■ Since the action of enzymes is affected by temperature, it is likely that the optimum temperature for the enzymes that function in humans would be body temperature. So, when the body temperature increases above normal, it is likely that our enzymes would not function normally with the result that normal enzyme-dependent reactions would be adversely affected.

3 Cyanide attaches to the active site of a cytochrome enzyme. How does cyanide cause death?

■ When cyanide attaches itself to the cytochrome enzyme in the mitochondria, it prevents the breakdown of sugars for energy. Without the needed energy to carry out active transport, the synthesis of proteins, the transport of nutrients, and the elimination of wastes, cells die almost instantly.

4 Use the metabolic pathway shown below to explain feedback inhibition.

■ As the final product is produced, it will begin to inhibit the action of enzyme A.

5 What is the source of the oxygen released during photosynthesis? What is the source of carbon dioxide fixed during photosynthesis?

■ Oxygen is produced as a byproduct from the photolysis of water. The carbon dioxide is produced from the cellular respiration of high-energy nutrients, such as sugars, which occurs in plants and animals.

6 How does light intensity affect the rate of photosynthesis?

■ As the light intensity increases, so does the rate of photolysis (the light-dependent reaction). With an increased amount of hydrogen available for carbon-fixation, photosynthesis can proceed at a faster rate.

7 A photosynthesizing plant is exposed to radioactively labelled carbon dioxide. In which compound would the radioactive carbon first appear? (Consider NADPH, PGA, RuBP, and glucose.) Explain your answer.

■ The carbon would appear first in glucose. The RuBP takes up the atmospheric carbon dioxide to become a six-carbon sugar.

8 The removal of carbon dioxide from pyruvic acid distinguishes fermentation from lactic acid anaerobic respiration. What would occur if an enzyme in your body removed the carbon dioxide from pyruvic acid before lactic acid is formed?

■ The pyruvic acid would produce ethanol (ethyl alcohol) instead—a highly intoxicating alcohol found in all alcoholic beverages.

9 Copy the chart below and place the correct answer in the appropriate column.

Characteristic	Anaerobic	Aerobic
1 Amount of ATP produced	2 ATP	36 ATP
2 Terminal electron acceptor	Pyruvic acid	Oxygen
3 Site of activity in cell	Cytoplasm	Mitochondria
4 Final products	Lactic acid	Water

10 Compare respiration and photosynthesis by completing a chart like the one below. Place a ✔ in the correct column.

Characteristic	Photosynthesis	Respiration
1 ATP used to initiate reaction		✔
2 ATP produced during reaction		✔
3 Oxidation reaction		✔
4 Reduction reaction	✔	
5 Oxygen is a product	✔	
6 Occurs in plant cells	✔	✔
7 Occurs in animal cells		✔
8 Carbohydrate is reactant		✔

CRITICAL-THINKING QUESTIONS

(pages 206–207)

1 In an attempt to study photolysis, a student uses a heavy isotope of oxygen in the form of water ($H_2{}^{18}O$). If the following apparatus were used, indicate where you would expect to find the heavy isotope of oxygen. Explain your answer.

■ The $^{18}O_2$ would first appear in the test tube above the water. As the aquatic plant carries on photolysis of the water, oxygen produced from that water would be collected above the water by displacing it.

2 The following graph shows the rate at which products are formed for an enzyme-catalyzed reaction. By completing the graph, predict how a competitive inhibitor added at time X would affect the reaction.

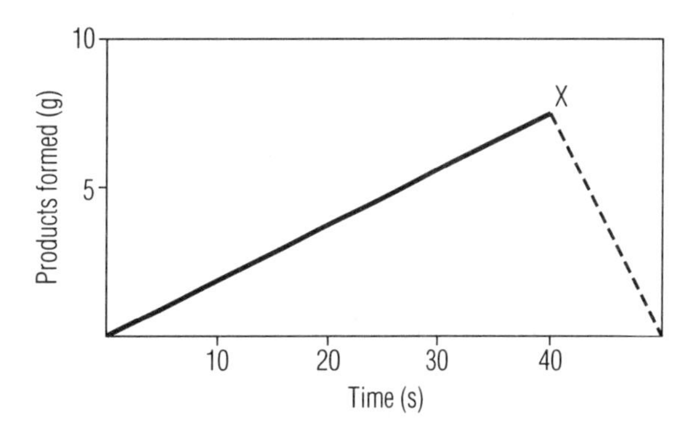

■ The line that would represent the production of products following the addition of a competitive inhibitor would begin to immediately drop off from point X and approach the horizontal axis.

3 Antibiotic drugs act as competitive inhibitors for metabolic pathways in invading microbes but do not interfere with the metabolic pathways in human cells. Using this information, explain how cancer-preventing drugs are designed to work. Why is it difficult to develop a cancer-preventing drug?

■ Cancer-preventing drugs must focus on cells that are reproducing or growing at extraordinarily rapid rates. Cancer is essentially a situation where a normal process like cell division has gone wild (out of control). The difficulty in producing effective cancer-preventing drugs lies in the fact that the drug must be able to prevent the cancer cell's metabolism but not interfere with normal cells.

4 The earth's early atmosphere was rich in carbon dioxide and low in oxygen. With the evolution of plankton, the composition of the atmosphere changed. Through photosynthesis, carbon dioxide in the atmosphere was fixed by the plankton, and oxygen was released. As plants became more abundant, carbon dioxide levels began to fall and oxygen levels began to rise. A number of scientists have suggested that increasing algae growth in the oceans may reverse increasing carbon dioxide levels and reduce atmospheric warming. One suggestion is to seed the oceans with iron and phosphates—nutrients that will increase plankton blooms. Comment on this strategy.

■ Certainly, adding nutrients to the ocean may produce algal blooms providing sufficient nutrient is added for the volume of the water. However, it is very difficult to predict all that might result from algal blooms. Answers to a variety of questions would need to be determined before such a step could be taken. For example: how might other organisms be affected? Would algal bloom increase the amount of dying plants as well? If the amount of dying material increased, would there be an increase in the amount of bacteria? Might such a step eventually lead to a decrease in the amount of oxygen produced by the marine plankton?

5 The following experiment was designed to demonstrate the relationship between plants and animals. Bromothymol blue indicator was placed in each of the test tubes. High levels of carbon dioxide will combine with water to form carbonic acid. Acids will cause the bromothymol blue indicator to turn yellow. The initial color of the test tubes is blue.

Predict the color change, if any, in each of the test tubes. Explain your predictions.

■ Contents of test tube A will remain blue, since no reactions occur here. Contents of test tube B will eventually turn yellow, since the snail is using oxygen and producing carbon dioxide. Carbon dioxide will combine with water to form carbonic acid. Contents of test tube C may turn yellow overnight, a time during which the plant is producing carbon dioxide, but will again turn blue during the day when the plant uses carbon dioxide and produces oxygen at a much faster rate than it produces carbon dioxide. Contents of test tube D will likely turn yellow during the night as both organisms carry on respiration, but during the day the plant will use carbon dioxide produced by itself and the snail at a fast enough rate to turn the color of the contents to blue. The plant will produce sufficient oxygen for itself and the snail to continue survival provided that the light is of sufficient intensity.

UNIT 3

Exchange of Matter and Energy in Humans

SUGGESTIONS FOR INTRODUCING UNIT 3

Chapters 8 to 13 examine how matter and energy are exchanged between humans and their environment during digestion, respiration, and excretion. The transport of nutrients is investigated in chapters 10 and 11. By examining the response of the human organism to the environment, Unit 3 provides a synthesis of the biological learning acquired in Unit 1, which investigates matter and energy at the level of the biosphere and ecosystems, and Unit 2, which investigates matter and energy at a cellular level.

Students might be encouraged to compare metabolic demands of unicellular and multicellular organisms. The following questions may assist students: How does the amoeba use matter and energy? What specializations have different cells in our bodies made to meet the same demands? The comparison can then be extended to larger systems such as ecosystems. The following chart may prove useful. Sample answers are provided.

Descriptor	Cellular level	Organ systems level	Ecosystems level
Energy input	Feeding and digestion (phagocytosis, pinocytosis, diffusion of materials through cell membranes, osmosis)	Feeding and digestion; circulatory system transports required nutrients	Solar energy
Energy transformations	Mitochondria, chloroplasts	Muscles	Food chains
Energy output	Synthesis of new proteins (ribosomes)	Movement, body heat	First and second order consumers

Note: The answers provided are only one way of expressing possible relationships and are not considered to be either exclusively correct or all-inclusive. It is more important for students to develop their own language and organizing system than to replicate any of the answers provided.

KEY SCIENCE CONCEPTS

Unit 3, Exchange of Matter and Energy in Humans, provides students with an opportunity to study the manner in which matter and energy are exchanged between organisms and the environment. Key concepts of **equilibrium** are introduced within the realm of metabolism. By using the human organism as a model, digestion, respiration, transportation, excretion, and defense are introduced through the study of body **systems.** The role of technology in maintaining equilibrium and related social issues are introduced in this unit. This unit provides a context in which **control systems** in Unit 4, Biology 20 of the Alberta curriculum can be introduced. ■

CHAPTER 8

Organs and Organ Systems

INITIATING THE STUDY OF CHAPTER EIGHT

Chapter 8 provides an overview of physiological systems by examining levels of organization (tissue, organ, and organ systems). The fetal pig dissection is designed to provide an overview of organs and organ systems. This should allow students to gain a better appreciation for the manner in which the form of organ systems ensures coordinated adjustments. Form is inextricably linked with function.

It should be noted that some students may object to using fetal pigs for dissection. Alternatives for the dissection are provided by a variety of videos which show dissection. A number of suggested video sequences from the laser disk are provided below.

The following suggestions provide one example of a way of beginning the chapter:

1. Ask students to read about the early work done on anatomy. Why did many of the scientists work in caves? What does this tell you about the way in which science was once regarded?
2. Collect several dissection guides of different mammals. Ask students to identify similarities and differences.

ADDRESSING ALTERNATE CONCEPTIONS

- Blood is usually not described by students as a tissue. Students can be asked to explain whether or not blood can be classified as a tissue.
- Grouping anatomical structures as organ systems, organs, and tissues can be confusing. Students may be asked to place various structures in the correct category. For example, is the brain an organ system, organ, or tissue? Is it correct to say brain tissue? Table 8.1 provides many excellent examples.
- Some students believe that X rays and other technological tools make diagnosis infallible. Students can be encouraged to speculate about the limitations of X-ray imaging or CAT scans. What types of things can't X rays do?

MAKING CONNECTIONS

- By examining the exchange of matter and energy in organ systems and organs in chapter 8, an association between Unit 1, which investigates matter and energy at the level of the biosphere and ecosystems, and Unit 2, which investigates matter and energy at a cellular level, is established.
- The information presented in chapter 8 serves as an overview to a more in-depth study of physiology, introduced in the succeeding chapters of Unit 3 and continued in Unit 4.

POSSIBLE JOURNAL ENTRIES

- Students might be asked to express their concerns before beginning the dissection and as the dissection proceeds.
- Students can be encouraged to develop a concept map which includes all key terms presented in the chapter.
- Group thinking and decision-making strategies may be recorded as each group prepares for the debate. Did the students change their minds during the preparation for the debate? Students may even be asked to record their initial positions about the social issue and to reflect upon these feelings after the debate has been completed. Did they change their minds after listening to opposing arguments? Does the use of artificial organs present ethical issues?

USING THE VIDEODISC

Section title	Page	Code	Frame (side)	Description
Importance of Organs and Organ Systems	210 – 211		3084 (all)	Organs
Homeostasis and Control Systems	223		2957 (all)	Homeostasis
Frontiers of Technology: Nuclear Medicine	214 – 215		20637 (1)	Movie showing 3-D X ray produced at the Mayo Clinic
Laboratory: Fetal Pig	217		2198 (all)	Fetal pig—external
Laboratory: Fetal Pig	217		2199 (all)	General view of fetal pig
Laboratory: Fetal Pig	218		2200 (all)	Female pig showing genital papilla
Laboratory: Fetal Pig	218		2201 (all)	Male pig showing scrotum
Laboratory: Fetal Pig	218		2202 (all)	Ventral view showing umbilical cord, mammary papilla
Laboratory: Fetal Pig	218		2203 (all)	Soft and hard palate
Laboratory: Fetal Pig	218		2204 (all)	Cross-section of mouth area
Laboratory: Fetal Pig	218		2205 (all)	Oral cavities and pharynx
Laboratory: Fetal Pig	218		2206 (all)	Glottis and epiglottis
Laboratory: Fetal Pig	218 – 219		2209 (all)	Liver and diaphragm
Laboratory: Fetal Pig	218 – 219		2210 (all)	Gallbladder
Laboratory: Fetal Pig	218 – 219		2211 (all)	Stomach and spleen labelled
Laboratory: Fetal Pig	218 – 219		2212 (all)	Pancreas
Laboratory: Fetal Pig	218 – 219		2213 (all)	Stomach showing pyloric sphincter
Laboratory: Fetal Pig	218 – 219		2214 (all)	Large intestine
Laboratory: Fetal Pig	221 – 222		2215 (all)	Kidney and ureters

Section title	Page	Code	Frame (side)	Description
Laboratory: Fetal Pig	221 – 222		2216 (all)	Ureters and bladder
Laboratory: Fetal Pig	221 – 222		2217 (all)	Kidney with peritoneum removed, showing ureter
Laboratory: Fetal Pig	221 – 222		2218 (all)	Female urinary system and reproductive system
Laboratory: Fetal Pig	221 – 222		2222 (all)	Male urogenital system
Laboratory: Fetal Pig	221 – 222		2223 (all)	Male urogenital system, closer look
Laboratory: Fetal Pig	221 – 222		2224 (all)	Scrotum and inguinal canal
Laboratory: Fetal Pig	221 – 222		2225 (all)	Testes, scrotum removed
Laboratory: Fetal Pig	221 – 222		2226 (all)	Vas deferens
Laboratory: Fetal Pig	221 – 222		2227 (all)	Testes
Laboratory: Fetal Pig	221 – 222		2228 (all)	Penis and testes

IDEAS FOR INITIATING A DISCUSSION

Table 8.1: Levels of Cell Organization

Ask students to locate the areas of the body which contain the various body systems and organs.

Figure 8.3: Human organ systems

Brainstorm to get as much information about the body systems as possible. Students likely know much more than they think they know. Ask students to compare the layouts of the various systems. They should note parallels among the circulatory, nervous, and skeletal systems.

Figures 8.4, 8.5, 8.6, and 8.7: X ray, CAT scan, NMR image, or NMI

Ask students if they have ever seen one of their own X rays, or had NMR or NMI. You may prefer to have them do a journal entry about their experiences if you feel anyone may not want to share his or her experiences with the class.

REVIEW QUESTIONS

(pages 216)

1 Differentiate between a tissue, an organ, and an organ system. Provide examples of each in your explanation.

■ Tissues are groups of similarly shaped cells which carry out the same function (e.g., muscle tissue). Organs are groups of tissues working together for a common function (e.g., the heart). Organ systems are composed of many organs which carry out a common function (e.g., the circulatory system).

2 List the four main groupings of tissue and provide an example of each.

■ The four main groupings of tissue are: epithelial (covering), connective, muscle, and nervous. The cells of the outer skin are epithelial tissue, for example. Tendons, cartilage, ligaments, bones, and blood are all examples of connective tissue. The heart, and all other muscles of the body are muscle tissue. The brain, spinal cord, and the sense organs are all composed of nervous tissue.

3 Why is blood considered to be a tissue?

■ Blood cells have a similar shape and carry out a similar function.

4 Why do different classification schemes for body systems exist?

■ Classification systems are a socially constructed way of organizing knowledge. Different scientists tend to classify things differently.

5 Discuss the advantages and disadvantages associated with X rays, CAT scans, nuclear imaging, and NMR.

■ X rays enable physicians to identify solid structures within the body. The morphology of tissues with different densities can be determined. CAT scans provide a 3-D view of organs or tissues by collecting a series of different X rays. Nuclear imaging employs radioisotopes, which allow organs to be studied as they function. Nuclear magnetic resonance is a technology which employs a magnetic field and radio waves to determine the behavior of molecules in soft tissue. NMR has the advantage of not using radioactive matter in the body.

6 Draw a diagram of the human body and label the following cavities: cranial, thoracic, and abdominal.

■ See Figure 8.8 for reference.

LABORATORY
FETAL PIG DISSECTION

(pages 217–219)

Time Requirements

Approximately three classes are required to complete the dissection.

Notes to the Teacher

- The fetal pig that you will be examining is somewhere between 112 and 115 days old.
- To ensure that your dissection provides your students with maximum benefit, make sure that you read the procedure carefully. The accompanying diagrams are designed to provide reference, but are not intended to direct the dissection.
- Students should be cautioned to follow proper safety procedures and should use the appropriate dissecting instruments. This and the following labs have been designed to minimize the use of a scalpel.

Observation Questions

a) Estimate and record the age of your fetal pig.

■ Usually 112 to 115 days will be the age of the pig.

b) What is the function of the umbilical cord?

■ It serves as a connection between the embryo and the placenta. Nutrients and wastes are exchanged between mother and fetus in the placenta.

c) What is the sex of your pig?

■ Answers will vary.

d) Indicate the position and the number of toes.

■ Students will likely see four toes. A vestigial toe is barely visible.

e) How many lobes does the liver have?

■ The liver has three lobes.

f) Describe the location of the gallbladder.

■ It is underneath the right lobe of the liver.

g) Describe the appearance of the pancreas.

■ It is white or a creamy color. The pancreas is narrow and held in place by the mesentery. Encourage descriptive language rather than providing a specific answer.

h) Describe the appearance of the inner lining of the stomach.

■ It is glistening; some students may see folds. Encourage descriptive language rather than providing a specific answer.

Laboratory Application Questions

1 State the function of the following organs.

Organ	Function
Stomach	Initial protein digestion
Liver	Multiple functions (glycogen storage)
Small intestine	Major area of digestion and nutrient absorption
Gallbladder	Stores bile
Pancreas	Digestive enzymes and insulin
Large intestine	Water reabsorption and waste storage
Spleen	House white blood cells

LABORATORY
FETAL PIG DISSECTION

(pages 220–222)

Observation Questions

a) What organs are found in the thoracic cavity?

■ The heart and lungs are found in the thoracic cavity.

b) Compare the size of the wall of a ventricle and an atrium.

■ The ventricle has a much thicker wall than the atrium.

c) Why does the left ventricle contain more muscle than the right ventricle?

■ It pumps blood much further.

d) Why do the lungs feel spongy?

■ It is the area of gas exchange; the tissue contains many air sacs.

e) Describe what happens.

■ The trachea returns to its original shape.

f) What function do the cartilaginous rings of the trachea serve?

■ They keep the trachea open.

g) Describe the shape and color of the kidneys.

The kidneys are kidney-bean-shaped; they appear darker in color, usually a deep red.

Laboratory Application Questions

1 State the function of the following organs.

Organ	Function
Heart	Pumps blood
Kidney	Filters impurities from the blood
Ureter	Conducts urine to the bladder
Urethra	Conducts urine from the bladder
Testes	Male reproductive organs
Ovaries	Female reproductive organs
Uterus	Site of the developing embryo or fetus

2 How is the abdominal cavity separated from the thoracic cavity?

■ The diaphragm is a sheet of muscle which separates the thoracic body cavity from the abdominal cavity.

3 Indicate why the male reproductive system is often referred to as the *urogenital* system while the female's is not.

■ In males, the sperm and urine are both conducted from the body by way of the urethra. The urethra of females conducts only urine.

APPLYING THE CONCEPTS

(page 227)

1 Describe the advantages associated with cell specialization.

■ Because cells specialize, they become more efficient at performing a limited number of tasks. Division of labor enables different cell types to work together to keep the entire organisms running efficiently.

2 Using epidermal tissue as an example, explain the relationship between cell shape and tissue function.

■ Squamous-shaped epithelial cells which cover tissue, are long and thin, providing maximum coverage. Columnar cells, cells which provide structural support, are pillar-shaped, providing maximum support.

3 List seven organ systems and provide at least two examples of organs that belong to those organ systems.

■ See Table 8.1 for reference.

4 Explain the advantages of the CAT scan over conventional X rays.

- CAT scans allow physicians to examine organs layer by layer, from a number of different angles. The computer assembles a number of individual X rays.

5 Explain the concept of homeostasis by describing how your body adjusts to cold environmental temperatures.

- Expect general answers; specific details will be provided in later chapters. If external temperatures are low and the body begins to cool, shivering will occur, thereby raising body temperatures. If your body begins to warm, perspiration occurs. The evaporation of perspiration causes cooling.

CRITICAL-THINKING QUESTIONS

(page 227)

1 The cost of nuclear medicine, CAT scans, and artificial body parts is extremely high and places a heavy financial burden on the health care system. Can we continue to support such expensive research projects?

- Expect a wide variety of answers. Although no one answer is correct, students should be encouraged to consider multiple points of view and reflect an answer which considers the question from alternate perspectives. Assessment can be provided if you consider the coherence of expression and the tolerance of alternate perspectives.

2 Liposuction is a fat-reducing technique in which fat cells are mechanically sucked out of the lower layer of the skin. Fees for this kind of surgery can range from between $750 to over $4000, depending upon the type of procedure. Discuss the moral and economic aspects of medical procedures that are designed to improve appearance.

- Expect a wide variety of answers. Although no one answer is correct, students should be encouraged to consider multiple points of view and reflect an answer which considers the question from alternate perspectives. Assessment can be provided if you consider the coherence of expression and the tolerance of alternate perspectives.

CHAPTER 9

Digestion

INITIATING THE STUDY OF CHAPTER NINE

1 Ask students to make a journal entry describing the function of the stomach, small intestine, and large intestine. They can check back on their initial ideas once they have completed the chapter.

2 Using the *Canada Food Guide*, students may be asked to record their daily energy intake. Students can make a daily record of all the foods and beverages they consumed over three days. They should record the type of food and quantity consumed, immediately, in a notebook. Caution them against relying on their memories.

Sample Record

Toast	1 slice
Coffee	1 cup
Milk	1 glass
Eggs	2 boiled

Instructions to Students

Prepare a table of your results. Calculate the energy value of the quantity of the food you consumed by referring to the food energy value table. If some of the foods you eat are not found on the table make approximations for their energy value by comparing them to similar foods. Indicate the approximation by placing an * above the food.

Food type	Quantity	Energy (kJ)	Daily energy

The table below outlines some common foods, and their energy value. For your convenience the table is divided into different groups. This table will enable you to calculate the amount of food energy that you take in each day. For a more complete guide see *Nutrient Value of Some Common Foods*, a supplement to the *Canada Food Guide*, from the Health and Welfare Department, Government of Canada. Each of the units is calculated on a normal serving.

Food	Measure	Weight (g)	Food energy (kJ)
Milk products			
Cheddar cheese	15 mL	7	710
Cottage cheese	250 mL	237	1020
Cream	15 mL	16	200
Powdered creamer	5 mL	2	40
Whipping cream	250 mL	252	3640
Milk (whole)	250 mL	257	660
2% milk	250 mL	258	540
Skim milk	250 mL	258	380
Buttermilk	250 mL	258	430
Evaporated milk	250 mL	356	1490
Ice cream	125 mL	70	590
Ice cream (soft)	125 mL	95	600
Yogurt with fruit	125 mL	74	530
Eggs			
Eggs (cooked in shell)	1 egg	50	330
Eggs scrambled in butter	1 egg	64	400
Fried egg	1 egg	46	350
Egg substitute	125 mL	126	790
Meats and meat products			
Wieners	1 wiener	50	520
Ground beef	1 pattie	90	1080
Roast beef (in oven)	2 pieces	90	1570
Steak (broiled)	1 piece	90	1330
Cod (broiled)	1.5 fillets	90	640
Cod (pan fried with butter)	1.5 pieces	90	960
Bacon	1 slice	15	380
Ham	2 pieces	90	1410
Pork chop	1 chop	98	1090
Spareribs		90	1660
Chicken breast (fried)	1 piece	76	650
Chicken (roasted)	1 piece	90	510
Turkey (roasted)	1 piece	90	720
Vegetables and related products			
Red kidney beans	250 mL	267	1190
Peas	250 mL	263	1260
Chick peas	250 mL	240	950
Asparagus	250 mL	153	130
Yellow beans	250 mL	132	130
Bean sprouts	250 mL	91	160
Beets (sliced)	250 mL	195	240
Broccoli	250 mL	164	180
Carrots (diced)	250 mL	153	200
Cauliflower	250 mL	127	110
Celery	250 mL	133	80
Corn	250 mL	175	620
Corn (creamed)	250 mL	243	760
Lettuce	250 mL	78	40
Mushrooms (canned)	250 mL	257	180
Mushrooms (fresh)	250 mL	257	120
Onions (raw)	250 mL	222	260
Onions (fried)	250 mL	184	1560
Parsnips	250 mL	169	440
Peas (cooked)	250 mL	169	510
Potatoes (medium baked in skin)	1	100	380
Potato (boiled)	1	136	440
French fries	10 pieces	57	650
Mashed potatoes (milk added)	250 mL	206	550
Mashed potatoes (butter added)	250 mL	206	820
Spinach (raw)	250 mL	32	40
Spinach (cooked)	250 mL	190	180
Tomatoes (raw)	1	150	35

Fruits and related products			
Apples (raw)	1	150	290
Applesauce (canned)	250 mL	269	1020
Bananas	1	175	420
Cantaloupe	1/2 melon	385	250
Cherries	250 mL	137	400
Cranberries	250 mL	100	190
Grapefruit (white)	1/2	241	190
Grapefruit (pink)	1/2	241	210
Grapefruit juice (canned concentrate)	1 can	207	1260
Grapes	250 mL	169	420
Grape juice (frozen concentrate)	1 can	216	1650
Lemon juice (fresh)	250 mL	257	260
Lemonade (frozen concentrate)	1 can	219	1800
Oranges (raw)	1	180	270
Orange juice (fresh)	250 mL	262	490
Orange juice (frozen concentrate)	1 can	213	1510
Peaches (whole)	1	114	150
Peaches (sliced)	250 mL	177	290
Peaches (canned and syrup added)	250 mL	271	880
Pineapples (raw)	250 mL	269	330
Pineapples (canned with heavy syrup)	250 mL	274	860
Raisins (seedless)	250 mL	174	2120
Raisins (25-mL pack)	1 package	14	170
Watermelon	1 slice	925	480

Breads and wheat products			
Bread (white enriched)	1 slice	30	340
Bread (60% whole wheat)	1 slice	30	300
Raisin bread	1 slice	25	270
Bread (French or Vienna)	1 slice	34	410
Cinnamon bun	1	50	660
Commercial hard roll	1	40	520
Hot-dog bun	1	50	570
Hamburger bun	1	60	690

Desserts			
Brownies	1	20	400
Chocolate chip	1 cookie	10	210
Chocolate marshmallow	1 cookie	19	310
Pancakes	1	27	250
Apple pies	1 sector	160	1720
Cherry pies	1 sector	160	1620
Lemon meringue pie	1 sector	140	1490
Pumpkin pies	1 sector	150	1330
Popcorn (with salt and oil)	250 mL	9	170
Potato chips	10 chips	20	480
Chocolate eclair (with custard filling)	1 piece	110	1320

Combination plates			
Cabbage rolls with meat	2 rolls	206	1090
Chili con carne	250 mL	264	1470
Chop suey with meat or poultry	250 mL	163	790
Chow mein chicken	250 mL	184	780
Egg rolls (pork)	2 rolls	146	2000
Irish stew	250 mL	211	1240
Fish stew	250 mL	237	600
Lasagna	1 slice	180	1100
Macaroni and cheese	250 mL	231	2080
Meat loaf	1 slice	70	110
Pizza, cheese	1 sector	75	740
Pizza, sausage	1 sector	105	1030
Spaghetti with meat balls	250 mL	260	1460
Tourtière (pork pie)	1 sector	139	1890

Beverages			
Cola	200 mL	197	320
Ginger ale	200 mL	195	260
Coffee (instant)	250 mL	235	20
Tea (instant)	250 mL	224	0
Orange juice (fresh)	250 mL	262	490
Orange juice (frozen concentrate)	1 can	213	1510
Grapefruit juice (canned concentrate)	1 can	207	1260
Milk (whole)	250 mL	257	660
2% milk	250 mL	258	540
Skim milk	250 mL	258	380
Buttermilk	250 mL	258	430

* Values taken from the *Nutrient Values of Some Common Foods*, Health and Welfare Department, Government of Canada, revised 1979. The classification of foods is provided by the *Canada Food Guide*, Health and Welfare Department, Government of Canada.

Although energy and nutrient requirements change with age, there are some general guidelines for a good healthy diet that can apply to all ages. These data below show the different energy needs by age group. Students can speculate about why different age groups require different amounts of food energy.

Sex	Age	Weight (kg)	Height (cm)	Energy per day (kJ)
Male	15 - 18	60	175	13 400
Female	15 - 18	54	165	8800
Male	19 - 22	60	175	13 400
Female	19 - 22	54	165	8800
Male	23 - 50	60	175	11 300
Female	23 - 50	54	165	7900

ADDRESSING ALTERNATE CONCEPTIONS

- Most students will identify the stomach as the major area of digestion, rather than the small intestine. Students can be encouraged to list the functions of the stomach and small intestine. Although both organs are vital, students will soon come to realize that the stomach is not the organ primarily responsible for digestion.
- Students often believe that gravity is primarily responsible for moving food along the gastrointestinal tract. Ask students to try drinking a glass of water from a straw while standing on their head.
- Students often do not think about the engineering problems facing the stomach. Ask them to imagine storing HCl and a protein-digesting enzyme in a protein container. Why doesn't the HCl denature the cell membranes of the cells that line the stomach? Cell membranes, as they already know, are composed of proteins. How does a parietal cell of the stomach, primarily composed of protein, store HCl, or how does the peptic cell store the protein-digesting enzyme?
- Many students may begin confusing liver function with that of the kidney, introduced in chapter 13. The term "excretion" is often incorrectly used to describe the functions of both organs. Students can begin discussing the often subtle difference between secretion and excretion in chapter 9. Although the liver secretes bile, it does not excrete it.

MAKING CONNECTIONS

- Any study developed around the themes of the exchange of matter and energy would be incomplete without a study of digestion. An understanding of anabolism and catabolism, as introduced in chapter 6, Chemistry of Life, requires support from this chapter on digestion. The acquisition of matter, as presented by the chapter on digestion, also sets the stage for investigating the circulatory system: the organ system responsible for the transport of digested matter.
- Enzyme function, as introduced in chapter 7, Energy within the Cell, is provided with a context for greater understanding in chapter 9. The fact that many potentially corrosive enzymes are stored in the inactive form helps explain many of the mysteries that enshroud digestion.
- The control of digestion by nerves sets the stage for investigating the nervous system and homeostasis, chapter 15, while hormonal controls for digestion provides a context for studying hormones and homeostasis, chapter 14.
- The knowledge presented in the chapter on digestion establishes an important bridge between food chains, food webs, and food pyramids, and the study of matter at the level of body systems.
- Liver function, as detoxifying potentially harmful substances, provides important bridges between physiological systems and environmental toxins. The linkage can be extended further when students begin investigating environmental carcinogens, as introduced in the heredity unit in chapter 23, Protein Synthesis. The fact that the actual substance may not be a carcinogen, but that it is broken down by the liver to carcinogenic metabolites can be presented and comprehended once students learn about liver function.

POSSIBLE JOURNAL ENTRIES

- Students might be asked to write a dialogue between the text and themselves, as they refute the idea that specialized organs for digestion are advantageous for multicellular organisms. By acting as a "devil's advocate," they can challenge their own learning and push understanding to a higher level.
- Ask students to view Figures 9.8 and 9.10, and draw a picture of what they might look like if they did not have intestinal villi. Without villi, the intestine would have to be 70 m long to maintain normal absorption. How would this alter waist size?
- Students can be encouraged to develop a concept map which includes all key terms presented in the chapter.
- Students can be encouraged to express the difficulties they experienced in organizing their learning of this challenging chapter. What things did they employ to aid them in constructing their knowledge? For example, some students may indicate that they were guided by Figures 9.5, 9.6, and 9.9 before re-reading the text. Other students may have attempted a series of their own drawings or constructed concept maps. The journal entry can provide students with more information on how they learn.
- Group thinking and decision-making strategies may be recorded as students prepare for the debate. Did the students change their minds during the preparation for the debate? Students may even be asked to record their initial positions about the social issue and to reflect upon these feelings after the debate has been completed. Did they change their minds after listening to opposing arguments? Why are people so concerned about diets? How can fad diets be harmful?

USING THE VIDEODISC

Section title	Page	Code	Frame (side)	Description
Organs of Digestion	229		3103 (all)	Peristalsis
Organs of Digestion	229		10159 (2)	Movie sequence showing swallowing and peristalsis
Stomach	230 – 231		2211 (all)	Stomach and spleen from fetal pig
Stomach	230 – 231		2213 (all)	Stomach showing pyloric sphincter
Stomach	230 – 231		3100 (all)	Peptide bond
Pancreas	232 – 233		2212 (all)	Pancreas
Liver and Gallbladder	235		2209 (all)	Fetal pig liver
Liver and Gallbladder	235		2210 (all)	Gallbladder from fetal pig
Colon	235 – 236		2214 (all)	Large intestine
Control of Digestion	239		11559 (2)	Movie sequence showing the action of gastric juices, time-lapse photography

IDEAS FOR INITIATING A DISCUSSION

Figure 9.3: Peristalsis

Ask students to speculate about whether astronauts experience difficulty drinking water in outer space. The same question can be approached by asking students if they can drink water while standing on their head. Most believe that gravity is what moves fluids to the stomach. Although gravity helps, peristaltic motions ensure that food reaches the stomach.

Figure 9.5: Synthesis of HCl in the lumen

Ask students to suggest a way of carrying strong acids and protein-digesting enzymes in a protein container. Because all cells in your body are made from protein, the stomach cells face much the same challenge.

REVIEW QUESTIONS

(page 231)

1 Provide examples of how the digestive system interacts with other organ systems.

■ Many possible examples may be provided. Cells of the digestive system require oxygen and nutrients carried by the circulatory system. The digestive system breaks down foods into components which can be transported to other cells of the body.

2 Define ingestion, digestion, absorption, and egestion.

■ Ingestion is taking in food. Digestion refers to a chemical and physical breakdown of larger molecules. Absorption involves the uptake and transport of nutrients. Egestion refers to the removal of undigested materials from the body.

3 Differentiate between physical and chemical digestion. Provide examples of each.

■ Physical digestion involves the breakdown of a large particle into smaller particles. Chewing and emulsification are examples. Chemical digestion involves the breaking of chemical bonds. Larger macromolecules are digested into component parts. Trypsin and pepsin are enzymes that regulate chemical digestion of proteins.

4 State the functions of the enzymes amylase, pepsin, and rennin.

■ Salivary amylase initiates the breakdown of carbohydrates. Pepsin initiates the breakdown of proteins. Rennin causes the coagulation of milk proteins.

5 What is the function of the mucous layer that lines the stomach?

■ It protects the cells of the lumen of the stomach from HCl.

6 What causes stomach ulcers?

■ Pepsin and stomach acids destroy the cells lining the stomach. This may occur if the mucous cells are impaired, or if an emulsifying agent reduces the protective covering.

(page 239)

7 Outline the mechanism by which stomach acids are buffered by pancreatic secretions.

■ Bicarbonate ions, released from the pancreas, travel to the small intestine by way of the pancreatic duct. HCl from the stomach is neutralized and the shape of the pepsin enzyme, coming from the stomach, is altered.

8 What is the function of bile salts?

■ Bile is an emulsifying agent. It breaks large fat droplets into smaller ones, thereby increasing the surface area.

9 What is cirrhosis of the liver?

■ It is a liver disorder characterized by the growth of fibrous connective tissue and fat in place of healthy liver cells.

10 Why is cellulose considered to be an important part of your diet?

■ Cellulose provides roughage, which stimulates egestion.

11 What are villi and what function do they serve?

■ They are small finger-like projections found inside the lumen of the small intestine. Villi increase surface area for absorption.

12 State the functions of gastrin and enterogasterone.

■ Gastrin is a hormone, produced in response to partially digested proteins appearing in the stomach, which increases gastric secretions. Enterogasterone is secreted in response to fats in the large intestine. It slows peristaltic movements, thereby allowing greater time for fat digestion.

LABORATORY
EFFECT OF pH ON PROTEIN DIGESTION

(page 237)

Time Requirements

Approximately 40 min are required to complete the setup for the laboratory. Results can be read the following class in approximately 10 min. Allow at least 10 min for cleanup.

Notes to the Teacher

- The activity presents some potential hazards because students are working with strong acids and bases. Both present problems in case of spillage. Safety procedures are included in the laboratory.
- Students should wear aprons and goggles during cleanup and observations.
- This activity may be well suited for a demonstration.

Observation Questions

a) Why was the graduated cylinder rinsed between adding the HCl and NaOH solutions?

■ Acids will neutralize bases.

b) Record the pH of each solution.

■ HCl is acidic and NaOH is alkaline.

c) Using tweezers or forceps, measure each of the egg white cubes.

■ Answers will vary.

d) Compare the amount of digestion in the test tubes.

■ Expect the HCl and pepsin solution to show the greatest digestion.

Laboratory Application Questions

1 Which test tubes served as controls for the experiment? Note: This experiment uses more than one control.

■ Test tubes 1, 3, and 5 served as controls because they did not contain any pepsin.

2 At what pH does pepsin work best?

■ Acid works best at acid pH; see test tube 4. It showed the fastest reaction rate—most digestion.

3 After interpreting the data from the experiment, a student concludes that HCl causes the digestion of proteins. Do you agree with this conclusion? Give your reasons.

■ No, because test tube 3, which also contained HCl, showed limited digestion. Pepsin and HCl must be present.

4 Using the data collected in the experiment, predict how an alkaline environment in the stomach would affect the digestion of carbohydrates, fats, and proteins.

■ Fats and carbohydrates are not digested in the stomach. The lab presents no evidence of fat or carbohydrate digestion. Protein digestion would be slowed. We assume that all bases work much the same way.

APPLYING THE CONCEPTS

(page 242)

1 Why are you able to drink water while standing on your head?

■ Peristalsis moves water down the gastrointestinal tract, by rhythmic contractions of smooth muscle.

2 Why does the stomach not digest itself? (Give two reasons.)

■ A protective layer made from mucous cells protects the stomach's cells. Pepsin, a protein-digesting enzyme, is stored in the inactive form. Cells are slowly replaced.

3 How is the duodenum protected against stomach acids? Why does pepsin not remain active in the duodenum?

■ $NaHCO_3$ from the pancreas buffers acids from the stomach. Once pepsin reaches the small intestine, the alkaline environment causes a change in the enzyme's geometry. The active site is altered and the enzyme, pepsin, no longer hydrolyzes proteins.

4 Why are pepsin and trypsin stored in inactive forms? Why can erepsins be stored in active forms?

■ If pepsin and trypsin were active they would begin digesting proteins and long-chain peptones. The cells would be digested. Erepsin only acts on partially digested proteins, therefore cells with intact proteins are not in danger.

5 Explain why the backwash of bile salts into the stomach can lead to stomach ulcers.

■ The bile salts can emulsify the stomach lining. Pepsin then gains access to cells in the lumen of the stomach.

6 Why do individuals with gallstones experience problems digesting certain foods?

■ Fatty food presents a problem, because the bile is released, but is unable to reach the small intestine. Emulsification of fats is all but eliminated and chemical digestion of fats is slowed. (Large fat droplets have much less surface area exposed to lipase enzymes. This slows the reaction rate.)

7 Why might individuals with an obstructed bile duct develop jaundice?

■ A product of red blood cell destruction, the heme portion of hemoglobin, causes the jaundiced condition. This metabolic product is toxic.

8 In cases of extreme obesity, a section of the small intestine may be removed. What effect do you think this procedure has on the patient?

■ This decreases absorption distance and more food is egested.

9 Trace the digestion of a spaghetti dinner from the mouth to the colon. Describe the enzymes and hormones involved in digestion.

■ Initial carbohydrate digestion begins in the mouth. Starches are broken down to smaller glucose-containing chains. Carbohydrate digestion continues once they enter the small intestine. Amylase enzymes from the pancreas and disaccharide enzymes from the small intestine and pancreas cleave more bonds, producing monosaccharides (mostly glucose). Assuming the spaghetti dinner contains some proteins, protein digestion is initiated by pepsin from the stomach. Proteins are hydrolyzed to long-chain peptones. This stimulates the release of the hormone gastrin, which accelerates gastric secretions. Protein digestion continues in the small intestine once trypsinogen, from the pancreas, is activated by enterokinase. Trypsin causes the breakdown of long-chain peptones into shorter-chain peptones. Eventually, the shorter-chain peptones are broken down to amino acids by erepsins from the small intestine. Meat fat is emulsified by bile salts in the small intestine. Pancreatic lipase causes chemical digestion of fats to fatty acids and glycerol in the small intestine. The presence of fats in the small intestine initiates the release of enterogasterone, which slows peristalsis, allowing greater time for fat absorption.

10 The following experiment was designed to investigate factors that affect lipid digestion. A pH indicator is added to three test tubes as shown below. (This indicator turns pink when the pH is 7 or above and clear when the pH is below 7.) The initial color of each test tube is pink. Predict the color changes for each of the test tubes shown in the diagram and provide your reasons.

■ Test tubes A and B remain pink. No chemical digestion occurred because no enzymes were present. Test tube C turns blue because fats were digested to fatty acids and glycerol. The fatty acids changed the pH.

CRITICAL-THINKING QUESTIONS

(page 242)

1 Coal-tar dyes are used to enhance the color of various foods. Some dyes, however, have been found to be carcinogenic (cancer causing). Many of the artificial food colors used in Europe are banned in North America, and many used in North America are banned in Europe. How is it possible that two groups of scientists could test a food dye and come to different conclusions?

■ Expect a wide variety of answers. Although no one answer is correct, students should be encouraged to consider multiple points of view and reflect an answer which considers the question from alternate perspectives. Scientific research is often influenced by a political agenda. This does not mean that data were falsified, but that the data were chosen.

2 The incidence of colon cancer is highest in countries where people eat the greatest quantities of animal proteins and fats. Individuals who live in countries where cereal grains form the basic diet have a much lower incidence of colon cancer. What conclusion might you draw from these data? Can colon cancer be eliminated by a change in diet?

■ Current research indicates that Canadians eat too many processed foods. Little waste is produced and wastes remain in the colon for extended periods of time causing a slow build-up of metabolic toxins. More frequent bowel movements, as would be stimulated by eating cellulose-based foods, would help reduce the problem. Roughage is an important part of the diet.

3 Comment on the research performed by William Beaumont on Alexis St. Martin. Would similar experiments be tolerated today? Give your reasons.

■ Students should indicate that the ethics which govern research are changing.

4 Salivary amylase breaks down starch to small-chain polysaccharides and disaccharides. The presence of these products can be detected by Benedict's solution. In the procedures shown on the next page, 5 mL of starch has been added to each of the test tubes. Benedict's solution changes color in the presence of reducing sugars.
Which of the procedures would best determine the optimal pH for the digestion of starch by salivary amylase? Critique the procedures not chosen.

■ Procedure 2 should be followed. Procedure 1 alters the amount of Benedict's solution and the amount of the enzyme, in addition to the pH (the independent variable). Procedure 3 also alters the amount of Benedict's solution. The amount of Benedict's solution must be constant.

5 Modify the experimental design provided above to investigate how pepsin concentration affects the rate of protein digestion.

■ Set up procedure 2, but use an acidic pH and alter the concentration of pepsin in each test tube. Only the concentration of the enzymes must be altered.

CHAPTER 10

Circulation

INITIATING THE STUDY OF CHAPTER TEN

1 Brainstorm about the importance of the circulatory system. Students might be asked to indicate why animals such as sponges and jellyfish do not have a circulatory system.

2 Ask students to estimate the amount of circulatory system they have in their body. Each estimate can be placed on the chalkboard. An estimate based on a 70 kg individual is about 96 000 km of blood vessels carrying blood to 60 trillion cells.

ADDRESSING ALTERNATE CONCEPTIONS

- It is often difficult for students to make a link between the circulatory system and metabolic rate. The greater the rate of metabolism, the more critical is the transport of oxygen and nutrients. Students may compare organisms shown in Table 10.1 and speculate about the demands placed on each circulatory system. The laboratory activity, Effects of Temperature on Peripheral Blood Flow, provides an important bridge between blood flow and body temperature.
- Some students assume arteries are blood vessels which carry oxygenated blood, while veins are blood vessels which carry deoxygenated blood. This idea can be confronted by examining the pulmonary artery and pulmonary veins. The pulmonary artery carries deoxygenated blood and the pulmonary vein carries oxygenated blood. Arteries should be defined as blood vessels that carry blood away from the heart. Veins are blood vessels that carry blood to the heart.
- The importance of capillaries is often understated. The entire circulatory system is designed to carry blood to and away from the capillaries, the functional part of the transport system. Ask students to speculate about a circulatory system without capillaries. (They may consider an open circulatory system, such as that of a crayfish.)
- The historical presentation of the circulatory system is provided to demonstrate how thinking has changed over time. As technological advancements improve the manner in which we view a functioning circulatory system, students can expect further changes in knowledge. The historical perspective is not intended to encourage a sense of superiority among the present generation. Past theories, like many student-constructed theories, were based on the best information available, and consistent within their own frame of reference.
- Some students have a tendency to equate terms such as cardiac output, stroke volume, and blood pressure. Asking students to perform calculations which address factors that affect cardiac output may help them discover how each term is connected, but distinct. Consider the following table:

Name	**Cardiac output (L/min)**	**Stroke volume (mL/beat)**	**Heart rate (beats/min)**
Jim (at rest)	5	70	(*71*)
Jim (exercising)	(*21*)	100	210
Jiri (at rest)	5	(*100*)	50
Jiri (exercising)	20	120	(*167*)

MAKING CONNECTIONS

- The link between digestion of food and transport of nutrients in a multicellular organism is made within the chapter.
- The development of the concepts associated with blood pressure provides a foundation for understanding capillary fluid exchange, developed later in the chapter, and the physiological adjustments associated with hemorrhage and allergic reactions which are introduced in chapter 11, Blood and Immunity. Blood pressure also provides needed information for a study of filtration, as introduced in chapter 13, Kidneys and Excretion.
- Counter-current circulatory systems, as adaptations for cold environmental temperatures, as presented in chapter 3, Aquatic and Snow Ecosystems, can be better understood by studying circulation. The laboratory activity, Effects of Temperature on Peripheral Blood Flow, provides an experiential context for formulated relationships between blood flow and body temperature.

POSSIBLE JOURNAL ENTRIES

- Students can be encouraged to develop a concept map which includes all key terms presented in the chapter.
- Students can be encouraged to express the difficulties they experienced in formulating an understanding of this challenging chapter. What things did they employ to aid them in constructing their knowledge? For example, some students may indicate that their understanding of one-directional blood flow was aided by Figure 10.18. Other students may have attempted a series of their own

drawings or constructed concept maps. The journal entry can provide students with more information on how they learn.

- Students might be asked to write a dialogue between the text and themselves, as they refute the idea that the transport of nutrients within a multicellular organism requires a circulatory system. By acting as a "devil's advocate," they can challenge their own learning and push understanding to a higher level.
- Group thinking and decision-making strategies may be recorded as students prepare for the debate. Did the students change their minds during the preparation for the debate? Students may even be asked to record their initial positions about the social issue and to reflect upon these feelings after the debate has been completed. Did they change their minds after listening to opposing arguments? Should people who refuse to alter lifestyles that are dangerous to their health be permitted equal access to health care?
- Students might record reasons why material learned in the chapter was particularly relevant to them. For example, students interested in athletics might indicate how they will use information in the section Adjustments of the Circulatory System to Exercise. Students may have relatives suffering from heart disease.

USING THE VIDEODISC

Section title	Page	Code	Frame (side)	Description
Blood Vessels	247 – 249		2745 (all)	Artery
Blood Vessels	247 – 249, 252 – 253		2470 (all)	Aorta
Blood Vessels	249 – 250		3282 (all)	Vein
Blood Vessels	247 – 249, 252 – 253		3283 (all)	Vena cava
Setting the Heart's Tempo	255 – 256		3094 (all)	Pacemaker
Setting the Heart's Tempo	255 – 256		12195 (2)	Movie sequence showing mammalian heart beat
Heart Sounds	258 – 259		12259 (2)	Blood flow and heart sounds are presented in a movie sequence
One-Way Blood Flow	252 – 254		13239 (2)	Movie sequence showing tricuspid heart valve
One-Way Blood Flow	252 – 254		13439 (2)	Pulmonary heart valve shown in a movie sequence
Blood Pressure	260 – 262		13747 (2)	Movie sequence showing vagus nerve stimulation on the beating heart
Blood Pressure	260 – 262		14051 (2)	Movie sequence showing the effects of an epinephrine injection into the bloodstream
Blood Pressure	260 – 262		14547 (2)	Movie sequence showing the effects of an acetylcholine injection into the bloodstream

Section title	Page	Code	Frame (side)	Description
Lymphatic System	266		2999 (all)	Lymphatic system
Lymphatic System	266		3000 (all)	Lymph
Lymphatic System	266		3001 (all)	Lymph node

IDEAS FOR INITIATING A DISCUSSION

Table 10.1: Various Types of Transport Systems

Ask students to speculate about why each animal uses oxygen at different rates. Would people use as much oxygen as mice? (We don't have as high a metabolic rate as the mouse. The mouse will lose body heat much faster than a human because it has a greater surface area to body volume ratio.)

Figure 10.5: Locating a pulse

Ask students to locate other areas of the body in which they can locate a pulse (e.g., ankle, neck, temporal lobe of skull).

Figure 10.13: X ray of obstructed artery

Ask students to locate the area of coronary obstruction.

Figure 10.17: Electrocardiogram

The chart shows atrial contractions, but only one ventricular contraction.

REVIEW QUESTIONS

(page 250)

1 Why do multicellular animals need a circulatory system?

- Single-cell organisms can rely on diffusion, but more complex organisms require a delivery system. Students might also suggest that larger organisms generally have a higher metabolic requirement and therefore require an efficient transport system.

2 Explain the importance of William Harvey's theory that blood circulated.

- Prior to Harvey's time many scientists did not identify the heart as a circulating pump. Blood was not thought of as carrying nutrients. Harvey's explanation provided a way of looking at the heart and blood within a system, the circulatory system.

3 How do arteries differ from veins?

- Arteries tend to have more elastic tissue and are able to withstand greater pressure. Lower pressure veins have a thinner middle layer.

4 What causes a pulse?

- The movement of blood through an artery which is close to the outer surface of the body causes a pulse. Ventricular contractions cause the pulsing of blood through arteries.

5 Why are aneurysms dangerous?

- The weakened wall of the blood vessel may eventually rupture, severely restricting blood flow. Without the delivery of oxygen and nutrients and the removal of wastes, cells will die. (The danger varies with the size and location of the blood vessel affected.)

6 Define vasodilation and vasoconstriction.

- Vasodilation is the widening of the diameter of a blood vessel. Vasoconstriction is the narrowing of a blood vessel.

7 Why are fat deposits in arteries dangerous?

- Blood vessels can narrow because of the fat deposits. This restricts blood flow to an organ.

8 What is the function of capillaries?

■ Diffusion of gases and nutrients between the blood and surrounding cells occurs in the capillaries.

9 Fluid pressure is very low in the veins. Explain how blood gets back to the heart.

■ Skeletal muscles massage blood back to the heart. Body movements such as stretching help the massaging action. In addition, the veins have a series of one-way valves which prevent the back flow of blood in veins.

(page 262)

10 Draw and label the major blood vessels and chambers of the heart. Trace the flow of deoxygenated and oxygenated blood through the heart.

■ Use blue and red pens. Blue arrows indicate the movement of blood from the right atrium to right ventricle and into the pulmonary artery. Red arrows indicate the movement of blood through the pulmonary vein into the left ventricle, into the left artery and out the aorta.

11 Differentiate between the systemic circulatory system and the pulmonary circulatory system.

■ The pulmonary circulation carries blood to and from the lungs. The systemic circulation carries blood to and from other organs. Arteries of the pulmonary circulatory system carry deoxygenated blood. Arteries of the systemic circulatory system carry oxygenated blood.

12 What causes the characteristic heart sounds?

■ The closing of the heart valves causes the heart sounds. The lubb sound is caused by the AV valves, while the dubb sound is caused by the closing of the semilunar valves.

13 What are coronary bypass operations and why are they performed?

■ Coronary bypass operations divert blood around an area of blockage to maintain adequate circulation for the heart muscle.

14 Explain the function of the sinoatrial node.

■ The sinoatrial node acts as a pacemaker to set the heart's beat rate. It initiates heart contractions.

15 What is an electrocardiogram?

■ An electrocardiogram is the graph made by tracings from an electrocardiograph, a technological device that monitors the heart's electrical activity.

Note: The electrocardiograph is the monitoring device, and the electrocardiogram is the graph.

16 Differentiate between systolic and diastolic blood pressure.

■ Systolic blood pressure is the blood pressure on an artery following ventricle contraction. Diastolic blood pressure is the blood pressure on an artery following ventricle relaxation.

17 Define cardiac output, stroke volume, and heart rate.

■ Cardiac output is the amount of blood pumped from each side of the heart every minute. Stroke volume is the quantity of blood pumped during each ventricular contraction. Heart rate is the number of times the heart beats per minute.

18 How do metabolic products affect blood flow through arterioles?

■ Metabolic products such as lactic acids cause arteriolar dilation, thereby increasing blood flow to local tissues.

(page 266)

19 How do blood pressure regulators detect high blood pressure?

■ Arteries stretch following ventricular contraction causing activation of baroreceptors. Nerves are stimulated when the pressure receptors stretch.

20 Outline homeostatic adjustment to high blood pressure.

■ A nerve message is sent from the blood pressure receptor to the medulla oblongata, the blood pressure regulator at the brain stem. Sympathetic nerves transmit nerve messages from the medulla oblongata to the heart, slowing heart rate and the force of heart contraction.

21 What two factors regulate the exchange of fluids between capillaries and ECF?

■ Fluid pressure forces fluid from the capillaries into the ECF and osmotic pressure, caused by proteins in the blood, draws fluid from the ECF into the capillaries.

22 Why does a low concentration of plasma protein cause edema?

- Plasma proteins establish osmotic pressure, the force which draws fluid into capillaries. If osmotic pressure is reduced, more fluid remains in the interstitial spaces, causing swelling.

23 What are lymph vessels and how are they related to the circulatory system?

- Lymph vessels are a system of open-ended vessels that collect proteins and debris from the ECF and return them to the circulatory system.

LABORATORY
EFFECTS OF TEMPERATURE ON PERIPHERAL BLOOD FLOW

(page 251)

Time Requirements

Approximately 30 min are required to complete the laboratory procedure.

Notes to the Teacher

- The peripheral blood flow kit can be ordered from Carolina Biological, order number 69-0975.
- Liquid crystal disks can be re-used. I have used some disks for more than 10 years.
- Students often experience difficulty using the arm-band designed to hold the thermography disks in place. I have had some success using plastic sandwich wrap or plastic bags. The disk must make complete contact with the skin for best results.

Observation Questions

a) Record the color displayed by the 34°C, 32°C, and 30°C disks.

- Answers will vary.

b) Record your results in a data table like the one below.

- Answers will vary, but generally the neck area is warmer.

c) Observe and record changes in the color of the disk for about 3 min or **for as long as the fingers can be comfortably immersed in the water.**

- Students will note cooling (color change from blue to colder colors such as brown or green) as the hand is immersed in cold water, indicating reduced blood flow.

d) Observe and record changes in the color of the disk for about 3 min.

- Students will note more blue and yellow colors on the disk as the hand is immersed in warm water, indicating greater blood flow.

Laboratory Application Questions

1 According to your results, which areas of the body have the greatest blood flow? Provide an explanation for your results.

- Answers will vary, but generally areas with arteries closest to the surface of the skin are the warmest areas.

2 How might your data have been affected if you had exercised before the experiment? Give your reasons.

- Subjects exercising will have greater heat production. Less sensitive disks might begin turning color during the exercise.

3 A subject places a sensitive thermography disk on her forehead. A bright blue appears. The subject then begins smoking and the color changes to a green, and then to yellow. What conclusions can you draw from this simple experiment?

- A variety of conclusions can be accepted. This question promotes lateral thinking. Possible answer: some factor in the cigarette smoke reduces circulation, thereby lowering the peripheral body temperature.

4 How does temperature affect peripheral circulation? Explain the homeostatic adjustment mechanism for immersion in both warm and cold water.

- Peripheral circulation in cold water is reduced. Students might speculate that this will help the body conserve heat. Conversely, peripheral circulation increases in warmer water. This increases the heat loss to the ambient air or surrounding water, thereby aiding cooling.

5 Scientists have long known that cancer cells are more active than normal cells. Cancer cells divide many times faster. Explain how liquid crystals have been used to diagnose cancerous tumors.

■ Cancerous tumors are hot spots. Metabolism occurs at a much faster rate in cancer cells.

LABORATORY
THE EFFECTS OF POSTURE ON BLOOD PRESSURE

(page 263)

Time Requirements

Approximately 50 min are required.

Notes to the Teacher

- Students often experience some difficulty hearing the low-pitched lubb, dubb sounds. A quiet room for the beginning of the lab is necessary.
- Teaching stethoscopes and electronic sphygmomanometers are available.
- Unusually high or low blood pressure readings are most often a sign of equipment malfunction. Aneroid sphygmomanometers, to operate accurately, must be calibrated. However, the intention of the laboratory is not to provide accurate readings but to present students with an opportunity to take blood pressure readings.

Observation Questions

a) Record the reading on the sphygmomanometer. This is the systolic blood pressure.

■ Answers will vary, but expect systolic blood pressure to read near 120 mm Hg. It should be noted that other values may indicate equipment problems and not blood pressure problems.

b) Record the reading on the dial of the sphygmomanometer when the sound disappears. This is the diastolic pressure.

■ Answers will vary, but expect diastolic blood pressure to read between 60 and 100 mm Hg.

c) Record the pulse rate while seated.

■ Answers will vary.

d) Record your results in a table like the one below.

■ Sample values are provided.

Position	Systolic BP (mm Hg)	Diastolic BP (mm Hg)	Pulse rate (beats/min)
Standing	120	80	70
Sitting	110	85	65
Lying	107	88	62

Laboratory Application Questions

1 Would you expect blood pressure readings in all of the major arteries to be the same? Explain your answer.

■ No, arteries further from the heart generally have lower blood pressure. Distance from the heart and diameter of the artery are two important factors.

2 Why should the lowest systolic pressure be recorded while you are lying down?

■ The heart is at the same level as the blood pressure receptors. Less pressure is needed to get blood to the head.

3 Atherosclerosis, or hardening of the arteries, is a disorder that causes high blood pressure. Provide an explanation for this condition.

■ Because the arteries have less stretch, blood is forced through a narrowed diameter.

4 Predict how exercise would affect systolic blood pressure. Provide your reasons.

■ The heart contracts with greater force, therefore systolic blood pressure increases. Diastolic blood pressure increases during exercise because muscles actively massage blood back to the heart; however, diastolic pressure drops following exercise if the subject is resting. The continued rapid heart rate means less filling time, and hence less filling pressure.
Note: Massaging action by skeletal muscles is reduced while relaxing.

5 Why might diastolic blood pressure decrease as heart rate increases?

■ Increased heart rate means less filling time between heart contractions and hence reduced filling and reduced cardiac output.

6 Design a procedure to investigate the role of exercise in influencing blood pressure.

- An investigation to study the role of exercise in influencing blood pressure should include the following steps.
  - Monitor blood pressure while resting.
  - Obtain readings 2 min, 4 min, etc. into the exercise.
  - Exercise must be at a constant rate.
  - Monitor blood pressure after exercise.

APPLYING THE CONCEPTS

(page 268)

1 Agree or disagree with the following statement and give reasons for your views: Oxygenated blood is found in all the arteries of the body.

- Disagree, because deoxygenated blood is found in the pulmonary artery.

2 Why does the left ventricle contain more muscle than the right ventricle?

- The right ventricle pumps blood to the lungs, organs close to the heart. The left ventricle pumps oxygenated blood to all parts of the body.

3 Why does blood pressure fluctuate in an artery?

- Blood pressure is greatest following ventricular contraction and lower while the heart relaxes.

4 Which area of the graph represents blood in a capillary? Explain your answer.

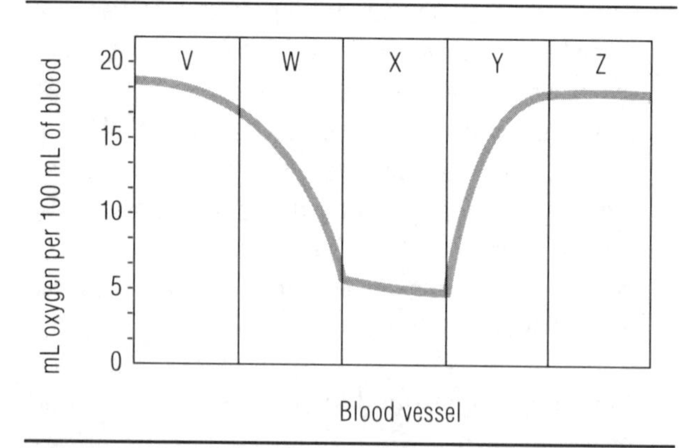

- Capillaries are represented by W and Y. Area W represents a body capillary, because oxygen diffuses from the blood to the surrounding cells.
  Note: Oxygen levels decrease.
  Area Y represents a pulmonary capillary.
  Note: Oxygen levels increase.

5 Using a capillary exchange model, explain why the intake of salt is regulated for patients who suffer from high blood pressure. (Hint: The salt is absorbed from the digestive system into the blood.)

- Salts increase the osmotic pressure in the capillaries, causing more fluids to move from the ECF into the blood. Increased fluid volume will elevate blood pressure.

6 Why do some soldiers faint after standing at attention for a long time?

- Inactivity of skeletal muscles decreases the amount of blood returning to the heart. Eventually, a lower diastolic pressure means that less blood will be pumped to the tissues. Blood pools in the lower extremities of the body and less blood is available to the brain.

7 Explain why someone who suffers a severe cut might develop rapid and weak pulse? Why might body temperature begin to fall?

- The cut reduces blood volume, which will manifest itself in reduced blood flow and a weaker pulse. In an attempt to compensate for lower blood pressure and decreased delivery of oxygen and nutrients the heart rate is increased and the pulse becomes rapid. Students should reason that because the circulatory system distributes heat throughout the body, a decreased blood flow means lower body temperatures. It should be noted that students might also reason that lower body temperatures might reflect slower rates of metabolism, a compensating mechanism for hemorrhage.

8 Why does the blockage of a lymph vessel in the left leg cause swelling in that area?

- Because proteins can no longer be removed from the ECF, an osmotic pressure is established outside of the capillary. The osmotic pressure inside of the capillary is opposed by the new osmotic pressure and fluids begin to accumulate in the ECF causing edema.

9 A fetus has no need for pulmonary circulation. Oxygen diffuses from the mother's circulatory system into that of the fetus through the placenta. Therefore, the movement of blood through the heart is highly modified: blood flows from the right atrium through an opening in the septum to the left atrium and then to the left ventricle. The opening between the right and left atria becomes sealed at birth. Explain why any failure to seal the opening results in what has been termed a "blue baby."

- Deoxygenated blood from the right side of the heart moves to the left side of the heart and begins to mix with the oxygenated blood. When the left ventricle contracts, some deoxygenated blood is pumped to the tissues along with the oxygenated blood, thereby lowering oxygen delivery and the efficiency of the circulatory system.

10 A person's blood pressure is taken in a sitting position before and after exercise. Compare blood pressure readings before and after exercise as shown in the following chart.

Condition	Systolic BP (mm Hg)	Diastolic BP (mm Hg)	Pulse rate (beats/min)
Resting	120	80	70
After exercise	180	45	160

a) Why does systolic blood pressure increase after exercise?

- The heart pumps with greater force during exercise in an attempt to deliver more blood to the tissues.

b) Why does diastolic blood pressure decrease after exercise?

- Heart rate is high and less filling time means that filling pressure is reduced.

CRITICAL-THINKING QUESTIONS

(page 269)

1 a) Nicotine causes the constriction of arterioles. Using the information that you gained about fetal circulation from question 9 in Applying the Concepts, explain why pregnant women are advised not to smoke.

- The placenta is the lifeline between the mother and the baby, supplying the developing fetus with nutrients and oxygen and removing wastes. Constriction of the arterioles caused by nicotine reduces the mother's blood flow to the placenta, in effect pinching the lifeline.

b) Mothers who smoke give birth to babies who are, on average, 1 kg smaller than normal. Speculate on the relationship between the effects of nicotine on the mother's circulatory system and the lower body mass of babies.

- Although the two circulatory systems remain separate, the fetus relies on the diffusion of nutrients from the mother's blood into that of the baby through adjoining capillary beds of the placenta. Restricted blood flow means less nutrient delivery and hence less chemical energy is made available to sustain life and growth, resulting in lower body mass.

2 Coronary heart disease is often related to lifestyle. Stress, smoking, alcohol consumption, and poor diet are considered to be contributing factors to heart problems. Should all people have equal access to heart transplants? Should people born with genetic heart defects be treated differently from people who have abused their hearts?

- Many opinions can be explored. Students should be encouraged to listen to other viewpoints. Most importantly, students should recognize that the answer to the question is found not within the discipline of science but that of ethics. The question illustrates the limits of using a scientific paradigm for problem solving.

3 Heart disease is currently the number-one killer of middle-aged males, accounting for billions of dollars every year in medical expenses and productivity loss. Should males be required, by law, to undergo heart examinations?

- Many opinions can be explored. Students should be encouraged to listen to other viewpoints. Most importantly, students should recognize that the answer to the question is found not within the discipline of science but that of ethics. The question illustrates the limits of using a scientific paradigm for problem solving.

4 Caffeine causes heart rate to accelerate; however, a scientist, who works for a coffee company, has suggested that blood pressure will not increase due to coffee consumption. This scientist states that homeostatic adjustment mechanisms ensure that blood pressure readings will remain within an acceptable range. Design an experiment that will test the scientist's hypothesis. For what other reasons do you think the scientist might have suggested that caffeine does not increase blood pressure?

- Many designs are possible; however, students should identify a control and attempt to limit the number of variables within their study. The methodology should attempt to uncover a cause and effect relationship.

5 It has been estimated that for every extra kilogram of fat a person carries, an additional kilometer of circulatory vessels is required to supply the tissues with nutrients. Indicate why obesity has often been associated with high blood pressure. Do only overweight people suffer from high blood pressure? Explain your answer.

- Speculation and prediction are promoted in this question. It is possible that the further you have to pump blood the greater is the force of heart contraction; however, many large people do not suffer from high blood pressure.

CHAPTER 11

Blood and Immunity

INITIATING THE STUDY OF CHAPTER ELEVEN

Have your students view the picture of David in Figure 11.1. Ask them to make a journal entry indicating how David's early life would be different from theirs. What types of things do your students believe that they would have missed most, if they had grown up inside the plastic bubble? Invite members of the class to share their journal entries.

ADDRESSING ALTERNATE CONCEPTIONS

- Students may be puzzled about why blood appears red in arteries and blue in veins. A variety of explanations are often formulated, many of which, although creative, do not agree with current medical information. The oxyhemoglobin complex gives blood its red color. (Actually the red blood cell appears pale orange. The composite of many red blood cells produces the red color.) Once oxygen is given up to cells of the body, the shape of the hemoglobin molecule changes, causing the reflection of blue light.
- We commonly speak of getting new blood into a company or organization. It was once assumed that younger people had more energy because they had younger blood. It was only logical, then, that transfusions with younger blood could provide more energy. Students may be asked to critique the long-established saying. Because the average blood cell only lives about 120 days, young people and elderly people both have blood cells that must be continuously replaced.
- Students who define blood antigens as foreign, invading substances may have difficulty understanding blood typing and transfusions. Blood antigen A, found on the red blood cells of people who have blood type A, is only a foreign invading substance when introduced into the blood of someone who does not have the A antigen, such as those with blood types B or O. The A antigen is not a foreign invading material for someone of blood type A. By defining the antigen as a substance that stimulates the production of antibod-

ies, some of the confusion can be eliminated. The text describes the A-marker as a glycoprotein, located on red blood cell membranes. The marker is present on red blood cells that are type A. People with blood type B have another marker, the B-marker. Blood type O has no markers, and hence is the universal donor.

- Some students believe that red blood cells move across the placental barrier, thereby carrying oxygen from the mother to the fetus. Misconceptions about placental function can be addressed by studying a rhesus-factor disorder called erythroblastosis fetalis. The fact that the disorder does not pose a problem for the firstborn, but for succeeding births, challenges the misconception that the placenta is permeable to red blood cells.
- Students concerned about getting HIV from toilet seats or casual contact are presented with information in the chapter that challenges this belief. As shown in Figure 11.14, the glycoprotein of the HIV virus only gains access to receptor sites on the T-cell lymphocytes. A lock-and-key geometry prevents the attachment of HIV to epithelial cells that cover the skin or lungs. HIV cannot be acquired by touch, nor is it an airborne infection.
- If students conceive of the terms antigen and antibody as specific things, rather than generalized groupings, they may conclude that any antibody provides immunity against any antigen. This is often expressed by the belief that any medicine will help a disease. The complementary geometry of glycoproteins of the invader and receptor sites challenges these views by providing counter-explanations about how antibodies work. Antibodies that attach themselves to the invading viruses (or other antigens) alter their shape, thereby preventing access to the entry ports. Misshapen viruses float around the body unable to find the appropriate entry port. This also provides clues about why different strains of influenza cause problems for people. Occasionally, the outer coat of the invader will change shape due to mutations. The mutated microbes are not tied up by the antibody and may gain access to the receptor site. This also provides a reference for understanding why exposing wet hair to low temperatures does not cause a cold.

MAKING CONNECTIONS

- The blood provides an interesting reminder of evolution. Blood has the same ions as the ancient seas, and in approximately the same relative concentration. In this manner, blood physiology provides a pathway to the study of evolution, as presented in chapter 4, Adaptation and Change, and developed further in chapter 24, Population Genetics.
- The subsection of the chapter entitled Allergies helps create a nexus between physiology and the environment; see Unit 1, Life in the Biosphere. Environmental pollutants are suspected for the ever increasing incidence of allergies. Global warming means more pollen, acid deposition is an irritant for epithelial cells of the respiratory tract, and toxic pesticides are amplified as they move through the food chain.
- The section, Antigen-Antibody Reactions, as presented in chapter 11, provides a model for understanding how the HIV virus infects T-cell lymphocytes. This knowledge is expanded in chapter 23, Protein Synthesis, in the case study, Human Immunodeficiency Virus. Students learn how the viral RNA inscribes information in the T-cell lymphocyte DNA and why AIDS is so difficult to cure.
- The Research in Canada section of the text describes Dr. Wanda Wenman's work on developing a vaccine for *Chlamydia trachomatis*, thereby creating an important bridge between immunity and chapter 17, The Reproductive System.
- The section, The Guided Missiles: Monoclonal Antibodies, provides an important connection between immunity and genetics. Information on monoclonal antibodies provides an important context for examining the mechanism of cloning, as addressed in chapter 18, Asexual Cell Reproduction.

POSSIBLE JOURNAL ENTRIES

- Students may write about their fears of HIV. Many of the fears may come from misconceptions that they hold about the virus. The misconceptions may be addressed in follow-up classes.
- Do any of the students have allergies? Do they believe that environmental pollutants can be linked to the ever-increasing incidence of allergies?

- Students can be encouraged to develop a concept map which includes all key terms presented in the chapter.
- Students can be encouraged to express the difficulties they experienced in formulating an understanding of this challenging chapter. What things did they employ to aid them in constructing their knowledge? For example, some students may indicate that their understanding of antigen-antibody formation was aided by constructing a model or doing a computer simulation. Other students may have attempted a series of their own drawings or have constructed concept maps. The journal entries can provide students with more information on how they learn.
- Decision-making strategies may be noted as each group prepares for the debate. Did the students change their minds during the preparation for the debate? Students may even be asked to record their initial positions about the social issue and to reflect upon these feelings after the debate has been completed. Did they change their minds after listening to opposing arguments? Should AIDS testing be compulsory? Who benefits and who could be harmed?

USING THE VIDEODISC

Section title	Page	Code	Frame (side)	Description
Components of Blood	271 – 272		2951 (all)	Hemoglobin
Components of Blood	271 – 272		2950 (all)	Heme
Components of Blood	273 – 274		2993 (all)	Leukocytes
Components of Blood	271 – 274		14805 (2)	Movie sequence showing blood cells and platelets moving through capillaries
Blood Clotting	276 – 277		3128 (all)	Platelets
Blood Clotting	276 – 277		16313 (2)	Movie sequence showing platelets and blood clotting
Blood Groups	277 – 278		18273 (2)	Movie sequence showing blood typing and agglutination
Immune Response	280 – 283		2739 (all)	Antigen
Immune Response	280 – 283		2736 (all)	Antibody
Antigen-Antibody Reactions	282 – 283		3109 (all)	Phagocytosis
How the Body Recognizes Harmful Agents	284 – 285		3238 (all)	T cell
How the Body Recognizes Harmful Agents	284 – 285		2756 (all)	B cell
How the Body Recognizes Harmful Agents	284 – 285		2758 (all)	Bacterium

Section title	Page	Code	Frame (side)	Description
How the Body Recognizes Harmful Agents	284 – 285		19769 (2)	Movie sequence showing destruction of a bacterium
Chemical Control of Microbes	289		2735 (all)	Antibiotic
Chemical Control of Microbes	289		19145 (2)	Movie sequence showing the effects of antibiotics on bacteria phagocytosis

IDEAS FOR INITIATING A DISCUSSION

Figure 11.5: Production of erythropoieten

Challenge students to identify factors which would stimulate the production of erythropoietin. (Living at high altitude, hemorrhage, and respiratory disorders which limit oxygen uptake or transport are but a few.)

Figure 11.7: Blood clotting

Ask students to view the picture and speculate about why blood banks remove calcium. (Sodium oxalate or potassium oxalate can be used to tie up free oxygen, thereby preventing blood clotting.)

Figure 11.9: Agglutination responses of blood types

Ask students to use the chart and determine the acceptability of various blood donors and recipients.

Figure 11.13: Model of antigen receptor sites

Ask students why some poisons only affect certain cells (e.g., muscle cells), but not other cells.

REVIEW QUESTIONS

(page 274)

1 Why is blood considered to be a tissue?

■ Blood is composed of a group of similarly shaped cells that carry out a similar function.

2 Name the two major components of blood.

■ Plasma and cellular components are the two major components of blood.

3 List three plasma proteins and indicate the function of each.

■ Albumins maintain osmotic pressure in capillaries; globulins produce antibodies to provide protection against invading microbes and parasites; and fibrinogens are important in blood clotting.

4 What is the function of hemoglobin?

■ The function of hemoglobin is to carry oxygen.

5 What is erythropoiesis?

■ The controlled production of red blood cells is called erythropoiesis.

6 List factors that initiate red blood cell production.

■ Any factor which will lower oxygen levels in the blood will stimulate red blood cell production. Exercise, moving to high altitudes, and hemorrhage are but a few.

7 What is anemia?

■ Anemia is a condition that results in the reduction of blood oxygen levels. It is usually associated with a reduced red blood cell production or lower than normal levels of hemoglobin.

8 How do white blood cells differ from red blood cells?

■ White blood cells contain a nucleus, are capable of some independent movement, and function as a defense against disease.

9 State two major functions associated with leukocytes.

■ Production of antibodies and phagocytosis are two major functions associated with leukocytes.

10 What is the function of platelets?

■ Platelets initiate blood-clotting reactions.

(page 287)

11 Outline the biochemical pathway involved in blood clotting.

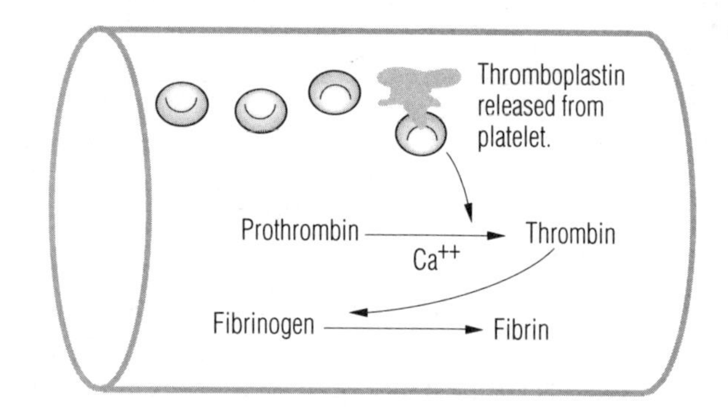

12 Define antigen and antibody.

■ Antigens are substances, usually proteins, that stimulate the formation of antibodies. Antibodies are formed to counter invading antigens.

13 Explain why type O blood is considered to be the universal donor. Why is type AB considered to be the universal acceptor?

■ Type O blood carries no A or B antigens and therefore no antibodies will be produced against type O blood. Type AB blood possesses both A and B antigens, and therefore will not produce antibodies against any other blood type.

14 How does Rh+ blood differ from Rh– blood?

■ RH+ blood contains an antigen for the Rhesus factor. RH– blood contains no Rhesus antigen.

15 Explain why erythroblastosis fetalis may affect a woman's second and third child, but not her first?

■ The mother will not have developed antibodies against the RH+ antigen with the first child because the mother's and baby's blood are separated by the placenta.

16 List the advantages and disadvantages associated with using artificial blood.

■ Artificial blood will not cause immune reaction, does not have to be screened for AIDS and other blood diseases, and requires no donor. However, the artificial blood is not very efficient at carrying oxygen and can only be used for short periods of time.

17 How do T-cell lymphocytes differ from B-cell lymphocytes?

■ T cells act as sentries which identify foreign invading antigens. B cells produce antibodies.

18 Differentiate between macrophages and lymphocytes.

■ Macrophages are phagocytotic white blood cells and lymphocytes are antibody-producing white blood cells.

19 Explain how B-cell, helper T-cell, and killer T-cell lymphocytes provide immunity.

Cell type	Function
Lymphocytes	Produce antibodies
Helper T cell	Act as sentries that identify foreign invading substances of cells
B cell	Produce antibodies
Killer T cell	Puncture cell membranes of infected cells, thereby killing the cell
Supressor T cell	Turn off the immune system

20 How do antibodies defeat antigens?

■ The antibodies attach themselves to the antigen. The antigens are often clumped together, making them more soluble and vulnerable. The antigen-antibody complexes may be engulfed and destroyed by macrophages.

21 How do memory T cells provide continuing immunity?

■ The memory T cell holds the imprint of the antigen on its outer surface.

22 Indicate why the suppressor T cells are important.

■ Suppressor T cells turn off the immune response, indicating that the danger has been controlled. Without the suppressor T cell, the immune response would continue and could become life threatening itself.

(page 290)

23 Comment on the statement that vaccines originated in the Far East.

- Ancient Chinese used scraped dried skin from smallpox patients.

24 Why did Edward Jenner inject cowpox viruses into a patient ?

- He believed that the cowpox would provide some type of immunity to smallpox.

25 What would have happened had Jenner first injected his patient with smallpox instead of cowpox?

- He would have most likely developed smallpox.

26 Why did Pasteur inject dead rabies into a patient who had contracted the living virus?

- Rabies will not immediately become active; an incubation period is required. By injecting the inactive virus, Pasteur hoped to stimulate antibody production. Pasteur hoped that by the time the virulent rabies became active, antibodies would be present. The antibodies should, in theory, prevent the antigens from gaining a foothold.

27 Why do people receive booster shots for polio?

- No antibody lives forever. Booster shots provide the continuous blueprint for the antigen.

28 Why do physicians not prescribe penicillin for the common cold?

- Penicillin only works on the cell walls of bacteria and bacteria-like organisms.

29 Why is sulfanilamide not classified as an antibiotic?

- It is not produced by another organism.

30 What chemical characteristics make sulfanilamide a desirable chemical therapy agent? What are its limitations?

- It blocks chemical reactions in the cells of invading bacteria, but not in the cells of humans.

LABORATORY

MICROSCOPIC EXAMINATION OF BLOOD

(page 275)

Time Requirements

Approximately 20 min are required.

Notes to the Teacher

- Figure 11.6 is useful for the identification of various leukocytes.
- Color and staining will vary according to technique.
- Use only prepared slides that are purchased from biological supply houses.

Observation Questions

a) Diagram the red blood cell of the fish.

- Answers will vary, but many students describe the network of blood vessels and the rapid movement of blood.

b) Estimate the size of the red blood cell.

- It is approximately 12 nm, but expect answers to vary.

c) Diagram a single human red blood cell.

- It is disk shaped. The nucleus is visible in the fish blood, but not in human red blood cells.

d) Estimate the size of the human red blood cell.

- It is approximately 9 nm, but expect answers to vary.

Chart called for in Procedure, step 4.
Approximate percentage values are provided.

Classification of leukocytes

Type	Description	Number	%
Granulocyte	Granular cytoplasm		
Neutrophil	Three-lobed nucleus, 10 nm (Wright's stain: purple nucleus, pink granules)	Will vary	65
Eosinophil	Two-lobed nucleus, 13 nm (Wright's stain: blue nucleus, red granules)		2 to 4
Basophil	Two-lobed nucleus, 14 nm (Wright's stain: blue-black nucleus, blue-black granules)		0.5

Agranulocyte	Nongranular cytoplasm	
Monocyte	U-shaped nucleus, 15 nm (Wright's stain: light bluish-purple nucleus, no granules)	4 to 7
Lymphocyte (small)	Large nucleus, 7 nm (Wright's stain: dark bluish-purple nucleus, no granules)	20 to 25
Lymphocyte (large)	Large nucleus, 10 nm (Wright's stain: dark bluish-purple nucleus, no granules)	2

Laboratory Application Questions

1 The red blood cells of fish contain a nucleus, while the human red blood cells do not. Indicate the advantage of the mammalian type of red blood cell over that of the fish.

■ Mammalian red blood cells have greater room for hemoglobin packaging and, therefore, greater oxygen-delivering capacity.

Blood tests are used to help diagnose different diseases. The chart below shows a few representative diseases. Use the chart to answer the questions 2 to 4.

Leukocyte change	Associated conditions
Increased eosinophils	Allergic condition, chorea, scarlet fever, granulocyte leukemia
Increased neutrophils	Toxic chemical, newborn, acidosis, hemorrhage, rheumatic fever, severe burns, acidosis
Decreased neutrophils	Pernicious anemia, protozoan infection, malnutrition, aplastic anemia
Increased monocytes	Tuberculosis (active), monocyte leukemia, protozoan infection, mononucleosis
Increased lymphocytes	Tuberculosis (healing), lymphocyte leukemia, mumps

2 Why would a physician not diagnose leukemia on the basis of a single blood test?

■ Many different diseases can be represented by similar symptoms.

3 What information might a blood test provide a physician about a patient being treated for the lung disease tuberculosis? Why would blood tests be taken even after the disease has been diagnosed?

■ The blood tests would provide additional information which may be used to confirm the diagnosis. By taking the blood tests after healing begins, the physician is able to trace the patient's recovery.

4 Leukemia can be caused by the uncontrolled division of cells from two different sites: the bone marrow or lymph nodes. Indicate how blood tests could be used to determine which of the sites harbors the cancerous tumor.

■ If the type of leukemia is associated with bone marrow tumors, granulocytes will be found in increased numbers. If the leukemia is associated with a tumor in lymph nodes, agranulocytes will be found in increased numbers.

CASE STUDY
DIAGNOSIS USING HEMATOCRITS

(page 276)

Time Requirements

Approximately 20 min are required.

Notes to the Teacher

Case studies provide an excellent opportunity for small group work.

Observation Questions

a) Calculate the hematocrit of the normal subject.

■ The hematocrit of the normal subject is approximately 45%.

b) Which subject do you believe has a low level of hemoglobin: A, B, C, or D?

■ Either C, because of lower red blood cell numbers, or D, because of the pale color, indicating a normal number of red blood cells, but less hemoglobin. The question is designed to raise some uncertainty. Generally lower levels of iron are manifest in a lower number of red blood cells, but not always.

Case-Study Application Questions

1 Cancer of the white blood cells is called leukemia. Like other cancers, leukemia is associated with rapid and uncontrolled cell production. Using the data in the case study, predict which subject might be suffering from leukemia. Give your reasons.

■ Subject B may be suffering from leukemia, because of higher than normal white blood cell numbers.

2 Although hematocrits provide some information about blood disorders, most physicians would not diagnose leukemia on the basis of one test. What other conditions might explain the hematocrit reading you chose in question 1? Give your reasons.

■ Many other infectious diseases such as mononucleosis can cause white blood cell numbers to increase.

3 Lead poisoning can cause bone marrow destruction. Which of the subjects in the case study might have lead poisoning? Give your reasons.

■ C might have lead poisoning because the red blood cell numbers are so low. Red blood cells are produced in the bone marrow.

4 Which subject lives at a high altitude? Give your reasons.

■ A, whose red blood cell numbers are unusually high, probably lives at a high altitude. This indicates a compensation for lower oxygen levels in the air.

5 Recently, athletes have begun to take advantage of the benefits of extra red blood cells. Two weeks prior to a competition, a blood sample is taken and centrifuged, and the red blood cell component is stored. A few days before the event, the red blood cells are injected into the athlete. Why would athletes remove red blood cells only to return them to their bodies later? What problems could be created should the blood contain too many red blood cells? Give your reasons.

■ Once red blood cells are removed, the body begins producing more. When the removed red blood cells are returned to the body, oxygen-carrying capacity is increased; however, some problems may result. The greater number of red blood cells means that the blood will have a lower fluid-to-cell ratio, and, therefore, will be more difficult to pump. The increased viscosity will place added strain on the heart. (The idea of pumping syrup makes a good visual analogy, if not a totally accurate scientific analogy.) The red blood cells also present a problem when they begin to break down, as the heme component of the hemoglobin is toxic.

APPLYING THE CONCEPTS

(page 294)

1 Sodium citrate and sodium oxalate are used by blood banks when storing blood. Both chemicals tie up calcium, which is present in the blood. Why do blood banks use these chemicals?

■ Platelets will rupture if they strike a rough object, such as a needle. Transportation of collected blood might also stimulate the rupture of the fragile platelets. Clotting requires calcium, and its removal by these anticoagulants ensures that transfusion recipients get blood, but not blood clots.

2 A physician notes fewer red blood cells and prolonged blood clotting times in a patient. White blood cell numbers appear to have increased, but further examination reveals that only the granulocyte numbers have increased, while agranulocytes have decreased. In an attempt to identify the cause of the anomaly, the physician begins testing the bone marrow. Why did the physician suspect the bone marrow? Predict what might have caused the problem. Provide the reasons behind your prediction.

■ Granulocytes are produced in the bone marrow. The problem might have been caused by a variety of factors which stimulate white blood cell production. Tumors, allergic reactions, chorea, scarlet fever, and rheumatic fever are a few possibilities.

3 What would happen if blood type A was transfused into people with blood types B, A, O, and AB? Provide an explanation for each case.

■ Type A blood would cause the formation of antibodies by types O and B blood. Blood types A and AB would not agglutinate donor type A blood.

4 The following illustrates how scientists determine blood type by cross matching. Heparin, an anticoagulant, is added to both a known and unknown blood sample to prevent blood clotting.

The samples are then placed in a centrifuge and separated into components. The red blood cells from the known sample, type A, are mixed with the serum of the unknown sample. (Plasma without clotting factors is called *serum*.) In the next step, the red blood cells from the unknown sample are mixed with the serum from type A blood. The data are provided in the following diagram. Predict the blood type of the unknown sample. Provide your reasons.

- Blood type AB would not form antibodies against type A blood (the A antigen is present in AB blood), but would stimulate the formation of antibodies in type A blood. (Type A does not have the B antigen found on AB red blood cells. Type A makes B antibodies in response to the foreign donor antigen.)

5 The serum containing antibody A will cause agglutination of donor blood that has antigen A. The serum containing antibody B will agglutinate donor blood that has antigen B. Use the data to the right to predict each of the blood types.

- Sample 1 = A; sample 2 = AB; sample 3 = O; sample 4 = B.

6 Why might people who have serious liver disease as a result of excessive alcohol consumption display prolonged blood clotting times?

- Liver tissue is replaced with fat and fibrous connective tissue. The fat cells cannot produce fibrinogen and prothrombin, proteins needed for blood clotting.

7 Why does exposure to the influenza virus not provide immunity to other viruses?

- Each virus has a distinctive outer coat (specific markers). Invading antigens are identified by the geometry of the outer coat.

CRITICAL-THINKING QUESTIONS

(page 295)

1 Individuals who work in a chemical plant are found to have unusually high numbers of leukocytes. A physician calls for further testing. Hypothesize about the physician's reasons for concern. Why might the physician check both bone marrow and lymph node areas of the body?

- The chemicals may have initiated allergic reactions or even caused leukemia. This question is designed to encourage logic-based predictions.

2 AZT (azidothymidine) is an experimental drug used against HIV (human immunodeficiency virus), the virus that causes AIDS. To date, the drug has been somewhat successful in arresting the infection, but is not considered to be a cure. A great deal of controversy has erupted over whether or not drugs that prolong life but do not cure the disease should be used, even experimentally. What is your opinion? Provide your reasons.

- Encourage different perspectives; there is no one answer.

3 The incidence of hepatitis and AIDS has increased the need for blood screening. Although all blood samples in Canada are screened for HIV and many other diseases, the same cannot be said about other countries. Do you think Canada should export screened blood to these countries? Give your reasons.

- Encourage different perspectives; there is no one answer.

4 Despite the availability of this information, many individuals still fear AIDS needlessly. Why do you think people fear AIDS? How can the fear of AIDS be reduced?

- Encourage different perspectives; there is no one answer.

CHAPTER 12

Breathing

INITIATING THE STUDY OF CHAPTER TWELVE

1 Ask students to describe adaptations that the astronaut would have to make to live on the moon. In what ways would those adaptations be similar to those that must be made for life at the bottom of the ocean?

2 Ask students why it is impossible to kill yourself by holding your breath.

ADDRESSING ALTERNATE CONCEPTIONS

- Students may use the terms breathing and respiration interchangeably; however, respiration is a much broader term than breathing. Respiration includes breathing movements, the transport of gases in the blood, and the energy-transforming processes of glycolysis and the Krebs cycle.
- Some students believe that all of the air inhaled is oxygen and all of the air exhaled is carbon dioxide. Approximately 20% of the air taken into the lung is not exchanged. Deoxygenated blood returning to the lung contains approximately 15 mL oxygen per 100 mL of plasma, while oxygenated blood contains approximately 20 mL of oxygen per 100 mL of plasma. Oxygen is exhaled and some carbon dioxide is inhaled; however, nitrogen is the most abundant gas both inhaled and exhaled.

MAKING CONNECTIONS

- The laboratory, Monitoring Lung Volume, provides students with added information that furthers their understanding of exercise physiology, which was introduced in chapter 10, Circulation.
- The section, Gas Exchange and Transport, provides a link between oxygen utilization and carbon dioxide production by the human body and the Carbon Cycle, introduced in chapter 1, Equilibrium in the Biosphere.
- The case study, Smoking and Lung Cancer, provides bridges between physiology, ecology, and genetics. Irritants from smoke create problems in artificial ecosystems, which were introduced in chapter 2, Energy and Ecosystems. In addition, the introduction of cancer, at a cellular level in chapter 12, provides a context for greater study in chapter 18, Asexual Cell Reproduction, when students study the section entitled Abnormal Cell Division: Cancer and chapter 23, Protein Synthesis, when students study the section entitled Oncogenes: Gene Regulation and Cancer.

POSSIBLE JOURNAL ENTRIES

- Students can be encouraged to develop a concept map that includes all key terms presented in the chapter.
- Students might record reasons why material learned in the chapter was particularly relevant to them. For example, students interested in athletics might indicate how they will use information on vital capacity, inspiratory reserve volume, and tidal volume.
- Students can be encouraged to express the difficulties they experienced in formulating an understanding of concepts presented in the chapter. What things did they employ to aid them in constructing their knowledge? For example, some students may indicate their understanding of carbon dioxide transport by viewing Figure 12.13. Other

students may have attempted a series of their own drawings or constructed concept maps. The journal entries can provide students with more information on how they learn.

- Decision-making strategies may be recorded as each group prepares for the debate. Did the students change their minds during the preparation for the debate? Students may even be asked to record their initial positions about the social issue and to reflect upon these feelings after the debate has been completed. Did they change their minds after listening to opposing arguments? Should governments ban the sale of tobacco products?

USING THE VIDEODISC

Section title	Page	Code	Frame (side)	Description
Importance of Breathing	296 – 297		31022 (4)	Movie sequence of shark breathing
Importance of Breathing	296– 297		31636 (4)	Movie sequence of frog breathing
The Human Respiratory System	297– 299		3259 (all)	Trachea
The Human Respiratory System	297– 299		2777 (all)	Bronchus
The Human Respiratory System	297– 299		2820 (all)	Cilium
The Human Respiratory System	297– 299		2720 (all)	Alveolus
The Human Respiratory System	297– 299		2783 (all)	Capillaries

IDEAS FOR INITIATING A DISCUSSION

Figure 12.2: Larynx showing vocal cords

Ask students if they have ever taken a drink of water or some other beverage and begun coughing. They may have had some fluid enter the trachea. Ask the students to examine the diagram and provide an explanation for what happened.

Figure 12.4: Position of diaphragm during inspiration and expiration

Ask students to explain why some women experience difficulty breathing during the later stages of pregnancy. (The baby pushes up against the diaphragm.)

REVIEW QUESTIONS

(page 301)

1 Which gases are exchanged during breathing?

■ Oxygen is inhaled and carbon dioxide is exhaled.

2 Differentiate between breathing and cellular respiration.

■ Breathing involves the intake and transport of gases to and from the cells of the body. Cellular respiration involves the oxidation of glucose within cells of the body to the lower-energy compounds, CO_2 and H_2O.

3 Trace the pathway of an oxygen molecule from the atmosphere to its attachment to a binding site on a hemoglobin molecule.

■ Oxygen moves through trachea, bronchus, and bronchiole into an alveolus. The oxygen molecule diffuses through the moist alveolar membrane into the plasma. Oxygen from the plasma attaches to the hemoglobin molecule.

4 Why does a throat infection cause your voice to produce lower-pitched sounds?

■ The swelling of the vocal cords in the larynx caused by the infection results in the production of lower-pitched sounds.

5 Describe the function of cilia in the respiratory tract.

■ The cilia are designed to remove foreign particles that become entrapped in the mucus that lines the respiratory tract.

6 Describe the movements of the ribs and the diaphragm during inhalation and exhalation.

■ During inhalation, the ribs move upward and the diaphragm moves downward, increasing the volume of the chest cavity. During exhalation, the ribs move downward and the diaphragm moves upward, decreasing the volume of the chest cavity.

(page 308)

7 How do CO_2 levels regulate breathing movements?

■ Carbon dioxide stimulates chemoreceptors in the medulla oblongata, the center of the brain which controls breathing.

8 Why does exposure to carbon monoxide (CO) increase breathing rates?

■ CO acts as a competitive inhibitor which attaches itself to the hemoglobin, thereby limiting the transport of oxygen. Low levels of oxygen are detected by the chemoreceptors in the carotid artery and aortic arch. The stimulated chemoreceptors send nerve messages to the medulla which in turn sends nerve messages to the diaphragm and rib muscles to increase breathing movements.

9 How does emphysema affect the lungs?

■ Inflammation of the bronchioles reduces air flow from the lungs. As pressure builds up in the lungs, alveoli begin to rupture. This reduces the surface area available for gas exchange in the lung.

10 How does partial pressure affect the movement of oxygen from the alveoli to the blood?

■ Gases diffuse from an area of high partial pressure to an area of lower partial pressure. Oxygen diffuses from air (partial pressure of 21 kPa) into the lungs (partial pressure 13.3 kPa). The largest difference in partial pressure is recorded between arteries (12.6 kPa) and veins (5.3 kPa).

11 How is CO_2 transported in the blood?

■ Approximately 9% of the CO_2 dissolves in plasma; 27% combines with hemoglobin to form carbaminohemoglobin; and 64% combines with water from the plasma to form carbonic acid (H_2CO_3).

12 Describe the importance of hemoglobin as a buffer.

■ H^+ ions formed from combining CO_2 and H_2O bind with hemoglobin, preventing fluctuations in pH.

LABORATORY
MONITORING LUNG VOLUME

(page 305)

Time Requirements

Approximately 40 min are required to complete the laboratory.

Notes to the Teacher

- Unless your school has a number of respirometers, it is recommended that students be assigned to groups and that respirometer time be scheduled. Wet or dry respirometers can be used.
- Disposable mouthpieces must be discarded with caution. It is recommended that disposable mouthpieces only be handled by students who use them.
- Most students exhale much harder than normal for tidal volume. Tidal volume measures a non-forced exhalation volume.

Observation Questions

a) Record the volume exhaled as *tidal volume*.

■ Expect answers to vary but normal tidal volume should be approximately 0.6 L to 0.9 L. Expect variation; body mass is a significant factor.

b) Record the value as *expiratory reserve volume.*

■ Approximately 3.8 L should be the expiratory reserve volume.

c) Record the value as *vital capacity.*

■ Vital capacity for males is approximately 5 L and females approximately 4 L. The difference is based largely on body mass.

Laboratory Application Questions

1 Predict how the tidal volume and vital capacity of a marathon runner might differ from that of the average Canadian.

■ Generally, the tidal volume is very close to average. Marathon runners do not breathe more deeply at rest; however, their vital capacity is generally greater than average.

2 How might bronchitis affect your expiratory reserve volume? Provide your reasons.

■ The swelling of the bronchioles will narrow the diameter and offer greater resistance to the movement of air during both inhalation and exhalation. However, forced exhalation will be affected more than inhalation. Forced exhalation allows less time to move the same quantity of air. Expiratory reserve volume will decrease and residual volume will increase, as more air remains in the lungs.

3 Predict how the respiratory volumes for a person with emphysema would differ from those collected during the laboratory.

■ The emphysema patient has difficulty exhaling. Therefore, less air is expelled, especially during exercise. (Also, higher levels of CO_2 might be found in the air.)

4 Why might respiratory volumes be measured during exercise? (Provide a list of things that could be investigated or diagnosed.)

■ Most individuals receive adequate ventilation during rest; however, exercise places greater demands on the system. Many symptoms such as low oxygen delivery become more pronounced during exercise, making diagnosis easier.

CASE STUDY
SMOKING AND LUNG CANCER

(page 310)

Time Requirements

Approximately 40 min are required to complete the assignment.

Notes to the Teacher

- Group work is recommended for case studies.
- Should students keep portfolios, case studies make appropriate submissions.

Observation Questions

a) How has cigarette smoke affected the mucous layer?

■ The smoke has increased the production of mucus.

b) Why does the build-up of tar in the bronchi limit air flow?

■ The tar causes the narrowing of the diameter of the bronchi and bronchioles.

c) In what area does the tumor begin to develop?

■ The tumor begins to develop in the basal layer of cells.

d) Why has the mucous layer in diagram C decreased in size?

■ Tumor cells have replaced many of the goblet cells, which secrete mucus.

e) Why does this characteristic make cancer especially dangerous?

■ The cancer is no longer confined to the lungs. It invades other tissues, disrupting their normal functioning.

Case-Study Application Questions

1 Based on the information that you gained in the case study, explain why smokers are often plagued by a cough.

■ The build-up of mucus caused by irritants in the smoke initiates a cough reflex.

2 Why is the slowing of the cilia lining the bronchi especially dangerous?

- A greater number of foreign particles enter the lungs. This increases the possibility of infections and toxic agents entering the respiratory tract.

3 Survey different smokers and calculate the amount of tar taken in each day. Most cigarettes contain about 15 mg of tar, but it should be noted that only 75% of the tar is absorbed. Show your calculations.

- Answers will vary.

4 Nicotine, one of the components of cigarettes, slows cilia lining the respiratory tract, causes blood vessels to constrict, and increases heart rate. Another component of cigarette smoke is carbon monoxide. As you have read, carbon monoxide competes with oxygen for binding sites on the hemoglobin molecule found in red blood cells. Analyze the data presented, and describe the potential dangers associated with smoking.

- This question allows students to construct many different answers. Students may indicate how smoking makes individuals susceptible to other infections. As described in question 3, the slowing of cilia may allow foreign particles to enter the lung. The lower oxygen levels associated with carbon monoxide may also be addressed. Some students may even consider how smoking leads to elevated blood pressure and use a capillary fluid exchange model to describe how smoking can lead to pulmonary edema.

APPLYING THE CONCEPTS

(page 312)

1 Why do breathing rates increase in crowded rooms?

- Carbon dioxide levels are usually elevated in crowded rooms. (Many students will suggest lower O_2 levels as a possible answer, but the CO_2 control is the most sensitive.) Elevated CO_2 levels stimulate the breathing center.

2 A patient is given a sedative that inhibits nerves leading to the pharynx, including those that control the epiglottis. What precautions would you take with this patient? Give your reasons.

- Eating and drinking could pose a danger. The trachea and bronchi are supported by cartilaginous rings, and, therefore, remain open. The epiglottis closes the opening of the trachea when a person swallows, but the sedative might cause it to stay open.

3 A man has a chest wound. The attending physician notices that the man is breathing rapidly and gasping for air.

a) Why does the man's breathing rate increase? Why does he gasp?

- His lung has collapsed and ventilation is severely hampered because the chest cavity can no longer establish pressure differences. To compensate, his respiratory rate increases. The gasp is a forced inhalation in an attempt to get more oxygen.

b) What could be done to restore normal breathing?

- The doctor could plug the hole in the chest cavity in an attempt to restore lower pleural pressure for inhalation.

4 During mouth-to-mouth resuscitation, exhaled air is forced into the victim's trachea. As you know, exhaled air contains higher levels of CO_2 than atmospheric air. Would the higher levels of CO_2 create problems or would they be beneficial? Provide your reasons.

- CO_2 in controlled amounts stimulates the breathing center. Even the CO_2 in exhaled air is reasonably low, usually less than 7%. Very high concentrations of CO_2 would be harmful, but this would not occur during mouth-to-mouth resuscitation.

5 Prior to swimming underwater, a diver breathes deeply and rapidly for a few seconds. How has the hyperventilating helped the diver hold her breath longer?

- The diver is saturating plasma and red blood cells with as much oxygen as possible, and removing the maximum amount of CO_2 by hyperventilating. Breathing impulses will not be signalled until blood CO_2 levels are elevated.

Use the following data table to answer questions 6, 7, and 8.

Individual	Breathing rate (breaths/min)	Hemoglobin (g/100 mL blood)	O_2 Content (mL/100 mL)
A (normal)	15	15.1	19.5
B	21	8	13.7
C	12	17.9	22.1
D	22	16.0	14.1

6 Which individual has recently moved from Calgary to Halifax? (Note: Halifax is at or near sea level, while Calgary is at a much higher altitude.) Give your reasons.

■ Subject C has just moved to Halifax; the higher hemoglobin levels were needed at higher altitude where the partial pressure of oxygen is lower. When the person moves to Halifax, the greater number of red blood cells and higher hemoglobin levels remain, at least for a while. Eventually, the red blood cells are destroyed and replaced with a number appropriate for living in Halifax, where O_2 levels are higher.

7 Which individual is suffering from dietary iron deficiency? Give your reasons.

■ Subject B is deficient in iron, and hemoglobin and blood oxygen levels are predictably lower. Iron is an important component of hemoglobin. The breathing rate has accelerated to compensate for reduced oxygen carrying capacity.

8 Which individual has been exposed to low levels of CO? Give your reasons.

■ It is subject D, The hemoglobin levels are normal, but blood oxygen levels are lower because CO interferes with the binding of oxygen to hemoglobin. The breathing rate has accelerated to compensate for reduced oxygen carrying capacity.

Use the graph to answer question 9. The graph compares fetal and adult hemoglobin.

9 Which hemoglobin is more effective at absorbing oxygen? What adaptive advantage is provided by a hemoglobin that readily combines with oxygen?

■ Fetal hemoglobin is more effective at absorbing oxygen. The fetus secures oxygen from the mother's blood by way of placental circulation.

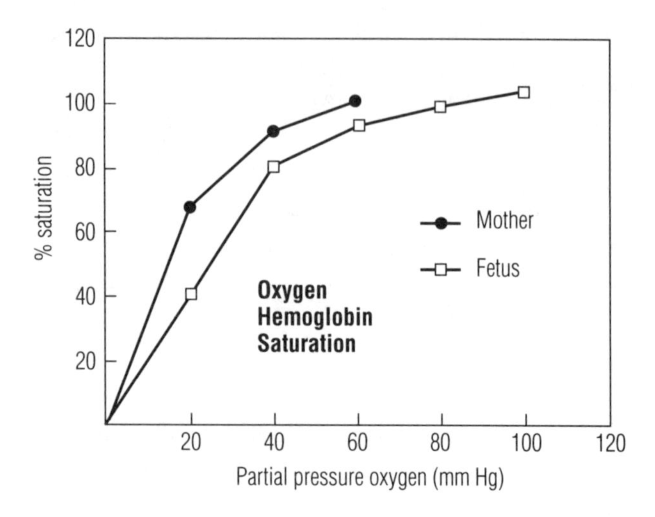

CRITICAL-THINKING QUESTIONS

(page 313)

1 A scientist sets up the following experimental design (shown on page 313).
Sodium hydroxide absorbs carbon dioxide. Limewater turns cloudy when it absorbs carbon dioxide.

a) Indicate the purpose of the experiment.

■ The purpose of the experiment is to determine whether CO_2 is a product of cellular respiration.

b) Which flask acts as a control?

■ The first flask containing limewater is the control, since it would detect any CO_2 in the air going to the animal.

c) Why is sodium hydroxide used?

■ It removes carbon dioxide from atmospheric air. This ensures that the CO_2 came from the animal.

2 The following chart shows a comparison of inhaled and exhaled gases.

	Percentage by volume	
Component	**Inhaled air**	**Exhaled air**
Nitrogen	78.62	74.90
Oxygen	20.85	15.30
Carbon dioxide	0.03	3.60
Water (vapor)	0.50	6.20

On the basis of the experimental data a scientist concludes that nitrogen from the air is used by the cells of the body. Critique the conclusion.

■ According to the data, the conclusion may indeed be possible. Although incorrect, the explanation is not without merit. Some students may indicate that nitrogen from the air is not used by the body, and, therefore, changes in concentration may reflect other chemical reactions that occur in the body. The elevated level of carbon dioxide is the most important key, because the products do not come from other gases but from carbohydrates. (Conservation of mass still occurs, but not all of the mass can be accounted for from the gases.)

3 Nicotine causes blood vessels, including those in the placenta, to constrict. Babies born to women who smoke are, on average, about 1 kg smaller than normal. This may be related to decreased oxygen delivery. Speculate about other problems that may face developing embryos due to the constriction of blood vessels in the placenta.

■ Less oxygen delivery means less available energy for cellular processes including cell growth and repair. Many other answers are possible.

CHAPTER 13

Kidneys and Excretion

INITIATING THE STUDY OF CHAPTER THIRTEEN

1 Ask students to view Figure 13.1, which shows a kidney transplant. Ask the students to make a journal entry indicating why they believe the kidney is an important organ. Encourage them to check their initial entry once the chapter has been completed.

2 Ask students to speculate why some beverages, such as those containing caffeine or alcohol, cause them to urinate more frequently.

ADDRESSING ALTERNATE CONCEPTIONS

- Kidney and liver functions are often expressed interchangeably. The term excretion is often incorrectly used to describe the functions of both organs. Students can re-address the often subtle difference between secretion and excretion introduced in chapter 9, Digestion. Although the liver is responsible for deamination and the secretion of urea, only the kidney excretes urea as a component of urine.
- Students may formulate the idea that the entire nephron is permeable to water. The case study, Comparing Solutes in the Plasma, Nephron, and Urine, should allow these students to confront this misconception. In the case study, students discover that the nephron is not only selectively permeable to different solutes, but that different regions of the nephron demonstrate selective permeability with respect to water.
- It is often difficult for students to differentiate between two regulating hormones of the kidney: ADH and aldosterone. ADH regulates water balance, and responds to any factor that increases

solute concentration. For example, increased exercise, fever, vomiting, diarrhea, and decreased fluid intake all increase blood solutes, causing the hypothalamus to shrink, thereby initiating the release of ADH. The ADH helps conserve body water by increasing water reabsorption from the nephron. By contrast, the aldosterone response is initiated by any factor which decreases blood pressure in the kidney. Although decreased fluid intake or perspiration brought on by extreme exercise could lower blood pressure as well as increase solute concentration, the two responses are not the same. Conditions such as hemorrhage will decrease blood pressure, causing the release of aldosterone, without activating ADH. Not every condition activates both regulatory mechanisms. ADH attempts to maintain a constant solute concentration, while aldosterone helps regulate blood pressure by maintaining body fluid volumes.

MAKING CONNECTIONS

- Any study developed around the themes of the exchange of matter and energy would be incomplete without a study of excretion. An understanding of anabolism and catabolism, as introduced in chapter 6, Chemistry of Life, requires support of this chapter on excretion. This chapter provides a missing piece to the puzzle which began with the acquisition of matter, as presented by chapter 9, Digestion, and the transport of digested matter as presented by chapter 10, Circulation.
- The hormonal control of solute concentration and fluid volume of the body provides a context for investigating hormones and homeostasis, presented in chapter 14.
- The knowledge presented in the chapter on excretion establishes an important bridge between the nitrogen cycle and the study of matter at the level of body systems.

POSSIBLE JOURNAL ENTRIES

- Students can be encouraged to develop a concept map that includes all key terms presented in the chapter.
- Students can be encouraged to express the difficulties they experienced in formulating an understanding of this challenging chapter. What things did they employ to aid them in constructing their knowledge? For example, some students may indicate that their understanding of water balance was aided by the flow charts presented in Figures 13.8 and 13.9. Flow charts are an effective way of organizing information and showing relationships. Other students may have attempted a series of their own drawings or constructed concept maps. The journal entry can provide students with more information on how they learn.
- Students might be asked to write a dialogue between the text and themselves, as they refute the idea that the hormonal control of water balance is essential. By acting as a "devil's advocate," they can challenge their own learning and push understanding to a higher level.
- Group thinking and decision-making strategies may be recorded as each group prepares for the debate. Did the students change their minds during the preparation for the debate? Students may even be asked to record their initial positions about the social issue and to reflect upon these feelings after the debate has been completed. Did they change their minds after listening to opposing arguments? Are mercury fillings harmful? Why do different groups of experts arrive at divergent conclusions?
- Students might record reasons why material learned in the chapter was particularly relevant to them. For example, students with diabetes mellitus might be interested to discover why the disorder, if untreated, will increase urine production. Students may have friends or relatives who have had kidney stones or renal dialysis.

USING THE VIDEODISC

Section title	Page	Code	Frame (side)	Description
Importance of the Kidneys	314 – 315		3272 (all)	Urea
Urinary System	315 – 316		2985 (all)	Kidneys
Urinary System	315 – 316		3273 (all)	Ureters
Urinary System	315 – 316		3274 (all)	Urethra
Nephrons	316 – 317		3055 (all)	Nephrons
Nephrons	316 – 317		2775 (all)	Bowman's capsule
Formation of Urine	317 – 319		2917 (all)	Filtration
Formation of Urine	317 – 319		3086 (all)	Osmosis / reabsorption
Formation of Urine	317 – 319		30353 (2)	Movie sequence showing how the nephrons function
Water Balance	320 – 322		2708 (all)	ADH
Water Balance	320 – 322		2709 (all)	Adrenal glands
Water Balance	320 – 322		2713 (all)	Aldosterone

IDEAS FOR INITIATING A DISCUSSION

Figure 13.2: Human urinary system

Ask students whether the renal vein or artery carries blood rich in nitrogen wastes. (The renal artery carries blood rich in wastes. The vein contains more CO_2 but lower concentrations of urea, because some of the urea has been removed by the kidney.)

Figure 13.4: Anatomy of the kidney

Have students compare the diagrams shown to 3-D models, if available.

Table 13.1: Comparison of Solutes

Why don't proteins and blood cells appear in Bowman's capsule? (They are unable to pass through the membrane, because the molecules are too large.)

REVIEW QUESTIONS

(page 319)

1 List the three main functions of the kidney.

■ Elimination of wastes, regulation of pH, and regulation of water balance are the three main functions of the kidney. The kidney also serves important functions in the regulation of blood pressure and produces renal erythropoietic factor, which is important in the production of red blood cells.

2 What is deamination and why is it an important process?

■ Deamination is the removal of an amino group from an organic compound. Excess amino acids are often converted into carbohydrates and other nutrient compounds.

3 How does the formation of urea prevent poisoning?

■ Nitrogen compounds such as ammonia are poisonous. Urea is much less toxic. Animals can store urea for short periods of time in their body. This permits controlled excretion of urine.

4 Diagram and label the following parts of the excretory system: kidney, renal artery, renal vein, ureter, bladder, and urethra. State the function of each organ.

■ See Figure 13.2 in text. The renal artery carries blood to the kidneys. The renal vein carries filtered blood away from the kidney. The ureter transports urine from the kidneys to the bladder. The urinary bladder stores urine. The urethra carries urine from the bladder outside the body.

5 Diagram and label the following parts of the nephron: Bowman's capsule, proximal tubule, loop of Henle, distal tubule, and collecting duct. State the function of each of the parts.

■ See Figure 13.4 in text. Bowman's capsule receives filtered blood from the glomerulus. The proximal tubule is the site of active transport of glucose, amino acids, and Na^+ ions. Eighty-five percent of the body's water is reabsorbed in this region. ADH acts on the upper part of the distal tubule and collecting duct to increase water reabsorption. If no ADH is present these areas only transport minerals such as salt.

6 List and describe the three processes involved in urine formation.

■ Filtration involves the movement of fluids from Bowman's capsule to the glomerulus. Reabsorption involves the movement of fluids from the nephron into the interstitial fluid and eventually the capillary net. Secretion involves the selective transport of fluids from the capillary net into the nephron.

(page 324)

7 What is ADH and where is it produced? Where is ADH stored?

■ ADH is antidiuretic hormone, a hormone that regulates water balance. ADH is produced in the hypothalamus and stored in the pituitary.

8 Describe the mechanism that regulates the release of ADH.

■ As solute concentration in the blood increases, cells of the hypothalamus shrink. This stimulates osmoregulators which send a nerve message to the pituitary to release ADH.

9 Where is the thirst center located?

■ The hypothalamus is the thirst center.

10 Describe the physiological adjustment to increased osmotic pressure in body fluids.

■ Osmoregulators are stimulated, and the pituitary releases ADH. ADH is carried in blood to the kidney. ADH enters the kidney and acts on the collecting duct and upper regions of the distal tubule, making them permeable to water. Greater volumes of water are reabsorbed. This helps prevent additional water loss.

11 Describe the behavioral adjustment to increased osmotic pressure in body fluids.

■ The shrinking of the cells of the hypothalamus initiates a thirst response and you begin drinking.

12 Discuss the mechanism by which aldosterone helps maintain blood pressure.

■ Low blood pressure in the kidneys or reduced blood flow to the kidneys causes the release of renin from the glomerulus area of the kidney. Renin activates angiotensinogen which in turn causes the release of aldosterone from the adrenal cortex. Aldosterone is released into the blood and causes an increased Na^+ reabsorption from the nephron. The Na^+ will increase osmotic pressure of the extracellular fluid and, therefore, increase water reabsorption.

13 What are kidney stones?

■ Kidney stones are caused by the precipitation of mineral solutes from the blood that may move into and/or become lodged in the renal pelvis, ureter, bladder, and urethra.

CASE STUDY

COMPARING SOLUTES IN THE PLASMA, NEPHRON, AND URINE

(page 320)

Time Requirements

Approximately 30 min are required to complete the case study.

Notes to the Teacher

- Group work is recommended for case studies.
- Should students keep portfolios, case studies make appropriate submissions.

Micropipettes were used to draw fluids from Bowman's capsule, the glomerulus, the loop of Henle, and the collecting duct. The data are displayed on the data table below. Unfortunately, some of the data were not recorded. The "no data" indicates the failure to record data.

Solute	Bowman's capsule	Glomerulus	Loop of Henle	Collecting duct
Note: quantities are recorded in grams per 100 mL				
Protein	8.0	0	0	0
Urea	0.05	0.05	1.50	2.00
Glucose	0.10	no data	0	0
Chloride	0.37	no data	no data	0.6
Ammonia	0.0001	0.0001	0.0001	0.04
Substance X	9.15	0	0	0

Case-Study Application Questions *(page 320)*

1 Which of the solutes was not filtered into the nephron? Explain your answer.

■ Proteins and substance X were not filtered. The chemicals appeared in the blood but not in the nephron.

2 Unfortunately, the test for glucose was not completed for the sample taken from the glomerulus. Predict whether or not glucose would be found in the glomerulus. Provide reasons for your prediction.

■ Yes, glucose should have been found. Glucose can pass through Bowman's capsule. The glucose has been actively transported from the proximal tubule and, therefore, does not show up in the loop of Henle.

3 Why do urea and ammonia levels increase after filtration occurs?

■ After water has been reabsorbed from the proximal tubule, the remaining solutes including ammonia become more concentrated.

4 Chloride ions (Cl^-) follow actively transported Na^+ ions from the nephron into the blood. Would you not expect the Cl^- concentration to decrease as fluids are extracted along the nephron? What causes the discrepancy noted?

■ Because filtration is an ongoing process, ions are continually being added to the nephron.

LABORATORY

DIAGNOSIS OF KIDNEY DISORDERS

(page 326)

Time Requirements

Approximately 40 min are required to complete the laboratory procedure. Allow another 30 min to complete the write-up.

Notes to the Teacher

- Combustrips can be purchased in most pharmacies.
- Preparation of samples:

Sample	Glucose (g)	Protein (gelatin) (g)	Water (L)
W	10	0	0.5
X	0	0	0.5
Y	0	0	2
Z	0	10	0.5

Add yellow food coloring to each of the samples. Sample X should be most concentrated (deepest yellow color) and sample Y least concentrated. Samples W and Z should be pale yellow, but contain more coloration than Y. Add enough sulfuric or hydrochloric acid to sample X to bring the pH near 5.

Observation Questions

a) Why was the graduated cylinder washed?

■ To prevent contamination, the graduated cylinder was washed.

b) Record the values for each of the samples in a table similar to the one below.

Sample	Glucose	Protein	pH
W	✓		about 7
X			about 5
Y			about 7
Z		✓	about 7

Laboratory Application Questions

1 Which subject do you suspect has diabetes mellitus? Provide your reasons.

■ Subject W probably has diabetes mellitus, because glucose was present in the urine. (We assume insulin injections are not properly regulating blood sugar levels.) This subject would be expected to have a high urine output.

2 Which subject do you suspect has diabetes insipidus? Provide your reasons.

■ Subject Y probably has diabetes insipidus, because urine output is high. The urine is very pale.

3 Which subject do you suspect has Bright's disease? Provide your reasons.

■ Subject Z probably has Bright's disease, because proteins are present in the urine.

4 Which subject do you suspect has lost a tremendous amount of body water while exercising? Provide your reasons.

■ Subject X has probably lost a tremendous amount of body water while exercising, because the urine is concentrated and acidic.

APPLYING THE CONCEPTS

(page 329)

1 Why is the formation of urea by the liver especially important for land animals?

■ Ammonia compounds released from the breakdown of amino acids are highly toxic. By combining the amino group with carbon dioxide to form urea, land animals produce a much less toxic compound which can be stored for a short period of time.

2 Predict how a drop in blood pressure would affect urine output. Give reasons for your prediction.

■ Decreased blood pressure will reduce filtration, thereby, decreasing the amount of urine produced.

3 A drug causes dilation of the afferent arteriole and constriction of the efferent arteriole. Indicate how the drug will affect urine production.

■ Blood will move into the glomeruli, but not move out as quickly. This should increase blood pressure in the glomeruli and this would increase filtration. As filtration increases, urine production should increase.

4 Why do the walls of the proximal tubule contain so many mitochondria?

■ Active transport occurs in the proximal tubule. The mitochondria supply ATP.

5 In an experiment, the pituitary gland of a dog is removed. Predict how the removal of the pituitary gland will affect the dog's regulation of water balance.

■ ADH is stored in the pituitary. Without ADH, urine production could be expected to increase.

6 How does excessive salt intake affect the release of ADH from the pituitary gland?

■ The salt would be absorbed and increase the osmotic pres-sure of the blood. Increased osmotic pressure would cause the hypothalamus to shrink, causing the release of ADH.

7 A drug that inhibits the formation of ATP by the cells of the proximal tubule is introduced into the nephron. How will the drug affect urine formation? Provide a complete physiological explanation.

■ Salt, amino acids, and glucose are actively transported from the nephron in the area of the proximal tubule. The reduced active transport capabilities would mean a higher solute concentration (increased osmotic pressure) in the nephron. Because of the higher osmotic pressure in the nephron, less water is reabsorbed. Urine production increases.

8 A blood clot lodges in the renal artery and restricts blood flow to the kidney. Explain why this condition leads to high blood pressure.

■ Reduced blood flow to the kidney is detected by the juxtaglomerular apparati. Renin is released which converts angiotensinogen to angiotensin. The angiotensin causes constriction of local arterioles (this increases blood pressure) and initiates the release of aldosterone from the adrenal cortex. Aldosterone, by increasing Na^+ transport from the nephrons, increases water reabsorption which helps elevate blood pressure.

9 For every 100 mL of salt water consumed, 150 mL of body water is lost. The solute concentration found in seawater is greater than that found in the blood. Provide a physiological explanation to account for the loss of body water. (Hint: Consider the threshold level for salt reabsorption by the cells of the nephron.)

■ Because the salt concentration in the nephron exceeds that which can be actively transported, a higher solute concentration remains in the nephron, creating high osmotic pressure. Water moves into the nephron from the extracellular fluid and this water is lost from the body along with the excess salt.

10 Explain why the presence of proteins in the urine can lead to tissue swelling, or *edema.*

■ The presence of proteins in the urine indicates either a breakdown of cells in the ureters, bladder, or urethra, or destruction of the glomerulus. Should the glomerulus break down, protein will enter the nephron, and create a high osmotic force in the nephron which will decrease water reabsorption. The loss of proteins with the urine also means that fewer proteins will remain in the blood. Students will have to think back to the circulation chapter to remember that proteins in the blood balance hydrostatic forces which push water from capillaries. By decreasing the osmotic force of the blood, more water moves out of the capillaries than into the capillaries. Hence, tissues swell.

CRITICAL-THINKING QUESTIONS

(page 329)

1 Alcohol is a diuretic, a substance that increases the production of urine. It also suppresses the production and release of ADH. Should individuals who are prone to developing kidney stones consume alcohol? Explain.

■ Increased fluid intake will increase filtration. By suppressing the release of ADH more, less concentrated urine will be produced. This indicates a potential benefit. However, students should also be encouraged to consider the long-term prospects of excessive alcohol consumption. Decreased liver function, associated with cirrhosis, will decrease important plasma proteins such as angiotensinogen and albumin which are important in regulating water balance and blood pressure.

2 In an experiment, four subjects each consumed four cups of black coffee per hour for a two-hour duration. A control group each consumed four cups of water per hour over the same two hours. It was noted that the group consuming coffee had a greater urine output than the control group. Provide a hypothesis that accounts for the data provided. Note: Formulate your hypothesis with an "If ... then ..." statement.

■ If coffee is consumed, then urine production will increase. Students may provide alternate suggestions for the data. Lateral thinking should be encouraged in this question. Urine production may not have been related to the coffee, but some other environmental factor.

3 Design an experiment that would test the hypothesis developed in question 2.

■ A variety of different experiments could be developed. Variables must be controlled and the collection period should be for at least 1 h following the consumption of fluids. The following provides a sample. Have 100 resting subjects drink only water for a 24 h period. Record urine output during that period and record the amount of water consumed. A few days later, have the same subjects drink the same amount of coffee at the same time of day and record urine output. The subjects should attempt to maintain the same amount of activity for both testing days.

4 Diseases such as syphilis, a venereal disease, can cause glomerular nephritis. (Other factors can also give rise to this disorder.) Glomerular nephritis is associated with the destruction of the high-pressure capillaries that regulate filtration. This occurs when proteins and other large molecules enter the filtrate. Once the filter is destroyed, it cannot be repaired. When a significant number of nephrons are destroyed, kidney function fails. Patients who suffer kidney failure can be treated by artificial dialysis or kidney transplants, two extremely expensive techniques. Should everyone who needs a kidney transplant be given the same priority?

■ Answers will vary. Students might be asked to consider a wide variety of situations. Should people who have knowingly abused their body have the same rights and privileges? This question allows students to explore many of the moral questions related to social decision-making.

UNIT 4

Coordination and Regulation in Humans

SUGGESTIONS FOR INTRODUCING UNIT 4

The human organism provides a model for studying the equilibrium between an organism's internal environment and its external environment. Beginning with the endocrine system students investigate how human body systems are regulated by chemical control agents, hormones. The integration of body systems depends on chemical controls which monitor and adjust physiological functions to maintain homeostasis. In chapter 15 students have the opportunity to study how the nervous system provides integrated control. Chapter 16 investigates how sensory nerve cells supply the central nervous system with information about the external environment and the quality of the internal environment.

In small brainstorming groups ask students to generate a list of physiological processes that must be regulated. The list can be checked once again at the end of the unit. A list of questions regarding the processes that are regulated can also be generated. How many of their initial questions have been answered?

KEY SCIENCE CONCEPTS

Unit 4, Coordination and Regulation in Humans, concentrates on equilibrium from the perspective of homeostasis. Technological devices used to assist control systems are introduced along with emerging social issues relating to their use. This unit provides a synthesis of the principles developed within Unit 1 of the Alberta Biology 30 Program of Studies. ■

CHAPTER 14

The Endocrine System and Homeostasis

INITIATING THE STUDY OF CHAPTER FOURTEEN

Chapter 14 provides an introduction to coordinating and regulating systems by studying the human endocrine system. The following suggestions provide examples for beginning the chapter:

1 Indicate that scientists can alter hormone levels in your body. Insulin and growth hormone can be injected. Challenge students to come up with

reasons why some hormone levels would be altered. Can they list practical applications?

2 Hormones are slow-acting control mechanisms, when compared with nerves. Why do humans have two control systems?

ADDRESSING ALTERNATE CONCEPTIONS

- The reasons for two control systems are not always obvious. Some students conclude that hormonal and nerve control systems act as backup regulators; however, each control system is adapted for different needs. The nervous system enables the body to adjust quickly to changes in the environment. The endocrine system is designed to maintain control over a longer duration.
- Many students assume that all hormones work in a similar manner, but two different types of hormones operate in the human body. Steroid hormones diffuse into the target cells where they combine with receptor molecules located in the cytoplasm. The hormone-receptor complex then moves into the nucleus and attaches to a segment of chromatin. The hormone activates a gene which sends a message to the ribosomes in the cytoplasm to begin producing a specific protein. Protein hormones exhibit a different action. Protein hormones combine with receptors in the cell membrane. Specific hormones combine at specific receptor sites. The hormone-receptor site complex activates the production of enzymes called adenyl catalase. The adenyl catalase causes the cell to convert ATP, the primary source of cell energy, into cyclic AMP. The cyclic AMP functions as a messenger, activating enzymes in the cytoplasm to carry out their normal functions.
- Some students may be aware of individuals who take pills for diabetes mellitus. They may assume the pills contain insulin; however, insulin cannot be taken orally. Protein hormones, such as insulin, are broken down by enzymes of the digestive tract. People suffering from maturity-onset diabetes can be helped by oral drugs known as sulfonamides. It is believed that the drugs, which are not effective for juvenile diabetes, stimulate the residual function of the islets of the pancreas in older people.

MAKING CONNECTIONS

- The information about cell membrane structure, gained in chapter 5, Development of the Cell Theory, and extended in chapter 11, Blood and Immunity, provides needed background to develop an understanding of negative feedback systems. The specificity of receptor site geometry helps students appreciate why some hormones affect certain organs but not others. The number of receptors found on individual cells may also vary. For example, liver cells and muscle cells have many receptor sites for the hormone insulin. Fewer receptor sites are found in less active cells such as bone cells and cartilage cells.
- The principles of homeostasis, developed by learning about the endocrine system in chapter 14, are directly applicable to the nervous system, studied in chapter 15. Knowledge gained because of recent technological breakthroughs has blurred the boundaries between the endocrine and nervous systems. Glands throughout the body are innervated by nerves. Some nerves produce chemical messengers that affect other cells. The injection of many hormones into the body produces a profound response in the nervous system. Nowhere is the division between the nervous system and endocrine system more obscure than in the hypothalamus.
- The concept of equilibrium, as presented within the context of regulating body systems, can be directly related to the Gaia hypothesis, introduced in chapter 1, Equilibrium in the Biosphere. Equilibrium is also dealt with at a cellular level in chapter 7, Energy within the Cell.
- The action of steroid hormones introduces the concept of protein synthesis, presented in greater depth in chapter 23, Protein Synthesis.

POSSIBLE JOURNAL ENTRIES

- Encourage students to comment on the award of the Nobel Prize to John MacLeod and not Charles Best.
- Students can be encouraged to express the difficulties they experienced in formulating an understanding of this challenging chapter. What things

did they employ to aid them in constructing their knowledge? For example, some students may indicate that their understanding of blood sugar balance was aided by the flow charts presented in Figures 14.12 and 14.14. Flow charts are an effective way of organizing information and showing relationships. Other students may have attempted a series of their own drawings or constructed concept maps. The journal entries can provide students with more information on how they learn.

- Group thinking and decision-making strategies may be recorded as students prepare for the debate. Did the students change their minds during the preparation for the debate? Students may even be asked to record their initial positions about the social issue and to reflect upon these feelings after the debate has been completed. Did they change their minds after listening to opposing arguments? Should anti-aging drugs be used? How would our society change because of a greater population of elderly people? Why do different groups of experts arrive at divergent conclusions?
- Students might record reasons why material learned in the chapter was particularly relevant to them. For example, students with diabetes mellitus might be interested to discover why the disorder can lead to a coma.

USING THE VIDEODISC

Section title	Page	Code	Frame (side)	Description
Importance of the Endocrine System	332		2889 (all)	Endocrine gland
Importance of the Endocrine System	332		2957 (all)	Homeostasis
Importance of the Endocrine System	332		2960 (all)	Hormone
Chemical Control Systems	333 – 334		2912 (all)	Feedback systems
Chemical Control Systems	333 – 334		2900 (all)	Equilibrium
Chemical Signals	334 – 335		2857 (all)	Cyclic AMP
Negative Feedback	335 – 336		3052 (all)	Negative feedback
The Pituitary: The Master Gland	336 – 337		3119 (all)	Pituitary
The Pituitary: The Master Gland	336 – 337		2967 (all)	Hypothalamus
Adrenal Glands	338 – 339		2709 (all)	Adrenal glands
Adrenal glands	338 – 339		3171 (all)	Releasing hormones
Insulin and the Regulation of Blood Sugar	340		2972 (all)	Insulin

Section title	Page	Code	Frame (side)	Description
Insulin and the Regulation of Blood Sugar	340		3301 (1)	Movie sequence on diabetes, showing anatomy of islets of Langerhans and historical research
Thyroid Gland	344 – 345		3256 (all)	Thyroid gland
Prostaglandins	346		3152 (all)	Prostaglandins

IDEAS FOR INITIATING A DISCUSSION

Figures 14.2 and 14.3: Action of protein and steroid hormones

Ask students to view the diagrams and compile a list which differentiates steroid and protein hormones.

Figure 14.4: Negative feedback of testosterone

Ask students to predict what would happen if the production of testosterone failed to be regulated. Would the male become more aggressive?

Figure 14.8: Effects of growth hormone

Growth hormone can be produced using genetic engineering techniques in which human genes that code for growth hormone are spliced into bacteria. The bacteria use the genetic information to construct human growth hormone, which can be harvested from the bacterial culture. Should growth hormone be supplied to individuals who wish to be basketball players or attain a height beyond what their genetic structure has determined?

Research in Canada: Discovery of insulin

Should John MacLeod have been given a Nobel Prize? Should Charles Best have been given a Nobel Prize?

REVIEW QUESTIONS

(page 338)

1 Define the term hormone.

■ Hormones are chemicals released by cells that affect cells in other parts of the body. Only a small amount of a hormone is required to alter a cell's metabolism.

2 What are target tissues or organs?

■ Target tissues have specific receptor sites that bind with hormones. These are the cells that hormones affect.

3 How are the nervous system and endocrine system specialized to maintain homeostasis?

■ The nervous system is specialized for quick adjustments to a changing environment, while the endocrine system maintains control over a longer duration of time.

4 Why is the pituitary called the "master gland"?

■ The pituitary regulates many other glands or hormones by way of tropic hormones.

5 Describe the signalling action of steroid hormones and protein hormones.

■ Steroid molecules diffuse from the capillaries into the interstitial fluid and into the target cells. The steroid hormones then combine with receptor molecules located in the cytoplasm. The hormone-receptor complex then moves into the nucleus and attaches to a segment of chromatin with a complementary shape. The hormone activates a gene which sends a message to the ribosomes in the cytoplasm to begin producing a specific protein. Protein hormones exhibit a different action. Protein hormones combine with receptors in the cell membrane. Specific hormones combine at specific receptor sites. The hormone-receptor site complex activates the production of enzymes called adenyl catalase. The adenyl catalase causes the cell to convert ATP (adenosine triphosphate), the primary source of cell energy, into cyclic AMP (adenosine monophosphate). The cyclic AMP functions as a messenger, activating enzymes in the cytoplasm to carry out their normal functions. (For example, when thyroid stimulating hormone (TSH) attaches to the receptor sites in the thyroid gland, cyclic AMP is produced in thyroid cells. Cells of the kidney and muscle have no receptors for thyroid stimulating hormone, and, therefore, they are unaffected. The cyclic AMP in the thyroid cell activates enzymes, which begin producing thyroxine, a hormone that regulates metabolism.)

6 What is negative feedback?

■ A negative feedback system is a control system designed to prevent chemical imbalances in the body. The body responds to changes in the external or internal environment. Once the effect is detected, receptors are activated and the response is inhibited, thereby maintaining homeostasis.

(page 341)

7 How would decreased secretions of growth hormone affect an individual?

■ Decreased secretions of growth hormone prior to puberty may cause dwarfism.

8 How would an increased secretion of growth hormone affect an individual after puberty?

■ Increased secretions of growth hormone cause the long bones and facial bones to widen. Often the bones will become more brittle.

9 Name two regions of the adrenal gland, and list two hormones in each area.

■ The adrenal medulla produces epinephrine and norepinephrine. The adrenal cortex produces cortisol and aldosterone (see Kidneys and Excretion).

10 How would high levels of ACTH affect secretions of cortisol from the adrenal glands? How would high levels of cortisol affect ACTH?

■ ACTH stimulates the release of cortisol. Therefore high levels of ACTH cause cortisol secretions to increase. The higher levels of cortisol cause ACTH secretions to drop. This is a negative feedback mechanism.

11 Where is insulin produced?

■ The islets of Langerhans cells in the pancreas produce insulin.

12 How does insulin regulate blood sugar levels?

■ When blood sugars are high, insulin is released. Insulin increases glucose utilization by cells and thus blood sugar levels are returned to a normal range.

13 How does glucagon regulate blood sugar levels?

■ When blood sugar levels are low, glucagon is released and liver glycogen is converted into glucose. Blood sugar levels rise as glucose leaves the liver.

(page 346)

14 How does thyroxine affect blood sugar?

■ Thyroxine increases metabolic rate. By increasing carbohydrate utilization, blood sugar levels can drop.

15 List the symptoms of hypothyroidism and hyperthyroidism.

■ Hypothyroidism causes lower metabolic rates, a tendency to gain weight, and lower levels of energy expenditure. People who have hypothyroidism tend to get cold easily and fatigue quickly. Hyperthyroidism increases metabolic rate, causing a tendency to reduced body fat, and higher levels of energy expenditure. People who have hyperthyroidism tend to remain warm even on cold days, and fatigue less easily.

16 How do the pituitary and hypothalamus interact to regulate thyroxine levels?

■ Should the metabolic rate decrease, receptors in the hypothalamus are activated. Nerve cells release thyroid releasing factor (TRF) which stimulates the pituitary to release thyroid stimulating hormone (TSH). Thyroid stimulating hormone is carried by way of the blood to the thyroid gland which in turn releases thyroxine. The hormone increases metabolism by stimulating increased sugar utilization by the cells of the body. Higher levels of thyroxine cause the pathway to be "turned off." Thyroxine inhibits the release of thyroid releasing factor from the hypothalamus, and thereby turns off the production of TSH from the pituitary.

17 What is a goiter?

■ An enlargement of the thyroid gland is called a goiter.

18 What are prostaglandins and what is their function?

■ Hormones that have a pronounced effect in a small localized area of the body are called prostaglandins.

LABORATORY

IDENTIFICATION OF HYPERGLYCEMIA

(page 343)

Time Requirements

Approximately 40 min are required.

Notes to the Teacher

- Preparation of solutions:

Solution code	Glucose (g)	Water (mL)	Food coloring yellow (drops)
A	0.0	100	10
B	10.0	100	5
C	2.5	100	6
D	1.5	100	10

- Students must wear eye goggles for both parts of the laboratory procedure and clean-up.
- Clinitest tablets must not be held by a bare hand. Always use forceps.
- Clinitest tablets can be purchased at any pharmacy. A color chart is provided in the package.

Observation Questions

a) Why must the medicine dropper be rinsed?

■ To prevent contamination, the medicine dropper must be rinsed.

b) Provide your data.

c) Provide your data.

Solution code	Benedict test	Clinitest
A	0	0
B	+2%	+2%
C	1.5% to 2%	+2%
D	1.0% to 1.5 % (will vary)	0.5% to 1.0% (will vary)

Laboratory Application Questions

1 Why is insulin not taken orally?

■ The protein hormone will be digested by protein digesting enzymes of the stomach and duodenum.

2 Explain why diabetics experience the following symptoms: low energy levels, large volumes of urine, the presence of acetone on the breath, and acidosis (blood pH becomes acidic).

■ Although blood sugar levels are high, the sugar does not enter nutrient-starved cells. Low energy is associated with lower levels of cell glucose. The large volumes of urine are produced because glucose remains in the nephron, creating an osmotic force which opposes the reabsorption of water. The acidosis occurs when fat metabolism is substituted for glucose metabolism. Intermediary metabolites of fat digestion cause the acidosis.

3 Why might the injection of too much insulin be harmful?

■ Blood sugars might drop too low, causing hypoglycemia. Few cells store glucose, therefore, low blood sugar will cause low supplies of sugar in cells. As blood sugar levels decline, so does the amount of energy available for cell metabolism. If energy reserves drop too low, cell metabolism slows or even stops.

4 How would you help someone who had taken too much insulin? Explain your answer.

■ Too much insulin lowers blood sugar. One obvious way of elevating blood sugar is to give the person some sugar. As blood sugar rises, cellular respiration makes energy available for metabolism.

CASE STUDY

THE EFFECTS OF HORMONES ON BLOOD SUGAR

(page 346)

Case-Study Application Questions

1 Which hormone injection did Bill receive at the time labelled X? Provide your reasons.

■ Bill may have received insulin. Insulin causes cells of the body to absorb glucose from the blood, lowering blood sugar levels. (Accept any condition which causes blood sugar levels to decline.)

2 What might have happened to Bill's blood sugar level if hormone X had not been injected? Provide your reasons.

■ Blood sugar levels would have remained high. Glucose would not be absorbed and utilized by body cells without insulin.

3 Explain what happened at time W for Bill and Farzin.

■ They may have eaten. Blood sugar began to increase at point W.

4 Explain why blood sugar levels begin to fall after time Y.

■ During exercise, increased energy demands accelerate the rate at which glucose is oxidized.

5 What hormone might Bill have received at time Z? Explain your answer.

■ Glucagon might have been received. Glucagon increases blood sugar levels by converting excess glycogen to glucose.

6 Why is it important to note that both Farzin and Bill had the same body mass?

■ Both would have similar energy demands. The experimenter is attempting to control variables.

7 What differences in blood sugar levels are illustrated by the data collected from Bill and Farzin?

■ Farzin, the nondiabetic subject, adjusts much faster to changes in blood sugar. Bill, the diabetic, relies on injections of insulin and other drugs for homeostasis.

8 Why do Bill and Farzin respond differently to varying levels of blood sugar?

■ Bill does not produce insulin. He must control blood sugar levels with diet, activity, and/or injections.

APPLYING THE CONCEPTS

(page 348)

1 Referring to the interaction between the hypothalamus and pituitary, indicate how the nervous system and endocrine system complement each other.

■ Releasing factors produced by the hypothalamus control pituitary function. The releasing factors are under nerve control. Endocrine hormones such as epinephrine complement nerve action. (This second point becomes more obvious after chapter 14.)

2 With reference to the adrenal glands, explain how the nervous system and endocrine system interact in times of stress.

■ Epinephrine is released from both nerves and adrenal glands. (Students have already been introduced to sympathetic nerves in the chapter on circulation. Their information base will be drastically increased in the chapter entitled, The Nervous System and Homeostasis.) Epinepherine increases the conversions of glycogen to glucose, thereby providing an increased energy source during times of stress. The nervous system increases heart rate, which speeds delivery of oxygen and glucose to the tissues.

3 A disorder called testicular feminization syndrome occurs when the receptor molecules to which testosterone binds are defective. Predict the effect of testicular feminization syndrome and explain how normal steroid hormone action is altered.

■ Steroid hormone molecules pass into a cell, combine with a receptor molecule, and then activate a gene in the nucleus. If the receptor molecule is faulty, the hormone has no way of activating the gene, and therefore, it is ineffective. The effects of the male sex hormone, in this case, are not exhibited.

4 A rare virus destroys cells of the anterior lobe of the pituitary. Predict how the destruction of the pituitary cells would affect blood sugar. Explain.

■ The anterior pituitary gland produces a variety of regulator hormones. TSH, a hormone of the anterior pituitary, initiates the release of thyroxine which regulates cell metabolism. As thyroxine levels decline, less blood sugar is oxidized. Impeding cellular metabolism slows the drop in blood sugar levels. Growth hormone is also produced in the anterior pituitary. Growth hormone increases fat metabolism, and switches fuels from glucose to fats. Without GH, the use of fat as a source of energy has declined. ACTH (also stored in the anterior pituitary) also stimulates the release of glucocorticoids from the adrenal glands. Cortisol, the most significant glucocorticoid, increases the level of amino acids in the blood. The amino acids are converted into glucose, which is the most usable source of cell energy. If cortisol levels drop, blood sugar levels may also fall.

5 Why do insulin levels increase during times of stress?

■ Epinephrine, a stress hormone, causes an increase in blood sugar levels. In turn, higher blood sugar levels initiate the release of insulin. This enables the excess blood sugar to enter the body cells.

6 Provide an explanation for the following symptoms associated with diabetes mellitus: lack of energy, increased urine output, and thirst.

■ Lack of energy occurs because less blood sugar enters the body cells. Insulin makes cell membranes permeable to glucose. Increased urine output occurs because sugar remains in the nephron. This increases osmotic pressure and reduces water reabsorption. Greater amounts of urine are produced. Thirst is associated with increased osmotic pressure of the blood (sugar is a solute) and the excessive water loss by urination.

7 A tumor on a gland can increase the gland's secretions. Explain how increases in the following hormones affect blood sugar levels: insulin, epinephrine, and thyroxine.

■ Increases in insulin and thyroxine would decrease blood sugar levels. Both hormones increase glucose utilization. Increases in epinephrine would increase blood sugar levels, because glycogen to glucose conversions are accelerated.

8 A physician notes that her patient is very active and remains warm on a cold day even when wearing a light coat. Further discussion reveals that even though the patient's daily food intake exceeds that of most people, the patient remains thin. Why might the doctor suspect a hormone imbalance? Which hormone might the doctor suspect?

■ The doctor might suspect a high level of thyroxine, which increases cell metabolism. If cell metabolism increases, more heat is liberated during accelerated cellular respiration.

9 With reference to negative feedback, provide an example of why low levels of iodine in your diet can cause goiters.

■ Iodine is an important component of thyroid hormones. Without iodine, thyroid production and secretion of thyroxine drops. This causes more TSH to be produced, and consequently, the thyroid is stimulated more and more. Under the relentless influence of TSH, the goiters continue to develop.

10 Three classical methods have been used to study hormone function:

- The organ that secretes the hormone is removed. The effects are studied.
- Grafts from the removed organ are placed within a gland. The effects are studied.
- Chemicals from the extracted gland are isolated and injected back into the body. The effects are studied.

Explain how each of the procedures could be used to investigate how insulin affects blood sugar.

■ By removing the gland, scientists were able to study how the animal adjusted to the environment without benefit of the hormone. By placing a graft back into the body, the hormone was reintroduced, but without benefit of nerve control. By using chemical extracts, different hormones could be isolated and studied separately. Some glands produce multiple hormones.

CRITICAL-THINKING QUESTIONS

(page 349)

1 A physician notes that individuals with a tumor on the pancreas secrete unusually high levels of insulin. Unfortunately, insulin in high concentrations causes blood sugar levels to fall below the normal acceptable range. In an attempt to correct the problem, the physician decides to inject the patient with cortisol. Why would the physician give the patient cortisol? What problems could arise from this treatment?

■ Cortisol causes the conversion of amino acids to glucose, thereby restoring glucose level. Accelerating the deamination process causes potential problems for both liver and kidney. Excessive amounts of nitrogenous wastes must be converted into the less toxic form, urea. Excessive amounts of urea alter the osmotic pressure of the blood and require elimination by the kidney. Excessively high concentrations of urea can also affect reabsorption from the nephron.

2 Some scientists have speculated that certain young female Olympic gymnasts may have been given growth hormone inhibitors. Why might the gymnasts have been given growth inhibitors? Do you think hormone levels should be altered to regulate growth patterns?

■ Shorter gymnasts have advantages, especially during tumbling routines. The taller the body, the more critical balance is. Students might speculate about what slowing the growth rate might do to other systems. No body system works in isolation. The ethical questions concerning drug-controlled athletic performances might also be addressed.

3 Cattle are given various steroid hormones to increase meat production. Recently some scientists have expressed concern that animal growth stimulators might have an effect on humans. Comment on the practice of using hormones in cattle. What potential problems might be associated with such procedures?

■ Students might indicate that these hormones might affect humans in similar ways. After all, cow insulin and pig insulin can be used by people. Humans might respond to these hormones in a similar or unpredictable manner. The question is designed to promote discussion but not to cause paranoia. Artificial steroid hormones, such as DES, are no longer used in Canada.

4 Negative-feedback control systems influence hormonal levels. The fact that some individuals have higher metabolic rates than others can be explained in terms of the response of the hypothalamus and pituitary to thyroxine. Some feedback systems turn off quickly. Sensitive feedback systems tend to have comparatively lower levels of thyroxine; less sensitive feedback systems tend to have higher levels of thyroxine. One hypothesis attempts to link different metabolic rates with differences in the number of binding sites in the hypothalamus and pituitary. How might the number of binding sites for molecules along cell membranes affect hormonal levels? How would you go about testing the theory?

■ The greater the number of binding sites, the more likely the cell would be to respond to the hormone. Individuals who have many binding sites would take up hormones more quickly. Fewer binding sites would mean that additional time would be required for the hormone to find the appropriate attachment site. The theory could be tested by adding controlled amounts of releasing factors to hormone-producing cells from two different tissue cultures. The amount of hormone produced by each tissue culture could be monitored. (Note: both tissue cultures must be the same size.) Students might suggest a variety of other possible solutions. There is no one correct answer, only a variety of problem-solving strategies to be evaluated.

CHAPTER 15

The Nervous System and Homeostasis

INITIATING THE STUDY OF CHAPTER FIFTEEN

Chapter 15 provides an introduction to coordination and regulation by studying the human nervous system. The following suggestions provide examples for beginning the chapter:

1 The nervous system is designed for rapid responses to environmental changes. The endocrine system studied in chapter 14 provides long-term sustained adjustments. Challenge small, brainstorming groups to identify situations in which hormonal control would be required and other situations in which nerve control is desired.

2 Ask students to estimate the number of nerve cells in the brain. There are 100 billion, greater than the number of visible stars in the Milky Way galaxy.

ADDRESSING ALTERNATE CONCEPTIONS

- Some students will imagine sensory nerve cells from the eye carrying a special light message to the brain. The idea that all sensory neurons work in essentially the same manner is difficult to imagine when related to experience. The fact that one nerve responds to light, while another responds to touch or sound can only be addressed by examining the specialization of sensory receptors. All neurons, providing they reach threshold potential, conduct impulses as waves of depolarization and repolarization.
- The concept of reflex arcs is not an easy one to understand. Many students believe that the brain controls reflexes. Ask students what happens when they touch a hot object with their finger. Do they remove their finger and then scream, or scream and then remove their finger? The fact that they remove their finger (it is part of the reflex arc) before screaming (it must be coordinated by the brain) provides evidence of reflex arcs. The laboratory, Reflex Arcs and Reaction Rate, provides examples of reflex arcs more appropriate for classroom demonstrations than touching hot objects.
- The difference between nerve impulses, which are electrochemical impulses, and electrical current is not easy to understand. Students often talk about nerves as if they were electrical wires; however, the differences between nerve action and electricity are important to developing a refined understanding of the nervous system. Current travels along a wire much faster than the impulse travels across a nerve. In addition, the cytoplasmic core of a nerve cell offers great resistance to the movement of electrical current. Unlike electrical currents, which diminish as they move through a wire, nerve impulses remain as strong at the end of a nerve as they were at the beginning. One of the greatest differences between the nerve impulse and electricity is that nerves use cellular energy to generate current. By comparison, the electrical wire relies on some external energy source to push electrons along its length. As early as 1900, Julius Bernstein suggested that nerve impulses were an electrochemical message created by the movement of ions through the nerve cell membrane. Evidence supporting Bernstein's theory was provided in 1939 when K.S. Cole and J.J. Curtis placed a tiny electrode inside the large nerve cell of a giant squid. A rapid change in charge was detected every time the nerve became excited.
- It is challenging for students to explain how animals detect the intensity of stimuli if nerve fibers either fire completely or not at all. Experience tells them that they are capable of differentiating between a warm object and one that is hot. To explain the apparent anomaly, they must examine the manner in which the brain interprets nerve impulses. Although stimuli above threshold levels produce nerve impulses of identical speed and intensity, variations with respect to frequency occur. The more intense the stimulus, the greater the frequency of impulses. Therefore, when a warm glass rod is placed on the hand, sensory impulses may be sent to the brain at a rate of, for example, one impulse every 30 s. A hot glass rod placed on the same tissue will also cause the nerve to fire, but the frequency of impulses is increased. The brain interprets the frequency of impulses. The different threshold levels of neurons provide a second way in which the intensity of stimuli can be detected. Each nerve is composed of many individual nerve cells or neurons. A glass rod at 40°C may cause a single neuron to reach threshold level, but the same glass rod at 50°C will cause two or more neurons to fire. The second neuron has a higher threshold level. The greater the number of impulses reaching the brain, the greater the intensity of the response. The brain interprets both the number of neurons excited and the frequency of impulses.

MAKING CONNECTIONS

- The principles of homeostasis, developed by learning about the endocrine system in chapter 14, are directly applicable to those applied to the nervous system, studied in chapter 15.
- The concept of equilibrium, as presented within the context of regulating body systems, can be directly related to the Gaia hypothesis, introduced in chapter 1, Equilibrium in the Biosphere. Equilibrium is also dealt with at a cellular level in chapter 7, Energy within the Cell.
- Nerve control, as presented in the section Homeostasis and the Autonomic Nervous System provides greater understanding of blood pressure regulation, as introduced in chapter 10, Circulation, or the manner in which peristaltic motions of the digestive tract are regulated, introduced in chapter 9, Digestion.

POSSIBLE JOURNAL ENTRIES

- Students can indicate why they believe that the work of Wilder Penfield was significant. Students may also comment on the promise presented by the current research of Dr. John McLean or Dr. Sergey Fedoroff.
- Students can be encouraged to develop a concept map that includes all key terms presented in the chapter.
- Students can be encouraged to express the difficulties they experienced in formulating an understanding of this challenging chapter. What things did they employ to aid them in constructing their knowledge? For example, some students may indicate that their understanding of nerve impulse transmission was aided by the model presented in Figure 15.9. Scientific models are an effective way of conceptualizing explanations. Other students may have attempted a series of their own drawings or constructed concept maps. The journal entry can provide students with more information on how they learn.
- Students might be asked to write a dialogue between the text and themselves, as they refute the idea that nerve impulses respond in an all-or-none fashion. By acting as a "devil's advocate," they can challenge their own learning and push understanding to a higher level.
- Decision-making strategies may be recorded as each group prepares for the debate. Did the students change their minds during the preparation for the debate? Students may even be asked to record their initial positions about the social issue and to reflect upon these feelings after the debate has been completed. How would a genetic link between chemical imbalances and drug dependency change the way in which we treat drug addicts? Why do different groups of experts arrive at divergent conclusions?
- Students might record reasons why material learned in the chapter was particularly relevant to them. For example, how can an understanding of neurotransmitters and synapses provide needed information about various illegal drugs?

USING THE VIDEODISC

Section title	Page	Code	Frame (side)	Description
Importance of the Nervous System	350 – 351		2912 (all)	Feedback systems
Vertebrate Nervous Systems	350		3059 (all)	Nervous system
Vertebrate Nervous Systems	350		3202 (all)	Somatic nervous system

Section title	Page	Code	Frame (side)	Description
Vertebrate Nervous Systems	350		2801 (all)	Central nervous system
Neurons	351 – 353		3063 (all)	Neuron
Neurons	351 – 353		3046 (all)	Myelin sheath
Neurons	351 – 353		3057 (all)	Nerve fiber
Neural Circuits	353		3192 (all)	Sensory neuron
Neural Circuits	353		2974 (all)	Interneurons
Electrochemical Impulse	356 – 361		2700 (all)	Action potential
Electrochemical impulse	356 – 361		3058 (all)	Nerve impulse
Electrochemical Impulse	356 – 361		20738 (2)	Movie sequence showing the neuron and the conduction of an impulse
The Synapse	362 – 364		3236 (all)	The synapse
The Synapse	362 – 364		2697 (all)	Acetylcholine
Central Nervous System	367 – 372		2776 (all)	Brain stem
Central Nervous System	367 – 372		2967 (all)	Hypothalamus
Central Nervous System	367 – 372		2804 (all)	Cerebellum
Central Nervous System	367 – 372		2805 (all)	Cerebral cortex
Central Nervous System	367 – 372		2806 (all)	Cerebrum
Central Nervous System	367 – 372		25028 (2)	Movie sequence showing voluntary movement and the sensory and motor activity of the cerebral cortex

IDEAS FOR INITIATING A DISCUSSION

Figure 15.4: Structure of nerves

Many students use the terms neuron and nerve interchangeably, but nerves are usually composed of many nerve cells or neurons. The diagram shows a bundle of axons from many neurons bundled into a nerve. Some nerves carry both sensory and motor neurons.

Figure 15.5: Reflex arc

Ask students to predict what would happen if the hand were removed, but the cell body from the sensory neuron remained. This could cause a phantom pain. Any stimulation of this neuron would cause a sensation to be felt in the hand which was no longer attached. Sensory information would be conducted to that part of the brain that registered sensory information for the hand. However, if the arm were cut above the cell body for this sensory neuron, the sensory neuron would die and no stimulus would be conducted to the brain.

Figure 15.14: Synapse

It was once thought that cholinesterase was held in vesicles found in the presynaptic neuron. Recent research has indicated that the cholinesterase actually is held along the postsynaptic membrane and is released during depolarization. Much of the information about nerve action has been derived from indirect evidence and, therefore, some will change as scientists gather additional explanations and substitute theories.

Figure 15.23: Action of opiates

The diagram shows how opiates have a similar geometry to that of endorphins. Does this provide any clues about why runners talk about a feeling of euphoria while and shortly after running?

REVIEW QUESTIONS

(page 354)

1 Differentiate between the PNS and CNS.

■ The central nervous system, or CNS, is composed of the nerves of the brain and spinal cord. The CNS serves as the coordinating center for incoming and outgoing information. The peripheral nervous system, or PNS, carries information between the organs of the body and the CNS.

2 What are neurons?

■ Neurons are the functional units of the nervous system. They conduct nerve impulses around the body.

3 Differentiate between sensory and motor nerves.

■ Sensory nerves are responsible for conducting information about the external environment to the central nervous system for processing. Motor nerves relay information from the CNS to the effectors, i.e., muscles and glands.

4 Briefly describe the function of the following parts of a neuron: dendrites, myelin sheath, Schwann cells, cell body, and axon.

■ Nerve cells are composed of many different parts. Dendrites are projections of cytoplasm that carry nerve impulses toward the cell body. The myelin sheath is a fatty covering over the axon of a nerve cell that speeds the rate of an impulse along a nerve. Schwann cells are the cells that comprise the myelin sheath. The cell body is the area of the nerve cell that contains the nucleus, and the axon is the extension of the cytoplasm that carries the nerve impulse to other nerves or effectors.

5 What is the neurilemma?

■ The neurilemma is a thin membrane that surrounds the axon of nerves in the PNS. The neurilemma promotes the repair of damaged axons, explaining how feeling can return to areas of the body with damaged peripheral nerves.

6 Name the essential components of a reflex arc and state the function of each of the components.

■ A reflex arc is the simplest nerve pathway. It has five essential components: the receptor, which detects the impulse; the sensory neuron, which conducts the stimulus; the interneuron, which transfers the impulse to the motor neuron; the motor neuron, which relays the impulse to the proper effector; and the effector, which acts upon the impulse.

(page 364)

7 What evidence suggests that nerve impulses are not electricity but electrochemical events?

■ Nerves have a high resistance to electrical current, which would reduce a current as it travelled along the nerve. Nerve impulses, however, remain constant along the length of the nerve, and also travel more slowly than electrical impulses. Electrodes placed inside nerves have detected electrochemical changes during the transmission of an impulse, confirming that nerve transmissions are electrochemical and not electrical.

8 Why was the giant squid axon particularly appropriate for nerve research?

■ The giant axon of a squid is useful for nerve research because of its large size. Electrodes can be inserted inside the axon. This allowed scientists to monitor changes in electrical charge across the nerve cell membrane.

9 What is a polarized membrane?

■ A polarized membrane has a charge that is caused by the unequal distribution of positively charged ions.

10 What causes the inside of a neuron to become negatively charged?

■ Positively charged ions are lost from inside of the resting membrane faster than they are added. This makes the inside of the membrane negative, relative to the outside of the membrane.

11 Why does the polarity of a cell membrane reverse during an action potential?

■ The action potential causes ion gates in the cell membrane to open, allowing ions to flow across the membrane, effectively reversing the polarity of the membrane.

12 What changes take place along a nerve cell membrane as it changes from a resting membrane to an action potential and then into a refractory period?

■ The resting cell membrane is more permeable to potassium than to sodium. Thus, the outside of the nerve is positive relative to the inside. An impulse causes sodium gates to open, allowing sodium ions to diffuse into the nerve, depolarizing the membrane. This depolarization causes the sodium gates to shut, and opens the potassium gates. Potassium diffuses out of the nerve. Sodium gates in adjacent areas open, and the action potential moves along the membrane. Sodium and potassium pumps then restore the polarization of the membrane.

13 Why do nerve impulses move faster along myelinated nerve fibers?

■ Nerve impulses move faster along myelinated nerves than they do along unmyelinated nerves because they jump along the nerve from one node of Ranvier to another, without passing through the rest of the nerve.

14 What is the all-or-none response?

■ A certain threshold stimulus is required to trigger an action potential along a nerve. However, once this threshold level is reached, further increase in the stimulus will not increase the response of the nerve; it is an all-or-none response.

15 Use the model of the synapse to explain why nerve impulses move from neuron A to neuron B, but not from neuron B back toward neuron A.

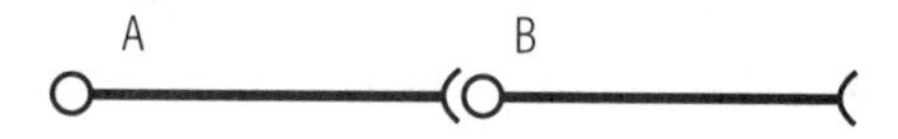

■ Acetylcholine is released at the end plate of neuron A. This causes depolarization of the dendrites of neuron B. If the nerve impulse moved from the axon of neuron B toward the dendrites it would not travel further. There is no transmitter chemical released from the dendrites of neuron B.

16 Explain the functions of acetylcholine and cholinesterase in the transmission of nerve impulses.

■ Acetylcholine is a neural transmitter found in the end plates of many nerve cells. When acetylcholine is released, it causes the postsynaptic membranes to become more permeable to sodium ions. The influx of sodium ions causes an action potential. Cholinesterase is released after the acetylcholine. It destroys the acetylcholine, which causes the sodium channels to close, and allows the nerve to begin its recovery phase.

17 Use a synapse model to explain summation.

■ Many neurons can meet at a synapse. In some cases, a single neuron is incapable of sufficiently depolarizing another neuron to cause an action potential. In cases such as this, chemicals released from two or more neurons may be required to initiate an action potential. This phenomenon is known as summation. See page 363 of the text.

(page 375)

18 Compare the structure and function of autonomic and somatic nerves.

■ The autonomic nerves consist of a pre- and a postsynaptic neuron. A single motor neuron is found in most somatic nerves. Autonomic nerves are nerves that regulate the organs of the body without conscious control. Somatic nerves lead to muscles and are regulated by conscious control.

19 State the two divisions of the autonomic nervous system and compare them by both structure and function.

■ The two divisions of the autonomic nervous system are the sympathetic and parasympathetic systems. The sympathetic system prepares the body for stress, while the parasympathetic system restores normal body conditions. Sympathetic nerves have short preganglionic and long postganglionic nerves; parasympathetic nerves have long preganglionic and short postganglionic nerves. Both release acetylcholine from the preganglionic nerve, but the postganglionic nerves of the sympathetic system release norepinephrine, while the postganglionic nerves of the parasympathetic system release acetylcholine.

20 What is the vagus nerve?

■ The vagus nerve is a major cranial nerve, and is a part of the parasympathetic nervous system. Branches of the vagus nerve regulate the heart, bronchi, liver, pancreas, and digestive tract.

21 State the function of the meninges and the cerebrospinal fluid of the brain.

■ The meninges of the brain are three protective membranes that surround the brain and spinal cord. Cerebrospinal fluid is a fluid that circulates between the innermost and middle meninges and through the central canal of the spinal cord. It acts as a shock absorber and as a nutrient and waste-transport system.

22 Describe the functions of the following: cerebrum, olfactory lobes, corpus callosum, cerebellum, thalamus, pons, and medulla oblongata

■ The brain is divided into many regions. The cerebrum is the largest and most highly developed area, and is responsible for storing sensory information and for initiating voluntary motor activities. The olfactory lobes receive and interpret information about smell. The corpus callosum allows communication between the two hemispheres of the brain. The cerebellum coordinates muscle movements. The thalamus coordinates and interprets sensory information. The pons relays information between the cerebellum and the medulla. The medulla oblongata acts as the connection between the peripheral and central nervous systems.

23 List the four regions of the cerebral cortex and state the function of each.

■ The four regions of the cerebral cortex are the frontal lobe, the temporal lobe, the parietal lobe, and the occipital lobe. The motor areas of the frontal lobe control movement of voluntary muscles. Association areas are linked to intellectual activities and personality. The sensory areas of the temporal lobe are associated with vision and hearing. Association areas are linked to memory and interpretation of sensory information. The sensory areas of the parietal lobe are associated with touch and temperature awareness. Association areas are linked to emotions and interpreting speech. Sensory areas of the occipital lobe are associated with vision. Association areas interpret visual information.

24 Explain the medical importance of brain-mapping experiments.

■ Brain-mapping experiments enabled scientists to identify the functions of the various areas of the brain.

25 What are endorphins? How do they work?

■ Endorphins are natural painkillers produced by the brain. They work by occupying sites in the substantia gelatinosa (SG) of the brain, which are normally used by pain signalling chemicals. The endorphins block the sites, reducing the intensity of the pain.

LABORATORY

REFLEX ARCS AND REACTION RATE

(page 355)

Time Requirements

Approximately 50 min are required.

Notes to the Teacher

- Demonstrate the knee-jerk reflex before beginning. Students who have knee problems should not be used as subjects.
- Encourage students to work in groups that revolve through stations. This permits supervision during the knee-jerk reflex. Some students have a tendency to strike the knee harder than necessary to evoke the reflex.

Observation Questions

a) Describe the movement of the leg.

■ The leg moves upward.

b) Compare the movement of the leg while the subject is clenching the book with the movement in the previous procedure.

■ The leg movement increases when the subject clenches the book.

c) Describe the movement of the foot.

■ The heel moves inward; the toes move outward.

d) Describe the movement of the toes.

■ The toes curl inward. Babies show the opposite reflex; their toes move outward in a grasping motion. Some evolutionary biologists have suggested that the early reflex can be related to arboreal life, however, this explanation is not well accepted.

e) Which pupil is larger?

■ The eye that was closed has the larger pupil.

f) Describe the changes you observe in the pupil.

■ The pupil of the eye that receives the light closes more quickly.

g) Record your data in a table similar to the one on the right.

■ Answers will vary.

Trial	Distance-right hand (cm)	Distance-left hand (cm)
1		
2		
3		
Average		

Laboratory Application Questions

1 A student touches a stove, withdraws his or her hand, and then yells. Why does the yelling occur after the hand is withdrawn? Does the student become aware of the pain before the hand is withdrawn?

■ The yelling is not part of the reflex. The sensory information must be processed through a myriad of neurons in the brain before yelling occurs. The hand is withdrawn before the student senses pain.

2 The neuron is severed at point X. Explain how the reflex arc would be affected.

■ The heat is detected, but the hand does not pull away from the heat source as part of the reflex. The person may still be able to consciously move the hand, but only after he or she notices that it is burning. The information would have to be processed by the CNS prior to any reaction.

3 Explain how the knee-jerk and Achilles reflexes are important in walking.

■ The stretching of the tendon in the knee causes the leg to straighten. The stretching of the Achilles tendon causes the foot to flex. Both motions are required in walking.

4 While examining the victim of a serious car accident, a physician lightly pokes the patient's leg with a needle. The light pokes begin near the ankle and gradually progress toward the knee. Why is the physician poking the patient? Why begin near the foot?

■ The physician is checking for nerve damage. By moving progressively closer to the upper leg, the physician is attempting to locate the position of the nerve damage. Nerves enter the spinal cord at various intervals. Those nerves that are found closer to the head are found in the upper areas of the spinal cord. Those found nearer the foot enter the lower areas of the spinal cord.

5 Why is the knee-jerk reflex exaggerated when the subject is clenching the book?

■ By clenching the book, muscles in various parts of the body are already pulling on tendons. By striking the knee, a greater number of stretch receptors are activated, causing a more exaggerated reflex. (Expect a number of novel answers for this question. The question is designed to promote thinking rather than providing a specific answer. The answer provided will not be the only possible answer.)

CASE STUDY
PHINEAS GAGE

(page 374)

Time Requirements

Approximately 20 min are required.

Case-Study Application Questions

1 Which lobe of Gage's brain had been damaged?

■ The frontal lobe had been damaged.

2 Provide a hypothesis that would explain why Phineas Gage's personality changed. How would you test your hypothesis?

■ Hypothesis: If the frontal lobe is damaged, then personality will change. The hypothesis could be tested by removing small sections of an animal's frontal lobe and studying its behavior. Changes in personality might be identified.

3 In one operation, first performed by Antonio Muniz, some of the nerve tract between the thalamus and the frontal lobes was severed. Why might a physician attempt such an operation?

■ The thalamus interprets sensory information and serves as a relay station to other parts of the brain. By cutting some of the nerve tracts leading from the thalamus and the frontal lobe, Dr. Muniz attempted to determine the manner in which patients responded to sensory information. This could be done to lower pain levels.

APPLYING THE CONCEPTS

(page 377)

1 How did Luigi Galvani's discovery that muscles twitch when stimulated by electrical current lead to the discovery of the ECG and EEG?

■ In 1791, an Italian scientist, Luigi Galvani, discovered that the calf muscle of frog could be made to twitch under electrical stimulation. Galvani concluded that the "animal electricity" was produced by the muscle. Although Galvani's conclusion was incorrect, it spawned a flood of research that led to the development of theories about how electrical current is generated in the body. The EEG and ECG monitor electrical fields in the brain and heart, respectively.

2 During World War I, physicians noted a phenomenon called "phantom pains." Soldiers with amputated limbs complained of pain or itching in the missing limb. Use your knowledge of sensory nerves and the central nervous system to explain this phenomenon.

■ If a leg had been severed, but the cell body of the sensory neuron remained intact, the neuron continued to send information to the CNS. The area of the CNS that was stimulated by the nerve was for the severed area of the leg. Although the section of the leg was removed, the sensory neuron continued to indicate that the area was irritated.

3 Explain the importance of a properly functioning myelin sheath by outlining the pathology of multiple sclerosis.

■ Multiple sclerosis is caused by the destruction of the myelin sheath that surrounds the nerve axons. The myelinated nerves in the brain and spinal cord are gradually destroyed as the myelin sheath hardens and forms scars or plaques. The destruction of the sheath results in short-circuits. Referred to as MS, multiple sclerosis often produces symptoms of double vision, speech difficulty, jerky limb movements, and partial paralysis of the voluntary muscles.

4 Explain why damage to the gray matter of the brain is permanent, while minor damage to the white matter may only be temporary.

■ Nerves within the brain that contain myelinated fibers and a neurilemma are called white matter. The neurilemma promotes regeneration of damaged axons. Some nerve cells within the brain and spinal cord, referred to as the gray matter, lack a myelin sheath. These nerve fibers will not be regenerated after injury. Damage to the gray matter is permanent.

5 Use information that you have gained about threshold levels to explain why some individuals can tolerate more pain than others.

■ Some individuals have nerves that have a higher threshold level and are not as easily activated. This means that it takes an impulse of greater intensity to activate these sensory neurons; hence, the person tends not to feel pain as easily.

6 Use the diagram below to answer the following question. The neurotransmitter released from nerve X causes the postsynaptic membrane of nerve Y to become more permeable to sodium. However, the neurotransmitter released from nerve W causes the postsynaptic membrane of nerve Y to become less permeable to sodium but more permeable to potassium. Why does the stimulation of neuron X produce an action potential in neuron Y, but fail to produce an action potential when neurons X and W are stimulated together?

■ Neuron X is an excitatory neuron. Its neuron transmitter causes the depolarization of neuron Y, making it easier to reach threshold level. Neuron W is an inhibitory neuron. Its neuron transmitter causes the hyperpolarization of the resting membrane Y, making it more difficult to achieve threshold level. When X and Y are stimulated together their effects on neuron Y cancel each other.

7 Botulism (a deadly form of food poisoning) and curare (a natural poison) inhibit the action of acetylcholine. What symptoms would you expect to find in someone exposed to botulism or curare? Provide an explanation for the symptoms.

■ Without acetylcholine being produced, postsynaptic membranes would not depolarize. Nerve transmission would be severely inhibited, because although sensory nerves would respond to stimuli, motor neurons could not be excited. Walking, sitting, and other movements that require motor nerves could not occur. Even breathing movements would stop.

8 Nerve gas inhibits the action of cholinesterase. What symptoms would you expect to find in someone exposed to nerve gas? Provide an explanation for the symptoms.

■ The postsynaptic membrane would not undergo a recovery period. Without a refractory period, the neuron would not be able to receive another impulse.

9 A patient complains of losing his sense of balance. A marked decrease in muscle coordination is also mentioned. On the basis of the symptoms provided which area of the brain might a physician look to for the cause of the symptoms?

■ The cerebellum is associated with these functions.

10 In a classical experiment performed by Roger Sperry, two patients viewed the word "cowboy." The first subject acted as a control. The second subject had had his corpus callosum severed. Note the right and left hemispheres have been supplied with different stimuli.

Predict the results of the experiment and provide an explanation for your predictions.

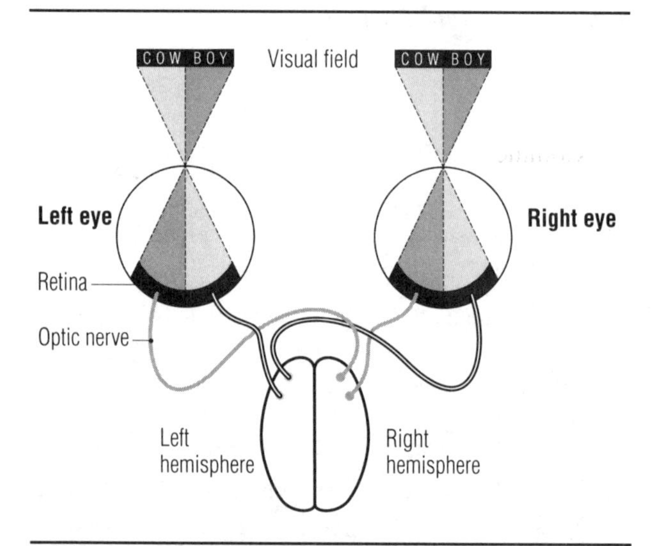

■ The control subject reported seeing the word "cowboy." For the second subject the word "cow" was transferred to the left visual field, the word "boy" to the right visual field. The left hemisphere, which received the word "cow," controls language. When asked what word he saw, the subject would say "cow." If the subject were to write a word with his left hand, he would write "boy," but would not be able to speak a word to indicate what he had written. When asked to write the word with the right hand, he would write "cow." The information from the right and left hemispheres would not be transferred. (Students should only be expected to indicate that the two hemispheres do not communicate information. The answer provided in the key gives a more complete explanation than could be expected from the students.)

CRITICAL-THINKING QUESTIONS

(page 378)

1 People with Parkinson's disease have low levels of the nerve transmitter dopamine. In 1982, a group of Swedish scientists grafted cells from a patient's adrenal glands into the patient's brain. The adrenal glands produce dopamine. In July 1987, Swedish scientists announced the first transplant of human fetal brain cells into other animals. In September 1987, officials at several hospitals in Mexico City announced the transplant of brain tissue from dead fetuses into patients with Parkinson's disease. Although this research is very new, it would appear that the fetal cells begin producing dopamine. Patients with Parkinson's disease who have received the transplanted tissue have demonstrated remarkable improvement. Comment on the ethics of this research.

■ Many opinions can be explored. Students should be encouraged to listen to other viewpoints. Most importantly, students should recognize that the answer to the question is found not within the discipline of science, but that of ethics. The question illustrates the limits of using a scientific paradigm for problem solving. It is important for students to recognize that the tissues are not taken from living fetuses. Have students considered that organs for transplants have been taken from other life forms? In chapter 18 they may have read about the transplant of liver cells from a mother to her child. A variety of religious concerns may be expressed. Although not all individuals in the class may agree with these beliefs, it is important that they respect and acknowledge them.

2 Research by Dr. Bruce Pomerantz of the University of Toronto has revealed a link between endorphins and acupuncture. Dr. Pomerantz believes that acupuncture needles stimulate the production of pain-blocking endorphins. Although acupuncture is a time-honored technique in the east, it is still considered to be on the fringe of modern western medical practice. Why have western scientists been so reluctant to accept acupuncture?

■ Western science is based on determining cause-and-effect relationships. Each step is broken down to causal relationships in an attempt to find an answer. Eastern science, by contrast, traditionally has looked at questions more holistically. Although we tend to be more comfortable with western science, eastern science has much to offer. However, not all of the answers for acupuncture are known (following western tradition—we don't know each component). Therefore, western scientists approach it cautiously.

3 Painkillers are big business. Television commercials tell us that the answer to our headaches, sore backs, or sore muscles is a pill. Should people be encouraged to take pills? What are the implications for individuals who refuse to give their children pills for pain?

■ A variety of opinions can be explored. Advertisers will indicate that their intention is to provide the public with needed information. Responsible advertising must be informative, but it must also be entertaining enough to capture the attention of the viewing audience. Students might also explore the implications of not taking pain-killing drugs. Do individuals build up a tolerance against the painkillers? Do these drugs cause side effects? What are the implications of not reducing pain levels for individuals who are suffering? Expect the solutions to differ widely.

4 Aspirin and Tylenol were both marketed long before their action was defined. Although both drugs are generally considered safe, they are not without side effects. Aspirin in high dosages can destroy the protective lining of the stomach. Tylenol in high dosages can cause liver damage. Should the biochemical mechanism of drug action be defined before a drug is marketed? Should drugs that produce side effects be sold over the counter? What tests would you like to see before drugs are approved for sale?

■ Many answers can be explored. This question helps students engage in a conversation that provides a better understanding of the nature of scientific research. Determining whether or not a drug should be taken requires careful examination of risks and benefits. It is important to note, however, that genetic diversity within a population makes analysis of risks very difficult to determine. A drug that might produce no measurable side effect in one person may cause considerable harm for another. Students should advocate testing of the drug for many individuals within a population, if they are willing to accept human testing. Random samples are most often suggested. Some students may suggest using double-blind studies in which one group is unknowingly given a placebo. Other students may reject human testing altogether.

5 The EEG has been used to legally determine death. Although the heart may continue to beat, the cessation of brain activity signals legal death. Ethical problems arise when some brain activity remains despite massive damage. Artificial resuscitators can assume the responsibilities of the medulla oblongata and regulate breathing movements. Feeding tubes can supply food, and catheters can remove wastes when voluntary muscles can no longer be controlled. The question of whether life should be sustained by artificial means has often been raised. Should a machine like the EEG be used to define the end of life?

■ Clearly this question extends beyond the boundaries of science, but requires an understanding of science and technology to frame the question and seek solutions. By approaching this question, students will gain an appreciation of how science and technology are changing the way in which we define life and death. Most student answers are rooted in personal, moral, and religious beliefs.

6 According to the study of craniometry, males were once considered more intelligent than females because they have larger skulls. Outline some of the ethical dangers associated with using skull size to define intelligence.

■ Skull size is commensurate with body size, not brain intelligence. If skull size marked intelligence, blue whales and elephants would dominate thinking.

CHAPTER 16

Special Senses

INITIATING THE STUDY OF CHAPTER SIXTEEN

Chapter 16 provides an introduction to coordination and regulation by studying the manner in which humans gather and assess sensory information. The following suggestions provide examples for beginning the chapter:

1 Try various sensory adaptations with the class. One of the adaptations is described in Figure 16.2. Students can be encouraged to provide their own examples of sensory adaptations. The sharp smell associated with formalin-preserved animals is often mentioned. After some time to adjust, most people indicate that there is only a faint smell.

2 Ask students to list sensory information that they are gathering as class begins. The feel of clothes, sounds coming from the hallway, and even the sensation of food in their stomach may be recorded.

ADDRESSING ALTERNATE CONCEPTIONS

- Students are provided with a second opportunity to confront the idea that the sensory neurons of the eye transmit light or that auditory neurons transmit sound. As mentioned in the previous chapter, a neuron that carries visual information from the eye functions in essentially the same manner as one that carries auditory information from the ear. Both carry electrochemical impulses. The ability to differentiate between visual and auditory information depends upon the area of the brain that receives the impulse. The brain interprets your reality. Visual information is stored in the posterior portion of the occipital lobe of the cerebrum, while auditory information is stored in the temporal lobe of the cerebrum.
- Students are often surprised to discover that despite an incredible collection of specialized sensory receptors, much of our environment is not detected. What we detect are stimuli relevant to our survival. Ask them to consider the stimuli from the electromagnetic spectrum. They experience no visual sensation from radio waves, infrared wavelengths, and ultraviolet wavelengths. Humans can only detect light with wavelengths between 350 nm and 800 nm. Our range of hearing, at least compared with many other species, is also limited.
- Although students are aware of blind spots when driving, most fail to realize a physiological connection. There are no rods and cones in the area in which the optic nerve comes in contact with the retina. Because of the absence of photosensitive cells, this area is appropriately called the blind spot. The demonstration of blind spot found in the laboratory, Vision, will help students link experiential learning with the knowledge of eye structure and function.
- Some students who listen to extremely loud music believe that hearing will be restored with rest. They are often surprised to find out that hearing loss associated with the destruction of the hair cells of the cochlea is permanent.

MAKING CONNECTIONS

- In many respects the laws of thermodynamics introduced in chapter 2, Energy and Ecosystems, and chapter 7, Energy within the Cell, provide a basis for understanding sensory receptors. A stimulus is a form of energy. Sensory receptors convert one source of energy into another. For example, taste receptors in the tongue convert chemical energy into a nerve action potential, a form of electrical energy. Light receptors of the eye convert light energy into electrical energy, and balance receptors of the inner ear convert gravitational energy and mechanical energy into electrical energy.

POSSIBLE JOURNAL ENTRIES

- Although his view was incorrect, Galen provided biologists of the second century with an insightful view of how the human eye worked. His research occurred centuries before physicists, such as Newton and Huygens, developed a comprehensive theory to explain light and provided a scientific model to begin thinking about vision. Ask students to comment on the contributions made to science by individuals who provided theories not currently accepted today.
- Students can be encouraged to develop a concept map that includes all key terms presented in the chapter.
- Students can be encouraged to express the difficulties they experienced in formulating an understanding of this challenging chapter. What things did they employ to aid them in constructing their knowledge? For example, some students may indicate that their understanding of the eye was aided by the Table 16.2, which summarized the structures and functions of the parts of the eye. Other students may have found the analogy of the eye and camera useful. The journal entry can provide students with more information on how they learn.
- Group thinking and decision-making strategies may be recorded as students prepare for the debate. Did the students change their minds during the preparation for the debate? Students may even be asked to record their initial positions about whether rock concerts are potentially harmful and to reflect upon these feelings after the debate has been completed.

USING THE VIDEODISC

Section title	Page	Code	Frame (side)	Description
Importance of Sensory Information	380 – 381		3192 (all)	Sensory neurons
What Are Sensory Receptors?	381 – 382		3150 (all)	Proprioceptors
Structure of the Eye	384 – 386		2837 (all)	Compound eye (insects)
Structure of the Eye	384 – 386		3177 (all)	Retina
Structure of the Eye	384 – 386		2923 (all)	Fovea
Structure of the Eye	384 – 386		2839 (all)	Cone
Structure of the Eye	384 – 386		3182 (all)	Rod
Structure of the Ear	384 – 386		2827 (all)	Cochlea

IDEAS FOR INITIATING A DISCUSSION

Figure 16.2: Sensory adaptation to temperature

Encourage various students to attempt the demonstration. The beaker containing hot water should be tested prior to immersing the hand. Use temperatures of about 50°C.

Figure 16.3: Taste receptors on the tongue

Encourage students to map their tongues at home. This demonstration is not recommended in schools.

Figure 16.6: Nerve pathway from the retina to the brain

Every year racquetballs detach people's retinas. Ask students to research racquetball injuries. Why should eye protection always be worn?

REVIEW QUESTIONS

(page 384)

1 Identify a sensory receptor for each of the following stimuli: chemical energy, mechanical energy, heat energy, light energy, and sound energy.

- Examples of a sensory receptor for each of the stimuli are: chemical energy—chemoreceptors in the carotid artery; mechanical energy—proprioceptors; heat—thermoregulators; light—eyes; sound energy—ears. A variety of other examples may be provided.

2 Do sensory receptors identify all environmental stimuli? Give examples to back up your answer.

- The sensory receptors in humans do not identify all environmental stimuli. For example, humans cannot detect radio waves, or indeed most of the electromagnetic spectrum apart from visible light.

3 Explain the concept of "sensory adaptation" by using examples of olfactory stimuli and auditory stimuli.

- Sensory adaptation occurs when sensory receptors become accustomed to an environmental stimulus, and fail to respond to it. For example, you quickly become accustomed to the smell of the ocean, and to the ticking of a clock in your bedroom.

4 Draw a diagram of the various chemoreceptors on the tongue.

■ See Figure 16.3 for reference.

5 Why are you less able to taste food when you have a cold?

■ Your sense of taste is reduced when you have a cold because your sense of smell and taste complement each other. A cold reduces your capacity to smell, thus reducing your sense of taste.

(page 393)

6 List the three layers of the eye and outline the function of each layer.

■ The three layers of the eye are the sclera, the choroid layer, and the retina. The sclera is a protective layer, the choroid layer focuses light on the retina, which captures the light and signals the brain.

7 Indicate the function of each of the following parts of the eye: vitreous humor, aqueous humor, cornea, pupil, iris, rods, cones, fovea centralis, and blind spot.

■ The eye contains these parts: the vitreous humor, which fills the large chamber behind the lens and maintains the shape of the eye; the aqueous humor, a transparent fluid behind the cornea which supplies the cornea with nutrients and refracts light; the cornea, which is a clear covering over the eye; the pupil, a hole in the iris through which light passes; rods, which respond to low-intensity light; cones, which respond to high-intensity colored light; the fovea centralis, a very sensitive part of the retina which contains only cones; and the blind spot, the area of the retina where the optic nerve is attached, and which does not contain any rods or cones.

8 How did the experiment performed by Wilhelm Kuhne in the 1880s provide an answer for the phenomenon of "positive afterimages"?

■ Positive afterimages are images that can still be seen after you close your eyes. In the 1880s, Wilhelm Kuhne demonstrated that the retina acts like a photographic film, storing images for a short period of time. Thus, strong images like bright lights will remain on the retina after your eyes are shut.

9 What are accommodation reflexes?

■ Accommodation reflexes are adjustments made by the lens and pupil to focus images of objects close to and far away from the eye.

10 Outline the mechanism by which rhodopsin converts light energy into a nerve impulse.

■ When photons of light strike molecules of rhodopsin, the rhodopsin divides into two components, retinene and opsin. This division produces an action potential, releasing transmitter substances from the end plates of the rods.

11 Why is vitamin A essential for proper vision?

■ Vitamin A is essential for proper vision because it is an essential part of the rhodopsin molecule. Without vitamin A, the molecule cannot form into its light-sensitive structure.

12 Why do the rods not function effectively in bright light?

■ The rods are very sensitive, and in bright light they break down faster than they can reform. Thus, they are most useful in situations where the supply of light is limited.

13 What three color-sensitive pigments are found in the cones?

■ The cones provide you with color vision. In the cones, opsin combines with one of three color-sensitive pigments, either red, blue, or green.

14 Identify the causes for each of the following eye disorders: glaucoma, cataract, astigmatism, nearsightedness, and farsightedness.

■ There are several common visual defects. These include glaucoma, which is caused by a build-up of pressure in the aqueous humor; cataract, which is a clouding of the lens or capsule; astigmatism, a condition where the lens or cornea is irregularly shaped; nearsightedness or myopia, where the eyeball is too long; and farsightedness or hyperopia, where the eyeball is too short.

(page 399)

15 Briefly outline how the external ear, middle ear, and inner ear contribute to hearing.

■ The external ear collects sound and carries it to the eardrum via the auditory canal. The middle ear starts at the eardrum and extends to the oval and round windows. Sound vibrations that strike the eardrum are amplified and concentrated as they are transferred to the oval window. The inner ear identifies sound waves and converts them into nerve impulses.

16 What function do the tympanic membrane, ossicles, and oval window serve in sound transmission?

■ The tympanic membrane, or eardrum, concentrates sounds to the ossicles. The ossicles then transfer the sound vibrations to the oval window. The oval window has a much smaller surface area than the tympanic membrane; thus, the sound is greatly amplified.

17 Categorize the following structures of the inner ear according to whether their functions relate to balance or hearing: organ of Corti, cochlea, vestibule, saccule, ampulla, semicircular canals, oval window, round window.

■ The inner ear has two functions: it is associated with both balance and hearing. The cochlea, organ of Corti, oval window, and round window are associated with hearing. The saccule, ampulla, and semicircular canals are associated with balance.

18 In 1660, Robert Boyle placed an alarm clock in a bell jar from which air was then removed. Why was Boyle not able to hear the alarm clock?

■ Boyle was not able to hear the alarm in the bell jar because sound requires a medium in order to propagate. The inside of the bell jar was a vacuum; thus, no sound was heard.

19 Explain how the ligaments, tendons, and muscles that join the ossicles of the middle ear provide protection against high-intensity sound.

■ As a safety mechanism, small muscles in the middle ear can partially immobilize the malleus and move the stapes away from the oval window, thus, reducing the intensity of sounds.

20 Differentiate between static and dynamic equilibrium.

■ Static equilibrium is movement of the head along one plane such as the horizontal or the vertical. Information is provided to indicate the position of the head. Dynamic equilibrium is the movement of the entire body, information is provided to maintain balance.

21 How do the saccule and utricle provide information about head position?

■ The saccule and utricle are two fluid-filled sacs. The jelly-like fluid contains calcium carbonate particles known as otoliths. Cilia in the sacs interpret the movement of the otoliths, providing information about the head position.

22 Describe how the semicircular canals provide information about body movement.

■ Three fluid-filled semicircular canals provide information about body movement. Rotational stimuli cause the fluid in the canals to move, which bends the cilia. The bent cilia initiate nerve impulses, which are carried to the brain and interpreted as body movements.

LABORATORY

VISION

(page 394)

Time Requirements

Approximately 40 min are required to complete the laboratory.

Notes to the Teacher

Snellen eye charts can be purchased from most biological supply houses.

Observation Questions

a) Indicate the visual acuity of your left eye.

■ Answers will vary.

b) Indicate the visual acuity of your right eye.

■ Answers will vary.

c) Measure the distance of the book from your left eye.

■ Answers will vary.

d) Measure the distance of the book from your right eye.

■ Answers will vary.

e) Describe what happened to the position of the pencil.

■ The pencil moved.

f) Describe what happened to the position of the pencil.

■ The pencil moved.

g) Which is the dominant eye? (The object appears to move the least in the dominant eye.)

■ Answers will vary.

h) Survey the people in your class to determine if right-handed people have dominant right eyes or if left-handed people have dominant left eyes.

■ Answers will vary.

i) Describe what you see.

■ The fingers look like little sausages.

j) Record your results.

■ Answers will vary.

k) Record your results.

■ Answers will vary.

Laboratory Application Questions

1 In an attempt to map the position of the blind spot, a student rotates a book 180° and follows the same steps described in part II. How would rotating the text help the student map the position of the blind spot?

■ The initial procedure helps locate the horizontal coordinate for the blind spot. By moving the book vertically, the vertical coordinate for the blind spot can be located.

2 In part III of the laboratory you discovered that the pencil moved when you viewed it with a different eye. Offer an explanation for your observation.

■ Each eye views the image from a different angle. Depth perception is provided because the brain is able to superimpose one image over another. Stereoscopic lenses work in much the same way.

3 Horses and cows have eyes on the sides of their heads; their visual fields overlap very little. What advantage would this kind of vision have over human vision?

■ This increases their field of view. It is more difficult for predators to sneak up on horses and cows.

4 Like humans, squirrels have eyes on the front of their face. The visual field of the squirrel's right eye overlaps with the visual field of its left eye. Why does a squirrel need this type of vision ?

■ Animals that live in trees need good depth perception. Estimating the distance of a branch is crucial for an animal living in a tree.

5 Students place a large piece of cardboard between their eyes while attempting to read two different books at the same time. Will they be able to read two books at once? Provide an explanation.

■ No, the two hemispheres do not work independently. One eye always wins out, most often the dominant eye.

LABORATORY
HEARING AND EQUILIBRIUM

(page 400)

Time Requirements

Approximately 40 min are required to perform the experiment.

Notes to the Teacher

Students will have a tendency to increase the number and speed of rotations during part II of the laboratory. Increasing the speed of rotations will cause motion sickness in some individuals.

Observation Questions

a) From which direction does the sound appear to be coming?

■ It appears to come from the right ear.

b) Describe any changes in the intensity of the sound.

■ The sound intensity seems to increase once the right ear is closed.

c) From which direction does the sound appear to be coming?

■ It appears to come from the left ear.

d) Describe any changes in the intensity of the sound.

■ The sound intensity seems to increase.

e) Describe any changes in the intensity of the sound.

■ Sound traveling through the meter stick seems more intense.

f) In which direction did the subject lean?

■ The subject leaned in the opposite direction to the spinning: to the left.

g) In which direction did the subject lean?

■ The subject leaned to the right.

h) Ask the subject to describe the sensation.

■ The subject usually feels as if he or she is falling.

Laboratory Application Questions

1 Provide explanations for the data collected in observation questions a, b, c, and d.

■ Sound waves are conducted faster through denser media. By closing the ear, vibrations travel through bone, which is denser than air. The sound intensity increases as sound waves move through bone. Much the same phenomenon occurs when sound waves travel through the wooden or plastic meter stick.

2 Using the data collected, provide evidence to suggest that sound intensity is greater in fluids than in air.

■ The denser the media are, the greater the sound intensity is. Although liquids are not used, solids have greater density than air.

3 Using the data collected, provide evidence to suggest that the fluid in the semicircular canals continues to move even after rotational stimuli cease.

■ The subject who was rotated in the chair continued to experience the rotational stimulus even after it had ended. Students also know that they feel dizzy even after they stop spinning. Once the fluid stops, the dizzy sensation stops.

4 What causes the falling sensation produced in step 8?

■ By placing the head on the shoulder, the fluid in the vertical semicircular canal is set into motion by a horizontal rotational stimulus. Once the head is lifted from the shoulder the inertia of the fluid causes it to continue moving; however, the movement is now vertical. The subject feels as if he or she is falling or dropping.

5 Describe the manner in which the semicircular canals detect changes in motion during a roller-coaster ride.

■ As you move downward along the roller coaster the fluid in the vertical semicircular canals moves upward. Actually the canal begins moving prior to the fluid. Once the fluid (endolymph) is set into motion, it continues in motion even after the subject changes direction. Eventually, the downward motion of the subject reverses the movement of fluids in the semicircular canal, causing a downward sensation. For a split second the visual information does not match the sensory information provided by the semicircular canals, causing motion sickness. Once again the subject is pushed upward.

APPLYING THE CONCEPTS

(page 403)

1 A person steps from a warm shower and feels a chill. Upon stepping into the same room another person says that the room is warm. What causes the chill?

■ Your temperature receptors are not designed to identify specific temperatures, but changes in temperature. Once your sensory receptors become accustomed to a stimulus, the neurons cease to fire. The phenomenon is referred to as sensory adaptation. This causes the body's thermostat to readjust. Once the subject steps out of the warm shower, the cooler, room-temperature air is detected by the cold temperature receptors of the body. Although the cold receptors have not identified something which is cold, they have identified a change in environment. The room is comparatively cold.

2 If a frog is moved from a beaker of water at 20°C and placed in a beaker at 40°C, it will leap from the beaker. However, if the temperature is slowly heated from 20 to 40°C, the frog will remain in the beaker. Provide an explanation for the observation.

■ The frog's nervous system adjusts to the gradual changes in temperature. The sensory nervous system is not designed to respond to specific temperatures, but to temperature changes.

3 The retina of a chicken is composed of many cones very close together. What advantages and disadvantages would be associated with this type of eye?

■ This enables the chicken to distinguish a seed on the ground. Subtle differences in color and texture can be detected. (The cones are high light intensity photoreceptors that detect color.) However, the chicken's vision would be poor in situations with low light intensity

4 Why do people often require reading glasses after they reach the age of 40?

■ The lens hardens as you age; extra protein layers are added. This decreases the near-point accommodation, and people tend to become unable to focus on objects near their eyes.

5 Indicate how the build-up of fluids in the eustachian tube may lead to temporary hearing loss.

■ The fluids may become packed against the tympanic membrane, slowing its vibrations. Occasionally, the fluid may even exert pressure on the middle ear itself.

6 A scientist replaces ear ossicles with larger, lightweight bones. Would this procedure improve hearing? Support your answers.

■ The lightweight bones might move more easily, but they would also be less dense. The increased amplification of the sound wave occurs because the ossicles are small and dense. Sound is concentrated from a large area into a much smaller area.

7 When the hearing of a rock musician was tested, the results revealed a general deterioration of hearing as well as total deafness for particular frequencies. Why is the loss of hearing not equal for all frequencies?

■ Only certain specialized hair cells along the basilar membrane of the cochlea have been destroyed. Each frequency or pitch excites a specific part of the cochlea. Violent vibration may dislodge the specialized hairs. See the Social Issue: Rock Concerts and Hearing Damage for a picture.

CRITICAL-THINKING QUESTIONS

(page 403)

1 One theory suggests that painters use less purple and blue as they age because layers of protein are added to the lens in their eyes and it gradually becomes thicker and more yellow. The yellow tint causes the shorter wavelength to be filtered. How would you test the theory?

■ Simple color discrimination tests might be suggested. It is important to consider the age of the subject, light intensity and distance from the object. A control should also be introduced.

2 Should individuals who refuse to wear ear protectors while working around noisy machinery be eligible for medical coverage for the cost of hearing aids? What about rock musicians or other individuals who knowingly play a part in the loss of their hearing?

■ Many opinions can be explored. Students should be encouraged to listen to all other viewpoints. Most importantly, students should recognize that the answer to the question is found not within the discipline of science, but that of ethics. The question illustrates the limits of using a scientific paradigm for problem solving.

UNIT 5

Continuity of Life

SUGGESTIONS FOR INTRODUCING UNIT 5

This unit provides a connection between human physiology and cellular reproduction. By beginning at the macroscopic level, students are provided with a personal reference for constructing their knowledge about the continuity of life in chapter 17. Chapter 18 examines asexual cell reproduction, in humans and other organisms. Chapter 19 examines reduction division and the formation of gametes.

Students may be asked to list occupations associated with the study of reproduction in their personal journals. Most will provide a very narrow list, which regards reproduction in personal terms. Few will consider cancer research as a study of cell reproduction, nor will many acknowledge aging as cell replacement and reproduction. Their list may be examined at the end of the unit. What additions would they make and why?

KEY SCIENCE CONCEPTS

Unit 5, Continuity of Life, links **equilibrium,** investigated through physiological systems, with the key concepts of **change,** introduced by the reproductive system. Students have the opportunity to follow the changes in the human reproductive structures through maturation and pregnancy to the non-reproductive years. They will also learn about the changes that occur during asexual cell reproduction and gamete formation. This unit reflects the principles developed in Units 2 and 3 of the Alberta Biology 30 Program of Studies. ■

CHAPTER 17

The Reproductive System

INITIATING THE STUDY OF CHAPTER SEVENTEEN

Chapter 17 provides an introduction to continuity of life. The following suggestions provide examples for beginning the chapter:

1 Ask students to begin by reading Research in Canada (page 426). Ask them about the significance of Dr. Martin's work.
2 Ask students to write a number of questions they have about reproduction and place them in a cardboard box. Selected questions will be

addressed at the end of the unit. Students can be asked to provide the answers.

ADDRESSING ALTERNATE CONCEPTIONS

- Students are often surprised to discover how similar males and females are. Male and female sex organs originate in the same area of the body, the abdominal cavity, and are almost indistinguishable until about the third month of embryonic development. At that time, genes cause differentiation. During the last two months of fetal development, the testes descend through a canal into the scrotum. In addition, males have low levels of estrogen and females have low levels of male sex hormones.
- Although females have a limited number of egg cells available for reproduction, and their reproductive years end at about 50 years of age, males' sperm cell production may occur daily and continue well into their 80s or 90s.
- The 28 day reproductive cycle presented in the text is the norm, but in no way reflects what occurs with every woman. Many females function quite normally with reproductive cycles of various lengths.
- Students often assume that day 14 is ovulation day, regardless of the length of the female's menstrual cycle. Day 14 represents what usually happens within a 28 day cycle; however, many women do not have a 28 day reproductive cycle. Should the cycle be longer, ovulation may occur at day 16, 18, 20, or even later.
- In many texts, examples of fetal distress are linked exclusively with drug or alcohol consumption by the mother. The work of Dr. Renee Martin, described in the section, Research in Canada, provides examples of how the father can also be associated with birth defects.

MAKING CONNECTIONS

- Reproduction insures the survival of a species. The diversity produced by new gene combinations provides a basis for natural selection, as introduced in chapter 4, Adaptation and Change, and presented in greater detail in chapter 24, Population Genetics. Only the best-adapted survive. Survival, then, is based on providing numerous and varied offspring.
- The reproduction chapter serves as a bridge between physiology and genetics. The physiological system provides an overview for sexual and asexual cell division introduced in chapters 18 and 19 and provides the foundation for understanding genes and gene combinations.

POSSIBLE JOURNAL ENTRIES

- Students might be asked to write a dialogue between the text and themselves, as they refute the idea that males are no longer necessary for human reproduction. By acting as a "devil's advocate," they can challenge their own learning and push understanding to a higher level.
- Students can be encouraged to develop a concept map that includes all key terms presented in the chapter.
- Students can be encouraged to express the difficulties they experienced in formulating an understanding of this challenging chapter. What things did they employ to aid them in constructing their knowledge? For example, some students may indicate that they were helped by the case study, Hormone Levels during the Menstrual Cycle. If completed by small groups, students might learn that active discussion is an effective way for them to learn. Other students may have attempted a series of their own drawings or constructed concept maps.
- Decision-making strategies may be recorded as each group prepares for the debate. Did the students change their minds during the preparation for the debate? Students may even be asked to record their initial positions about the social issue and to reflect upon these feelings after the debate has been completed. Did they change their minds after listening to opposing arguments? Should laws be instituted to prevent pregnant women from drinking excessively?

USING THE VIDEODISC

Section title	Page	Code	Frame (side)	Description
Importance of Reproduction	406 – 407		39043 (2)	Movie sequence showing echinoderm development; time-lapse photography of the egg cell dividing following fertilization
The Male Reproductive System	407 – 408		2665 (all)	Human sperm
The Male Reproductive System	407 – 408		2668 (all)	Male reproductive system
Testes and Spermatogenesis	408 – 409		3205 (all)	Sperm cells
Testes and Spermatogenesis	408 – 409		3208 (all)	Spermatocytes
Testes and Spermatogenesis	408 – 409		3209 (all)	Spermatogenesis
Testes and Spermatogenesis	408 – 409		3249 (all)	Testis
Testes and Spermatogenesis	408 – 409		33467 (2)	Movie sequence showing spermatogenesis
Seminal Fluid	410		3191 (all)	Seminal vesicles
Seminal Fluid	410		3153 (all)	Prostate gland
Hormonal Control of the Male Reproductive System	410 – 411		3250 (all)	Testosterone
The Female Reproductive System	412 – 413		2669 (all)	Female reproductive system
The Female Reproductive System	412 – 413		2884 (all)	Egg
The Female Reproductive System	412 – 413		3091 (all)	Ovum
The Female Reproductive System	412 – 413		34543 (2)	Movie sequence showing human ovary and uterus
Menstrual Cycle	415 – 416		2893 (all)	Endometrium
Menstrual Cycle	415 – 416		2902 (all)	Estrogen
Menstrual Cycle	415 – 416		3145 (all)	Progesterone
Menstrual Cycle	415 – 416		35139 (2)	Movie sequence showing ovulation and transport of the egg in the Fallopian tube

Section title	Page	Code	Frame (side)	Description
Fertilization and Pregnancy	420 – 421		2670 (all)	Fertilization
Prenatal Development	421 – 422		2909 (all)	Extra embryonic membranes
Prenatal Development	421 – 422		2722 (all)	Amnion
Prenatal Development	421 – 422		2815 (all)	Chorion
Prenatal Development	421 – 422		2914 (all)	Fetus
Prenatal Development	421 – 422		3120 (all)	Placenta

IDEAS FOR INITIATING A DISCUSSION

Figure 17.3: Human sperm cell

Ask students to indicate how the sperm cell is specialized for movement. The fusiform shape, reduced cytoplasmic mass, and the placement of mitochondria next to the flagellum might be mentioned.

Figure 17.10: Sperm cell attached to ovum

Students might be referred to the June 1992 issue of *Discover* magazine. The entire issue is devoted to reproduction.

REVIEW QUESTIONS

(page 411)

1 Name the primary male and female reproductive organs.

■ The primary male reproductive organs are the testes and the primary female reproductive organs are the ovaries.

2 What would happen if the testes failed to descend into the scrotum?

■ If the testes did not descend into the scrotum, the male would not be able to produce viable sperm. Sperm do not develop at body temperature; they require an environment several degrees cooler.

3 Describe the function of the following structures: Sertoli cells, seminiferous tubules, and epididymis.

■ Sertoli cells are support cells that nourish the developing sperm. Seminiferous tubules are long, thin tubes about 250 m in length where the immature sperm cells divide and differentiate. The epididymis is a coiled tube found along the outer edge of the testis where the sperm mature.

4 What is semen? Where is it found? What function does it serve?

■ Semen is a male ejaculatory fluid which provides a swimming medium for the sperm cells. Semen is composed of fructose and prostaglandins from the seminal vesicles, mucus-rich fluids from Cowper's gland, and an alkaline buffer from the prostate gland.

5 What is spermatogenesis?

■ Spermatogenesis is the formation of mature sperm cells from spermatogonia.

6 Outline the functions of testosterone.

■ Testosterone is a male sex hormone produced in the interstitial cells of the testes. It is responsible for stimulating spermatogenesis, regulating and influencing the development of secondary male sexual characteristics, and is associated with sex-drive levels.

7 How do gonadotropic hormones regulate spermatogenesis and testosterone production?

■ (While sperm cells develop inside the seminiferous tubules, testosterone is produced in the interstitial cells of the testes. As the name suggests, the interstitial cells are found between the seminiferous cells. Both functions of the testes are directly under the control of the hypothalamus and the pituitary gland. Negative-feedback systems ensure that adequate numbers of sperm cells and constant levels of testosterone are maintained.) The gonadotropic hormones are produced and stored in the pituitary glands. Male follicle stimulating hormone or FSH stimulates the production of sperm cells in the seminiferous tubules. Male luteinizing hormone or LH promotes the production of testosterone by the interstitial cells.

8 Using examples of LH and testosterone, explain the mechanism of negative feedback.

■ At puberty, the hypothalamus secretes gonadotropin-releasing hormone (GnRH). GnRH activates the pituitary gland to secrete and release FSH and LH. The FSH acts directly on the sperm-producing cells of the seminiferous tubules, while LH acts on the interstitial cells to produce testosterone. In turn the testosterone itself increases sperm production. Once high levels of testosterone are detected by the hypothalamus, a negative-feedback system is activated. Decreased GnRH production slows the production and release of LH and leads to less testosterone production. Testosterone remains in check. The feedback loop for sperm production is not well understood. It is believed that FSH acts on Sertoli cells, which produce a peptide hormone that sends a feedback message to the pituitary, inhibiting production of FSH.

(page 418)

9 Diagram the female reproductive system and label the following parts: vagina, ovaries, cervix, oviducts, uterus, and endometrium.

■ See Figure 17.6 for reference.

10 Can a woman who has reached menopause ever become pregnant? Explain your answer.

■ Under normal circumstances, a woman who has reached menopause can no longer become pregnant because she is no longer capable of producing eggs.

11 What is menstruation? Why is it important?

■ Menstruation is the flow phase of the menstrual cycle. In this phase, the uterus sheds the endometrium. Students should infer that if the endometrium were not shed, it could serve as a nutrient for any microbe that invaded the uterus. Also, an increasingly thick endometrium would distend the uterus.

12 Explain why ectopic pregnancies are dangerous.

■ An ectopic pregnancy is one in which the embryo embeds in the oviduct. This type of pregnancy is dangerous because the amounts of nutrients and glandular tissues are limited and because the oviduct is not able to stretch to accommodate the developing embryo.

13 What is a Pap test?

■ A Pap test is a cell sample taken from the cervix that is used to diagnose cervical cancer.

14 Describe the process of ovulation. Differentiate between primary oocytes, secondary oocytes, and mature ova.

■ Ovulation is the release of an egg from a follicle held within the ovary. The follicles are made of two types of cells, the primary oocyte and granulosa cells. The granulosa cells provide nutrients for the oocyte. The primary oocyte undergoes meiosis I, splitting into two cells. One of the cells receives the majority of the nutrients and cytoplasm, and becomes the secondary oocyte. The other cell forms a polar body and soon dies. The follicle pushes outward against the wall of the ovary, eventually bursting it, and releasing the oocyte, which then enters the oviduct. In the oviduct, the oocyte undergoes meiosis II, forming an ovum and a polar body which soon dies.

15 Describe how the corpus luteum forms in the ovary.

■ The corpus luteum is formed by follicle cells surrounding the dominant follicle cell. The corpus luteum secretes the hormones estrogen and progesterone, which are essential for pregnancy. These hormones maintain the endometrium.

16 Describe the events associated with the flow phase, follicular phase, and luteal phase of menstruation.

■ The flow phase of the menstrual cycle is marked by the shedding of the endometrium. This process takes about

5 days. This next phase is know as the follicular phase, during which time the follicle develops within the ovary. This phase lasts between 6 and 13 days, and is also characterized by high concentrations of the estrogen hormone. This phase is followed by ovulation, and finally by the luteal phase. The start of the luteal phase is signalled by the development of the corpus luteum.

17 Outline the functions of estrogen and progesterone.

■ Two female sex hormones are estrogen and progesterone. Estrogen promotes the development of female secondary sex characteristics, including the development of body hair, breasts, and the thickening of the endometrium. Progesterone also stimulates the endometrium, and inhibits ovulation and uterine contractions.

18 How do gonadotropic hormones regulate the function of ovarian hormones?

■ The gonadotropic hormones FSH and LH regulate the production of estrogen and progesterone. FSH stimulates follicle development in the ovary. The follicle secretes estrogen. After ovulation, LH stimulates the transformation of the remaining follicles into the corpus luteum. The corpus luteum secretes both estrogen and progesterone.

19 Predict how low secretions of GnRH from the hypothalamus would affect the female menstrual cycle.

■ Low secretion of gonadotropin-releasing hormone (GnRH) from the hypothalamus may affect the release of FSH and LH from the pituitary gland. If these hormones are not re!eased, the menstrual cycle may not be initiated.

20 With reference to the female reproductive system, provide an example of a negative-feedback control system.

■ A negative-feedback control system is one in which the products of the reaction inhibit the initial stimulating agents. In the case of the female reproductive system, FSH stimulates the production of estrogen. As the estrogen concentration rises, the production and secretion of FSH is inhibited.

DEMONSTRATION
MICROSCOPIC VIEW OF THE TESTES

(page 412)

Time Requirements

About 10–15 min are required.

Observation Questions

a) Estimate the number of seminiferous tubules seen under low-power magnification.

■ Answers will vary.

b) Estimate the size of the seminiferous tubule and the interstitial cell.

■ It is approximately 70 μm, but diameter will vary depending upon location.

c) Diagram five different cells viewed in the seminiferous tubule.

■ See Figure 17.2 for reference.

d) Would you expect to find mature sperm cells in the seminiferous tubule? Give your reasons.

■ Although sperm cells are produced in the testes, they mature in the epididymis, a compact, coiled tube attached to the outer edge of the testes. Sperm cells develop their flagella in the epididymis and begin swimming motions within 4 days.

CASE STUDY
HORMONE LEVELS DURING THE MENSTRUAL CYCLE

(pages 419–420)

Time Requirements

Approximately 20 min are required.

Observation Questions

a) Identify as w, x, y, or z, the two gonadotropic hormones represented in the diagram.

■ FSH or follicle stimulating hormone is represented by x and LH or luteinizing hormone is represented by z.

b) Identify the ovarian hormones shown in the diagram.

- Estrogen is represented by w and progesterone and some estrogen are represented by y.

c) Which two hormones exert negative-feedback effects?

- Estrogen and progesterone exert negative-feedback effect.

d) Graph the data provided. Plot changes in temperature along the *y*-axis (vertical axis) and the days of the menstrual cycle along the *x*-axis (horizontal axis).

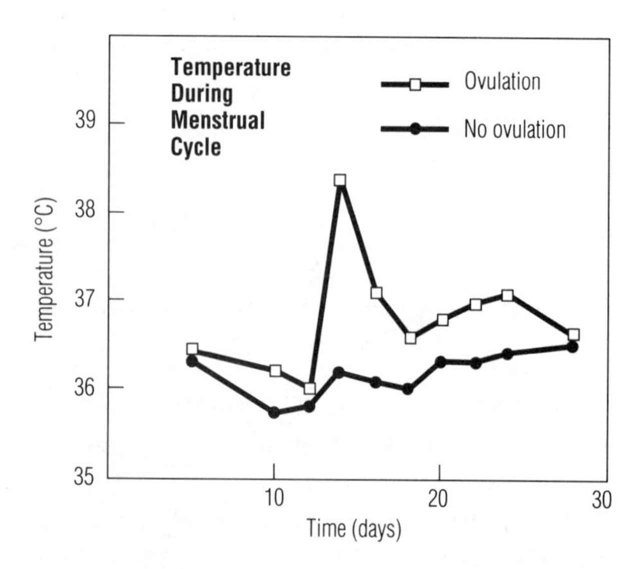

e) Assuming the menstrual cycle represents the average 28-day cycle, label the ovulation day on the graph.

- Ovulation occurs on the 14th day.

f) Describe changes in temperature prior to and during ovulation.

- The temperature decreases prior to ovulation and rises above normal body temperature during ovulation.

g) Compare body temperatures with and without a functioning corpus luteum.

- Body temperatures with a functioning corpus luteum are higher than those without.

h) Identify the events that occur at times *X* and *Z*.

- Ovulation occurs at *X* and menstruation occurs at *Z*.

i) Identify by letter the time when follicle cells produce estrogen.

- Estrogen is being produced at *W* and *Y*.

j) Identify by letter the time when the corpus luteum produces estrogen and progesterone.

- Estrogen and progesterone are produced by the corpus luteum at *Y*.

k) How does LH affect estrogen and progesterone?

- LH increases progesterone and estrogen levels by stimulating the formation of the corpus luteum. The corpus luteum produces both female sex hormones.

Case-Study Application Questions

1 Explain why birth control pills often contain high concentrations of progesterone and estrogen.

- High concentrations of these hormones prevent the release of ova from the ovaries.

2 Why would a woman not take birth control pills for the entire 28 days of the menstrual cycle? On which days of the menstrual cycle would the pill not be taken?

- High levels of estrogen and progesterone would maintain a thickened endometrium. The pills would not be taken during the period of menstruation.

APPLYING THE CONCEPTS

(page 428)

1 Predict the consequences if the testes failed to descend from the abdominal cavity during embryo development.

- It is likely that the sperm which would be produced would not be viable because they would be at a higher temperature than when the testes are found within the scrotum.

2 Anabolic steroids act in a similar fashion to testosterone by turning off secretions of gonadotropic hormones. What effects do anabolic steroids have on secretions of testosterone? Explain your answer.

- The anabolic steroids would provide negative feedback to the pituitary gland resulting with a decrease in the secretion of gonadotropin-releasing hormone (GnRH). Decreased GnRH production slows the production and release of LH, leading to lower testosterone production.

3 The seminal vesicle, prostate gland, and Cowper's gland contribute secretions to the semen. How would reproduction be affected should secretions from these glands be inhibited?

■ Semen provides nutrients for the sperm cell and contains buffers that protect sperm cells from the acidic environment of the vagina. Without the semen, sperm could not survive for any length of time in the vagina and would likely die before they could fertilize the ova.

4 An experiment was performed in which the circulatory systems of two mice with compatible blood types were joined. The data collected from the experiment are expressed in the table to the right. (Note: + indicates found; - indicates not found.) Explain the data collected.

■ All the necessary hormones for sperm production are present in both mice because they are carried in the bloodstream. The pituitary gland in one mouse produces a sufficient amount of hormones to stimulate the secretion of hormones by the testes of the other mouse. Sperm, being produced by the testes, can only be present in the mouse which still has its testes.

5 Draw a diagram of the female reproductive system. Label the sites of ovulation, fertilization, and implantation.

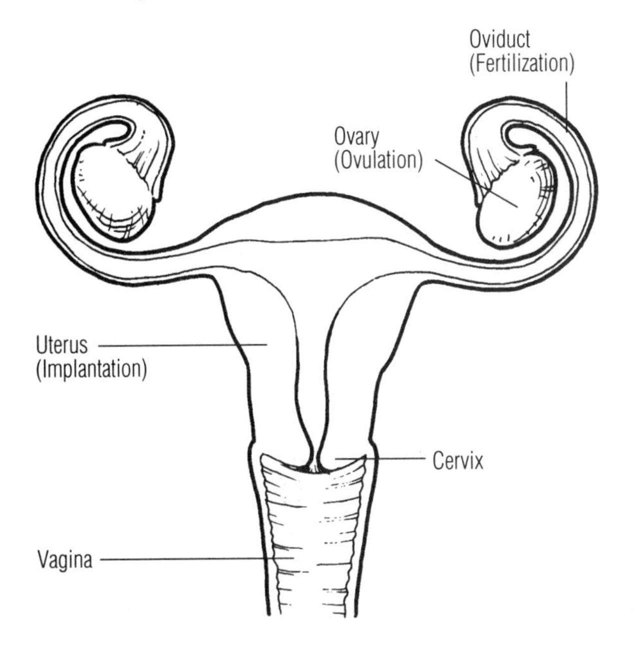

6 The graph below shows estrogen and progesterone levels during a menstrual cycle.

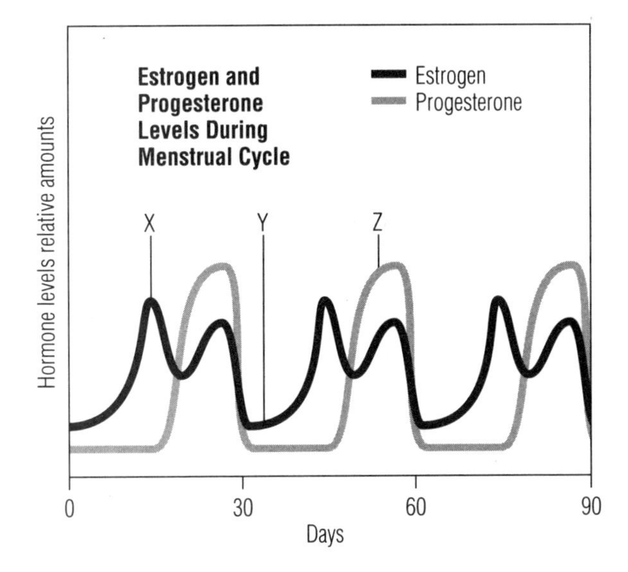

a) On which day (*X*, *Y*, or *Z*) would ovulation occur? Explain your answer.

■ Ovulation would occur at *X* when the secretion of estrogen is at a peak with little presence of progesterone.

b) On which day (*X*, *Y*, or *Z*) would you expect to find a functioning corpus luteum. Explain your answer.

■ A functioning corpus luteum would be present at *Z*, when both estrogen and progesterone are present.

7 The following graph shows hormone levels during the 40 weeks of pregnancy.

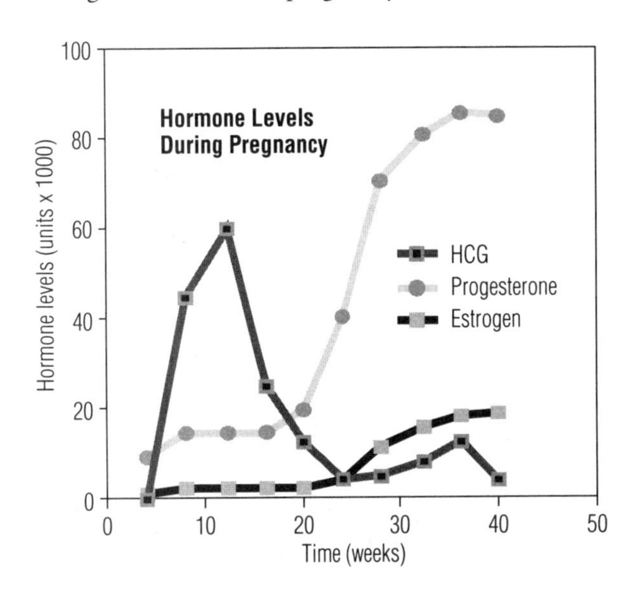

Explain the changes in HCG, estrogen, and progesterone levels during pregnancy.

■ At pregnancy, the outer layer of cells from the zygote forms a hormone, HCG (human chorionic gonadotropic hormone), which maintains the corpus luteum. The functioning corpus luteum keeps producing progesterone and estrogen, which in turn maintain the endometrium. The endometrium and embryo remain in the uterus. Cells from the embryo and endometrium combine to form the placenta. Materials are exchanged between the mother and developing baby through the placenta. After about four months of pregnancy, the placenta begins producing HCG, estrogen, and progesterone. High levels of progesterone prevent further ovulation. This means that once a woman is pregnant, she will not become pregnant once again until after the baby is born.

8 During the eighth month of pregnancy, the ovaries of a woman are surgically removed. How will the removal of the ovaries affect the fetus? Explain your answer.

■ Since the ovaries (corpus luteum) produce the hormones (estrogen and progesterone) which maintain the endometrium, the woman would go into labor, and the baby would be born immediately, if the ovaries were removed.

9 Explain why only one corpus luteum may be found in the ovaries of a woman who has given birth to triplets.

■ The triplets may have come from a single egg. In this case, the triplets must be identical. If the triplets were fraternal, three separate corpus lutei would be found.

CRITICAL-THINKING QUESTIONS

(page 429)

1 Design an experiment to demonstrate the independent functions of the interstitial cells and seminiferous tubules of the testes.

■ The action of the interstitial cells might be blocked by interfering with the production of male FSH. A competitive inhibitor might attach to FSH, or a chemical, similar in shape to FSH, might exert a negative-feedback effect on FSH. The experimenter could then check to see if lower testerone levels affected sperm cell production. Other experiments may be suggested.

2 Design an experiment to show how female gonadotropic hormones are regulated by ovarian hormones.

■ Levels of FSH and LH could be increased or decreased. Ovulation, the thickness of the uterine lining, and hormone (estrogen and progesterone) levels could be monitored. The experimental design should attempt to control variables. A great variety of experiments can be attempted. The three methods for hormonal research, extirpation (removal of the gland), graft, and chemical isolation and injection may be employed. See chapter 14 for further information on hormone research.

3 A 1988 article reported that prenatal testing had been banned in South Korea following a rise in male births. Seoul hospitals noted the change in the ratio of male to female births from 102.5 males for every 100 females, to 117 males for every 100 females. One statistician claimed that the change in ratio could be attributed to selection of male fetuses following prenatal testing. In India, doctors have been told not to disclose the sex of a child following amniocentesis testing. Consider the implications for a society that attempts to select the sex of unborn children.

■ The answers for this question could vary considerably. They would depend on a number of variables including the different beliefs that students may bring with them from their family, cultural, and religious orientations. It may be quite appropriate to discuss the variety of answers given. At any rate, the answers would probably allude to the fact that a change in the ratio of male and female births would produce an unbalanced population which would become even more unbalanced over time.

4 The embryo is sensitive to drugs, especially during the first trimester. In the 1950s a drug called thalidomide was introduced in Europe to help prevent morning sickness. Unfortunately, the drug altered the genes regulating limb bud formation in developing embryos. Before the drug could be withdrawn from pharmacies, children were born with lifelong disabilities. Although drugs such as thalidomide are no longer available, a great debate centers around drugs that may have a less pronounced effect on the fetus. Some evidence suggests that tranquilizers may cause improper limb bud formation. Even some acne drugs have been linked to facial deformities in newborns, and the antibiotic streptomycin has been associated with hearing problems. Explain why the embryo is so sensitive to drugs.

- Even a few cells in the embryo may represent organs or organ systems in the adult organism. So, when genes are altered in a cell of an embryo, such alterations may produce serious effects in organs or organ systems in the fully formed fetus.

CHAPTER *18*

Asexual Cell Reproduction

INITIATING THE STUDY OF CHAPTER EIGHTEEN

Chapter 18 provides an introduction to cell division. The following suggestions provide examples for beginning the chapter:

1. Ask students to imagine how their lives would change if cell division ever stopped. Imagine their appearance if new cells did not replace dead skin cells. Without cell division, scratches would never heal and blemishes would never disappear. Ask students to draw a picture of what they believe they would look like, if skin cell replacement stopped.
2. Ask students to hypothesize about aging. Do older people just have older cells? Are all of the cells in your body approximately the same age? Students can enter their initial thoughts in a journal and check them at the end of the chapter. Have they changed their thinking about aging?

ADDRESSING ALTERNATE CONCEPTIONS

- Interphase is not part of mitosis. Interphase describes the processes of cell activity between cell divisions. Most cells, even the rapidly dividing immature red blood cells, spend the majority of their time in this phase. Cells grow during interphase. They carry out the chemical activities that sustain life during this period. The single-stranded chromosomes are returned to double strands during interphase. If the cell did not duplicate its chromosomes during interphase it would never again be able to divide. Each new cell requires one of the strands of information to carry out its functions.
- Although there are many science-fiction movies and books about cloning people, no evidence exists suggesting that a human has ever been cloned in a laboratory. (Identical twins are nature's clones.) Initial cloning experiments were directed at plant tissue clones, a distinctly different process than those of nuclear transplants used on animals.

- Contrary to what many students believe, cancer is a broad group of diseases. Unlike most diseases which can be described in terms of cell or tissue death, cancer can be described in terms of too much life. Much more active than normal cells, cancer cells divide at rates that far exceed those of their ancestors.
- Cancer is not exclusive to humans. It is found in plants and animals alike. Sunflowers and tomatoes often display a form of cancer called a gall. These plant tumors are caused by invading bacteria, fungi, or insects. Evidence of cancerous tumors has been found in dinosaur bones and even in the cells found in the wrapped linen of ancient mummies.

MAKING CONNECTIONS

- Cell division reinforces the principles of the second part of the cell theory, introduced in chapter 5, Development of the Cell Theory. The first part of the cell theory states that all living things are made up of one or more cells. The second part of the cell theory states that cells come from pre-existing cells by cell division. Prior to this time it was believed that some single-celled organisms developed from nonliving objects. The linking of life to reproduction was an important step in the development of modern biology.

POSSIBLE JOURNAL ENTRIES

- Ask students to uncover many open-ended questions presented in the chapter. Students will be surprised to discover that, although scientists can describe the events of cell division, why cells divide remains a mystery. We don't understand why some cells divide and others do not, what causes aging, whether aging can be slowed or prevented, and why highly specialized cells tend to reproduce more slowly than unspecialized cells. Chapter 18 is more about scientific questions and proposed explanations than about irrefutable answers.
- Students can be encouraged to develop a concept map that includes all key terms presented in the chapter.
- Students can be encouraged to express the difficulties they experienced in formulating an understanding of this challenging chapter. What things did they employ to aid them in constructing their knowledge? For example, some students may indicate that they were helped by the laboratory, Mitosis. Other students may have attempted a series of their own drawings or constructed concept maps.

USING THE VIDEODISC

Section title	Page	Code	Frame (side)	Description
Importance of Cell Division	430 – 431	[barcode]	41291 (2)	Movie sequence showing frog development, cleavage to neurulation; time-lapse photography and animation of cleavage, gastrulation, and neurulation
Principles of Mitosis	431– 432	[barcode]	8398 (2)	Movie sequence of mitosis in the African blood lily
Principles of Mitosis	431– 432	[barcode]	9510 (2)	Movie sequence of cytokinesis in plant and animal cells
Stages of Cell Division	432	[barcode]	2561 (all)	Interphase

Section title	Page	Code	Frame (side)	Description
Stages of Cell Division	432		2562 (all)	Early prophase
Stages of Cell Division	432		2563 (all)	Late prophase
Stages of Cell Division	433		2564 (all)	Early metaphase
Stages of Cell Division	433		2565 (all)	Late metaphase
Stages of Cell Division	433		2566 (all)	Early anaphase
Stages of Cell Division	433		2567 (all)	Late anaphase
Stages of Cell Division	433		2568 (all)	Telophase
Stages of Cell Division	432 – 433		1651 (all)	Whitefish blastula
Stages of Cell Division	432		1652 (all)	Prophase—whitefish
Stages of Cell Division	433		1653 (all)	Metaphase
Stages of Cell Division	433		1654 (all)	Anaphase
Stages of Cell Division	433		1655 (all)	Early telophase
Stages of Cell Division	433		1656 (all)	Late telophase
Stages of Cell Division	432 – 433		1657 (all)	Plant cell mitosis—onion root tip
Stages of Cell Division	432 – 433		1658 (all)	Multiple phases
Stages of Cell Division	432		1659 (all)	Prophase
Stages of Cell Division	433		1660 (all)	Metaphase
Stages of Cell Division	433		1661 (all)	Anaphase
Stages of Cell Division	433		1662 (all)	Telophase
Cloning	437 – 439		13433 (1)	Movie sequence showing plant tissue techniques used in cloning rubber plants

IDEAS FOR INITIATING A DISCUSSION

Figure 18.9: Cloning a carrot

Tissue cloning techniques have tremendous commercial value. A group of cells grown in tissue culture can give rise to identical offspring which can be grown worldwide. Discuss the possible advantages of tissue culture techniques. The social issue provides an excellent follow-up.

Figure 18.11: Liver-tissue transplant patient

Discuss the idea of using tissue transplants. Although few people objected to using liver tissue transplants, much greater concern was expressed about brain tissue transplants.

Figure 18.12: Cloning mammals

Mammal cloning techniques have produced special strains of mice used in cancer research. Some biologists and nonbiologists have expressed concerns that as the techniques develop, people may one day be cloned. Consider the potentials of cloning organs for people who need transplants, or cloning exact duplicates.

REVIEW QUESTIONS

(page 434)

1 Compare the daughter cells with the original mother cell.

■ The daughter cells are identical to each other and identical to the mother cell.

2 Describe the structure and function of the spindle fibers.

■ Spindle fibers are microtubules that align and direct chromosomes during cell division.

3 During interphase, what event must occur for the cell to be capable of undergoing future divisions?

■ Single-stranded chromosomes must become double-stranded chromosomes before they can divide again. The DNA must be duplicated.

4 Describe the events of prophase, metaphase, anaphase, and telophase.

■ During prophase, the nuclear membrane dissolves, chromosomes shorten and thicken, centrioles divide, and spindle fibers form. The chromosomes also become attached to the spindle fibers by means of the centromere. Metaphase follows prophase. The chromosomes line up along the equatorial plate and the cell prepares for nuclear division. The chromosomes divide and the single-stranded chromatids move to opposite poles during anaphase, the next phase of division. During telophase, the cell will re-organize the nucleus.

5 Compare and contrast cell division in plant and animal cells.

■ The animal cell has asters and centrioles, but plant cells do not. A division plate also helps separate plant cells, while animal cells show a cytoplasmic furrow during cytoplasmic division.

6 What would happen if you ingested a drug that prevented mitosis?

■ Dead cells would not be replaced. Because red blood cells live on average about 120 days, the individual could expect the number of red blood cells to fall significantly. Bone cells, skin cells, and the lining of the digestive tract could not be replaced.

(page 440)

7 What is cloning?

■ Cloning is the making of exact duplicates of cells or organisms.

8 Discuss the economic importance of plant tissue cloning.

■ Cloning permits reproduction without the fusing of sex cells. Seedless grapes, oranges, and other fruits can be propagated by cloning. Because the fruits do not contain a seed they are incapable of reproducing sexually. The process also permits the duplication of desired characteristics without the surprises brought about by the new gene combinations characteristic of sexual reproduction. Many flowers are produced by this method. Each flower is almost a mirror image of the parents and each other.

9 What is a nuclear transplant?

■ It is the removal of a nucleus from one cell, and its subsequent transfer into the cytoplasm of another cell.

10 Define totipotency.

■ It is the ability to support cell division that will carry a cell from the fertilized egg to a multicellular animal.

11 Why is it difficult to clone human cells?

■ Cells lose totipotency once they mature. Brain cells, blood cells, and other mature cells have differentiated. Human cloning techniques therefore must be concentrated at the earliest stages of cell development.

12 Why are identical twins often called "nature's clones"?

■ The cells during the earliest stages of development, the blastula stage, separate from a rapidly dividing mass and each new cell mass develops into a new human. Because each cell mass carries the same DNA, each of the twins is identical.

(page 443)

13 Do all the cells of your body divide at the same rate? Explain.

■ No, some cells divide faster than others. Mature red blood cells do not divide, while the immature red blood cells found in the bone marrow divide at very fast rates. Cells of the skin and cells that line the digestive system divide very quickly. Nerve cells, by comparison, do not divide very quickly.

14 What evidence suggest that cells have a "biological counter"?

■ Leonard Haflick found that some cells extracted from the heart tissue of a developing embryo could divide 50 times. After the 50th division the cell died. Haflick then froze cells before the 50th division. No matter how long the cell was frozen, it would only divide until 50. The cell seemed to keep track of the number of divisions, no matter how long it was placed in suspended animation.

15 What evidence supports the theory that the human body is only capable of living about 115 years, even if disease were eliminated?

■ Cells are capable of a maximum number of divisions. As cells die through the normal wear and tear of life, no new cells will be available for replacement.

16 In what ways does cellular communication regulate the division of normal cells?

■ Cells divide only when needed. The replacement of cells after a sunburn provides an example. Cells of the body do not divide when separated from surrounding cells even when grown in tissue culture.

17 How rapidly can a cancer cell grown in tissue culture divide? Is it likely that the cancer cell could divide that quickly in the human body? Explain.

■ Cancer cells grown in tissue culture can divide every 24 h. At this rate the tumor would produce more than one billion cells in a single month. The tumor would reach a mass of 10 kg in only six weeks.

LABORATORY

MITOSIS

(page 436)

Time Requirements

Approximately two 40 min classes are required to complete the assignment, if the students have acquired some proficiency with the microscope.

Notes to the Teacher

- The diagram and photomicrographs within the text provide an excellent visual reference; however, some students will be tempted to construct their diagrams from either the pictures or diagrams provided in the text, unless otherwise instructed.
- We recommend taking a few minutes to explain the importance of making biological drawings, and emphasizing that students are to draw only what they see.

Observation Questions

a) How do the cells of the meristematic area differ from the mature cells of the root?

■ Cells of the meristematic region are shorter and more compact.

b) Draw, label, and title each of the phases. It is important to draw and label only the structures that you can actually see under the microscope.

■ See Figure 18.5 for reference.

c) Compare the appearance of the animal cells with that of the plant cells.

■ Answers will vary, but students may indicate that the plant cells are larger. The chromosomes of the animal cells are shorter, but more difficult to identify.

d) Cells in interphase = ___
Cells actively dividing = ___

■ Answers will vary.

e) Calculate the percentage of cells that are undergoing mitosis:

$$\frac{\text{Number of cells dividing}}{20} \times 100 = \_\_\% \text{ dividing}$$

■ Answers will vary.

f) Cells in interphase = ___
Cells actively dividing = ___

■ Answers will vary.

g) Calculate the percentage of cells that are undergoing mitosis:

$$\frac{\text{Number of cells dividing}}{20} \times 100 = \% \text{ dividing}$$

■ Answers will vary.

h) Compare the percentage of animal cells that are undergoing mitosis with the percentage of plant cells that are undergoing mitosis.

■ Answers will vary.

Laboratory Application Questions

1 Predict what will happen if both sister chromatids move to the same pole during mitosis.

■ The cell without some of the needed genetic information will likely die. The linking of gene information to the functions of a cell is important. The prediction for the cell that has received too much gene information will vary. At this point in the course, the student should be given credit for a variety of different predictions about what will happen if a cell receives too much genetic information. (Although nondisjunction is often thought of in meiosis, it occurs in mitosis.)

2 A cell with 10 chromosomes undergoes mitosis. Indicate how many chromosomes would be expected in each of the daughter cells.

■ Each daughter cell would have 10 chromosomes.

3 Predict what will happen if a small mass of cells breaks off from a human blastula.

■ An identical twin will be produced.

4 Herbicides like 2,4-D and 2,4,5-T stimulate cell division. Why does the stimulation of cell division make these chemicals effective herbicides?

■ The cells divide at such a fast rate that they do not specialize into different tissues. Without specialized tissues, plant cells are not provided with adequate levels of nutrients to support their growth. Eventually the plant dies.

LABORATORY
IDENTIFICATION OF A CANCER CELL

(page 444)

Time Requirements

Approximately 40 min are required.

Notes to the Teacher

- See Figure 18.16 for reference.
- Epithelial cells taken from Pap smears or the buccal cavity make ideal slides. In most cases the cells are separated and students only have a single layer of cells to view. Slides can be purchased from various biological supply houses.

Observation Questions

a) Locate the dermal and epidermal layers. Draw a line diagram showing the position of the epidermal and dermal cell layers.

■ The diagram should show an irregular basal layer dividing the epidermal and dermal layers of the skin.

b) Are the cells of the epidermis invading the dermis ?

■ Yes they are.

c) Estimate the size of a cancerous cell in micrometers (μm).

■ Answers will vary depending on the slides used.

d) Estimate the size of the nucleus of a cancerous cell in micrometers (μm).

■ The nucleus of the cancerous cell is larger than the nuclei of noncancerous cells.

e) Determine the nucleus-to-cytoplasm ratio:

$$\text{Ratio} = \frac{\text{nucleus } (\mu m)}{\text{cytoplasm } (\mu m)}$$

■ Answers will vary depending on the slides used.

f) Record the nucleus-to-cytoplasm ratio for the normal cell.

■ The normal cell has more cytoplasm and generally a smaller nucleus.

g) Sample answers:

	Cell size	Nuclear shape	Nuclear size
Normal cell	2 μm	round	0.5 μm
Cancerous cell	1.5 μm	irregular	0.8 μm

Laboratory Application Questions

1 Cancerous cells are often characterized by a large nucleus. Based on what you know about cancer and cell division, provide an explanation for the enlarged nucleus.

■ The cancer cell is much more active than the normal cells. The enlarged nucleus may be explained by the increased activity.

2 Why are malignant (cancerous) tumors a greater threat to life than benign tumors?

■ The malignant tumor can metastasize. This means that some of the cells from the tumor may dislodge and re-establish themselves in a new area. Malignant tumors can invade other tissues, spreading the cancer.

3 Provide a hypothesis that explains why the skin is so susceptible to cancer.

■ The outer tissues of the body come in contact with many environmental agents that can cause cancer. Students should be encouraged to present different hypotheses.

4 A scientist finds a group of irregularly shaped cells in an organism. The cells demonstrate little differentiation, and the nuclei in some of the cells stain darker than others .

a) Based on these findings, would it be logical to conclude that the organism has cancer? Justify your answer.

■ The organism may be cancerous. Cancer cells divide so rapidly that they do not have time to differentiate. The nuclei may absorb more stain because they are more active.

b) What additional tests might be required to confirm or disprove the hypothesis that the cells are cancerous?

■ The rate of cell division would be monitored. The nuclei would have to be measured and the cell would be studied in tissue culture.

APPLYING THE CONCEPTS

(page 447)

1 Consider what would happen if you remained a single cell from the onset of life. How would your life as a single-cell organism differ from that of a multicellular organism?

■ The single cell would have limited size. As cells increase in size, volume will increase faster than surface area. Because a cell secures nutrients through the cell membrane, the poorer surface area-to-volume ratio limits a cell's growth. A single cell lacks specialization and must do all of the work required to maintain life processes by itself. The lack of specialization will decrease the efficiency of a cell as it attempts to do every job. The lack of specialization may also limit the number of jobs performed by a single cell, and thereby limit the potentials for life.

2 How does cell division in plants and animals differ? What is responsible for the differences?

■ Plant cells do not have centrioles or asters. Plant cells form a division plate during telophase in order to separate the dividing cells, while animal cells create a furrow between dividing cells.

3 How does the mechanism of mitosis support the theory that many living things come from common ancestors?

■ The mechanism of mitosis is very similar among all living things. The more similar the organisms, the greater the number of similarities in cell division.

4 What evidence suggests that one of your nerve cells carries the same number of chromosomes and the same genetic information as one of your muscle cells?

■ Both cells originated from the same mother cell. Each mitotic division produces identical, or nearly identical cells.

5 Explain why orchids cloned from tissue cultures are so similar?

■ Clones are identical. All descend from the same mother cell.

6 Discuss some of the economic benefits of plant tissue cloning.

■ Cloning techniques prevent change that occurs with new gene combinations. After a parent with preferred traits is identified, cloning techniques produce identical offspring.

7 Explain why a better understanding of the mechanism of cell division may enable scientists to regenerate limbs.

■ The limbs of salamanders and starfish will regenerate, if severed. The cells of these lower forms of life are totipotent. By understanding why these cells continue to divide even after they specialize, scientists might be able to induce specialized human cells to begin dividing.

8 How might the puzzle of aging be unlocked by an understanding of the mechanisms of cell division?

■ If aging is to be understood as the end of a cell's potential to divide, continuing mitosis by controlling the cell's ability to replace its worn-out predecessors may prevent aging.

9 How might the puzzle of cancer be unlocked by an understanding the mechanisms of cell division?

■ Cancer is uncontrolled cell division. By understanding why specialized cells have a limited ability to divide, scientists may be able to slow the rate at which abnormal cells divide.

10 Hypothesize about why some types of cancer are more difficult to detect than others.

■ Cancers which are found inside the human body, such as liver tumors cannot be detected externally. Cancers such as lung cancer tend to metastasize. Also, cells in some organs divide faster than others. The rate of cell division alone will not signify cancer.

CRITICAL-THINKING QUESTIONS

(page 447)

1 X rays and other forms of radiation break chromosomes apart. Using the information gained in this chapter, assess some of the medical problems caused by the atomic bombs dropped on the Japanese cities of Hiroshima and Nagasaki in World War II.

■ Mitosis will be impaired. Unequal numbers of chromosomes or chromosome fragments will locate themselves in the daughter cells. It is quite likely that neither daughter cell will survive. This means that new cells will not replace worn-out cells. In some cases, radiation has been noted to cause certain forms of cancer. Many citizens of Hiroshima and Nagasaki developed leukemia well after the atomic blast. Students may be asked to predict some of the disorders that may arise because cells are not replaced.

2 Predict some of the potential problems that might arise if all plant reproduction were controlled by tissue cloning. Would the same problems exist if all domestic animals were restricted to cloning? Explain your answer.

■ Little variation would exist in the plant world. All orchids, for example, might come from the same parent. Any new disease that affects one orchid would have a similar impact on all orchids. Sexual reproduction causes variation. Variation, caused by new combinations of genes, might ensure that some plants are more resistant to the disease, and the species may avoid extinction. Variation may even produce orchids with even more desirable traits. The same problems would exist with animals.

3 A girl suffering from malignant melanoma—a cancer of the pigment cells of the skin—had one of her cancer cells grafted into the skin of her mother. In theory, scientists believed that the invading cancer cell from the dying child would stimulate the production of antibodies within the mother. Antibodies would destroy the foreign invading cancer cells. Scientists hoped to harvest the antibodies from the mother and transfuse some of them into the blood of the dying girl. The experiment did not work, and the young girl died a few days after the procedure. The true failure of the experiment was not known until some months later. The mother was diagnosed as having skin cancer. The melanoma also killed the mother. This incident raises moral and ethical questions. Should untried procedures like this be attempted on humans? Explain.

■ Clearly this question extends beyond the boundaries of science, but requires an understanding of science and technology to frame the question and seek solutions. By approaching this question, students will gain an appreciation of how science and technology are changing the way in which we define life and death. Most student answers are rooted in personal, moral, and religious beliefs.

4 Chemotherapy treatments given to cancer patients are designed to destroy the rapidly dividing cancer cells. From what you know about normal cell division, predict what other body cells could be affected by the chemotherapy. What side effects might you expect such a treatment to produce?

■ Students are most aware of problems with hair. Follicle cells that produce hair are extremely active. Other active cells such as those found in the skin and digestive tract may also slow. Dry skin and stomach upset may also be indicated.

CHAPTER 19

Sexual Cell Reproduction

INITIATING THE STUDY OF CHAPTER NINETEEN

Chapter 19 introduces gamete formation as reduction division. An understanding of chapter 19 provides the framework for the next sections of the text, Unit 6, Heredity, and Unit 7, Change in Populations and Communities.

- Introduce the social issue at the beginning of the chapter. Ask students to record their initial thinking. Their initial reactions can be reflected upon once they have gathered more information. Has the thinking of your students changed as information is gathered?
- Bring in a picture of the Royal family (or some other family that is well known) and ask students to identify some of their traits. Which traits appear to be traced from one generation to the next? What type of variation do they observe? Now show students a picture of animals which reproduce asexually. Ask them the same questions. Clearly, organisms that reproduce asexually have much greater similarity. Use the opportunity to link the discussion of sexual reproduction with genetic variation.

ADDRESSING ALTERNATE CONCEPTIONS

- Students often confuse meiosis and mitosis. The demonstration, Comparing Meiosis and Mitosis, should provide them with an opportunity to actively sort out the quandary.
- The terms diploid and haploid chromosome number can create confusion for some students. The example provided in the text indicates that the human diploid chromosome number is 46 and that the haploid chromosome number is 23. Some students will assume that 46 represents the diploid chromosome number, regardless of the organism. Other examples can be introduced, such as *Drosophila*, with a diploid chromosome number of 8 and haploid chromosome number of 4, or mice, with a diploid chromosome number of 22 and haploid chromosome number of 11.
- Some students will assume that because they look more like one parent than the other, they contain more genetic information from that parent. However, we obtain an equal amount of nuclear genetic material from each parent. The text indicates that each of the 23 chromosomes received from the father is matched by 23 chromosomes from the mother. The paired chromosomes are called homologous chromosomes because they are similar in shape, size, and gene arrangement. The student's appearance is determined by the manner in which the genes from homologous chromosomes interact. In chapter 24, students discover that some genetic information comes from the mitochondria of the egg cell, therefore, the mother actually contributes more genetic information, regardless of the sex of the child. The mitochondria of the sperm cell do not enter the egg. We recommend not introducing this information until students gain a framework for understanding Mendelian genetics. Although factual, the added information blurs the meaning of homologous chromosomes and may lead to confusion as students attempt to construct initial meanings.

MAKING CONNECTIONS

- Meiosis serves as a conceptual bridge between reproductive physiology and genetics. In animals, meiosis takes place in the testes and ovaries, described in chapter 17, The Reproductive System. The testes produce sperm cells and the ovaries produce egg cells. Plants also form sex cells, or gametes, by meiosis. The pollen cells of flowering plants are male sex cells, described in greater detail in chapter 20, Genes and Heredity. Plant gametes, like animal sex cells, have a haploid chromosome number. The fusion of male and female gametes restores the diploid chromosome number.
- Crossing-over, introduced within the context of meiosis, serves as a prologue to Mapping Chromosomes and Gene Recombinations in Nature introduced in chapter 21, The Source of Heredity. In turn, these sections serve as background for understanding the section, Gene Splicing Techniques and Gene Mapping, and the Research in Canada segment, The Cystic Fibrosis Gene, also in chapter 21.
- Meiosis provides a context for understanding the basis of natural selection, variation. Natural selection is presented in chapter 4, Adaptation and Change, and again in greater detail in chapter 24, Population Genetics. By understanding the principles of meiosis, students recognize why abnormal meiosis is associated with nondisjunction disorders. In turn, this section of the chapter acts as a springboard to developing an understanding of gene selection and changes in gene frequencies, which are introduced in chapter 24. The case study, Genetic Disorders as Models for Evolution, presented in chapter 24, establishes a bridge between meiosis (chapter 19), genetics (chapters 20–23), and evolution (chapter 24).
- The section Research in Canada, Down Syndrome, establishes additional links between chapter 17, Reproduction, and chapter 19, Sexual Cell Reproduction. Segments entitled Reproductive Technology, Redefining Motherhood, and the social issue Limits on Reproductive Technology, also create a nexus between sexual cell division and reproductive physiology.

POSSIBLE JOURNAL ENTRIES

- Students might be asked to write a dialogue between the text and themselves, as they refute the idea that virgin births are possible. They may consider what might happen if a primordial egg cell ovulates prior to meiosis. What would happen, should such a 2n-egg cell implant in the uterus. Some geneticists have indicated that the event, common in *Daphnia* and lower invertebrates that undergo parthenogenetic reproduction, might even occur in higher mammals. By acting as a "devil's advocate," the students can challenge their own learning and push understanding to a higher level.

- Students can be encouraged to develop a concept map that includes all key terms presented in the chapter.
- Students can be encouraged to express the difficulties they experienced in formulating an understanding of this challenging chapter. What things did they employ to aid them in constructing their knowledge? For example, some students may indicate that the laboratory, Human Karyotypes, aided their understanding of nondisjunction. Other students may have attempted a series of their own drawings.
- Decision-making strategies may be recorded as each group prepares for the debate. Did the students change their minds during the preparation for the debate? Students may even be asked to record their initial positions about the social issue and to reflect upon these feelings after the debate has been completed. Did they change their minds after listening to opposing arguments? Should limits be placed on reproductive technology?
- Students may identify open-ended questions that are not answered in the chapter. For example, why do sperm cells produce four cells following meiosis, compared with only one viable egg cell? Although the text provides a description of the differences, no mechanism for the differences in cell division is offered. Why are cytoplasmic divisions different from female cytoplasmic divisions during meiosis? What advantages are served by restricting the number of egg cells produced by meiosis?

USING THE VIDEODISC

Section title	Page	Code	Frame (side)	Description
Importance of Meiosis	448 – 449		2926 (all)	Gametes
Importance of Meiosis	448 – 449		2873 (all)	Diploid chromosome number
Importance of Meiosis	448 – 449		2958 (all)	Homologous chromosomes
Stages of Meiosis	449 – 451		2553 (all)	Early stages of meiosis phase I showing diploid chromosome number
Stages of Meiosis	449 – 451		2554 (all)	Shows synapsis of homologous chromosomes
Stages of Meiosis	449 – 451		2555 (all)	Each paired chromosome splits into sister chromatids
Stages of Meiosis	449 – 451		2556 (all)	Sister chromatids begin to split from the pair
Stages of Meiosis	449 – 451		2557 (all)	Chromatids are now shorter and thicker
Stages of Meiosis	449 – 451		2558 (all)	Metaphase I: homologous chromosomes line up along the equatorial plate

Section title	Page	Code	Frame (side)	Description
Stages of Meiosis	449 – 451		2559 (all)	Late anaphase I
Stages of Meiosis	449 – 451		2560 (all)	Anaphase II
Stages of Meiosis	449 – 451		2958 (all)	Homologous chromosomes
Stages of Meiosis	449 – 451		3251 (all)	Tetrad
Stages of Meiosis	449 – 451		2855 (all)	Crossing-over
Development of Male and Female Gametes	442 – 453		33467 (2)	Movie sequence showing spermatogenesis
Development of Male and Female Gametes	452 – 453		38275 (2)	Movie sequence showing fertilization of a mouse egg; polar bodies and first four stages of zygote division are also shown
Development of Male and Female Gametes	452 – 453		3194 (all)	Sex chromosomes
Abnormal Meiosis: Nondisjunction	454 – 455		3070 (all)	Nondisjunction
Frontiers of Technology: Amniocentesis	458		2722 (all)	Amnion
Frontiers of Technology: Amniocentesis	458		2888 (all)	Embryo

REVIEW QUESTIONS

(page 453)

1 A muscle cell of a mouse contains 22 chromosomes. Indicate the number of chromosomes you would expect to find in the following cells of the same mouse:

a) daughter cell formed from mitosis
b) skin cell
c) egg cell
d) fertilized egg cell

■ The answers are: **(a)** 22; **(b)** 22; **(c)** 11; and **(d)** 22.

2 Differentiate between haploid and diploid cells in humans.

■ The haploid chromosome number in humans is 23 and the diploid chromosome number is 46. The haploid chromosome number is characteristic of sex cells, while all the other cells of the body are diploid. Haploid cells have one member of each of the homologous pairs of chromosomes.

3 In what ways does the first meiotic division differ from the second meiotic division?

■ During the first division, the homologous chromosomes line up and move to opposite poles. The first meiotic division is a reduction division. During the second division the double-stranded chromosomes line up and chromatids move to opposite poles.

4 What is a tetrad?

■ A tetrad occurs when homologous, double-stranded chromosomes come together during prophase I.

5 Explain why synapsis may lead to the exchange of genetic information between chromosomes.

- Crossing-over may occur. Chromosome segments intertwine and break. Occasionally genetic material is exchanged between homologous chromosomes.

6 What are homologous chromosomes?

- They are chromosomes that are identical in shape and gene arrangement. One of the homologous pairs of chromosomes comes from the mother, the other comes from the father.

7 Do homologous chromosomes have the same number of genes? Explain.

- Yes, the homologous chromosomes must have the same number of genes. Your father and mother contribute the same number of genes to all of the chromosomes except the sex chromosomes.

8 Do homologous chromosomes have identical genes? Explain.

- No, not all of the genes contributed from your mother are identical to those contributed by your father. Although they have genes coding for the same characteristics, they do not necessarily have the same genes. For example, both your father and mother have genes for eye color, although it is unlikely that they have the same eye color.

9 Compare the mechanisms of gametogenesis in males and females.

- The male process, spermatogenesis, produces four haploid sperm cells that have a reduced amount of cytoplasm compared with the mother cell. The female process, oogenesis, produces one viable ootid, which has approximately the same amount of cytoplasm as that found in the mother cell. Both gametes undergo two divisions and divide in a similar fashion.

(page 462)

10 What is nondisjunction?

- Nondisjunction can occur in either meiosis or mitosis. In meiosis, nondisjunction results from either homologous chromosomes moving to the same pole in meiosis I, or both chromatids moving to the same pole, during meiosis II.

11 Use a diagram to illustrate how nondisjunction in meiosis I differs from nondisjunction in meiosis II.

- The diagram should show that meiosis I involves a reduction division, while meiosis II involves the division of chromatids.

12 Differentiate between monosomy and trisomy.

- Trisomy occurs when a zygote has three homologous chromosomes. A single member from the homologous pair is inherited for monosomic conditions.

13 What is Down syndrome?

- It is a genetic disorder involving trisomy of chromosome pair number 21.

14 What is a karyotype?

- It is a pictorial representation of homologous chromosomes.

15 What is Turner syndrome?

- It is a genetic disorder involving a monosomy XO of sex chromosomes.

16 Indicate some of the benefits of amniocentesis.

- Amniocentesis permits prenatal diagnosis of genetic disorders.

17 What is *in vitro* fertilization?

- It is fertilization in glass, or out of the body of a female.

DEMONSTRATION
COMPARING MEOISIS AND MITOSIS

(page 454)

Time Requirements

Approximately 20 min are required if it is done as a demonstration.

Notes to the Teacher

The yarn can be placed on an overhead projector.

Observation Questions

a) What cell structures do the strands of yarn and the paper clips represent? What area of the cell does the center line represent?

■ The strands of yarn represent chromatids and the paper clips represent the centromeres. The center line represents the equatorial plate.

b) What is the diploid chromosome number of this cell?

■ The diploid number is four.

c) What structure do the single strands of yarn represent in a true cell?

■ The strands represent chromatids.

d) How many chromosomes are in each of the daughter cells?

■ There are four chromosomes in each of the daughter cells.

e) Compare the daughter cells with the mother cell.

■ The cells are identical.

f) On what basis are the simulated chromosomes considered to be homologous?

■ The two strands of yarn are the same length.

g) What is the diploid chromosome number?

■ The diploid chromosome number is four.

h) How does metaphase I of meiosis differ from metaphase I of mitosis?

■ During meiosis I, the homologous chromosomes line up along the equatorial plate; however, during mitosis the double-stranded chromosomes line up along the equatorial plate.

i) What important event occurs during synapsis?

■ Crossing-over, in which genetic information is exchanged, may occur during synapsis.

j) What is the haploid chromosome number?

■ The haploid chromosome number is two.

k) Compare the resulting daughter cells of mitosis and meiosis.

■ Mitosis always produces two identical daughter cells, while meiosis produces four haploid daughter cells.

LABORATORY
HUMAN KARYOTYPES

(page 459)

Time Requirements

Approximately 45 min are required to complete the assignment.

Notes to the Teacher

To use the karyotype chart provided on page 187, you may photocopy one copy per student group.

Observation Questions

a) On what basis are the chromosomes arranged in pairs?

■ They are arranged by the size, shape, and banding of the chromosomes.

b) In what ways does chromosome pair 2 differ from pair 20?

■ The chromosomes are longer and have a different banding arrangement. (Note that the bands indicate genes.)

c) What is the sex of the individual shown in the chart above?

■ He is male.

d) How many chromosomes are found in the individual?

■ Forty-seven chromosomes are found in the individual.

e) What is the sex of the individual?

■ She is female.

f) On the basis of the information shown in the karyotype chart, provide a diagnosis.

■ She has Edward's syndrome.

Laboratory Application Questions

1 Would the diagnosis of Turner syndrome in a single cell necessarily mean that every cell of the body would contain 45 chromosomes? Explain your answer.

■ Yes, the single egg cell divides many times to produce duplicate cells.

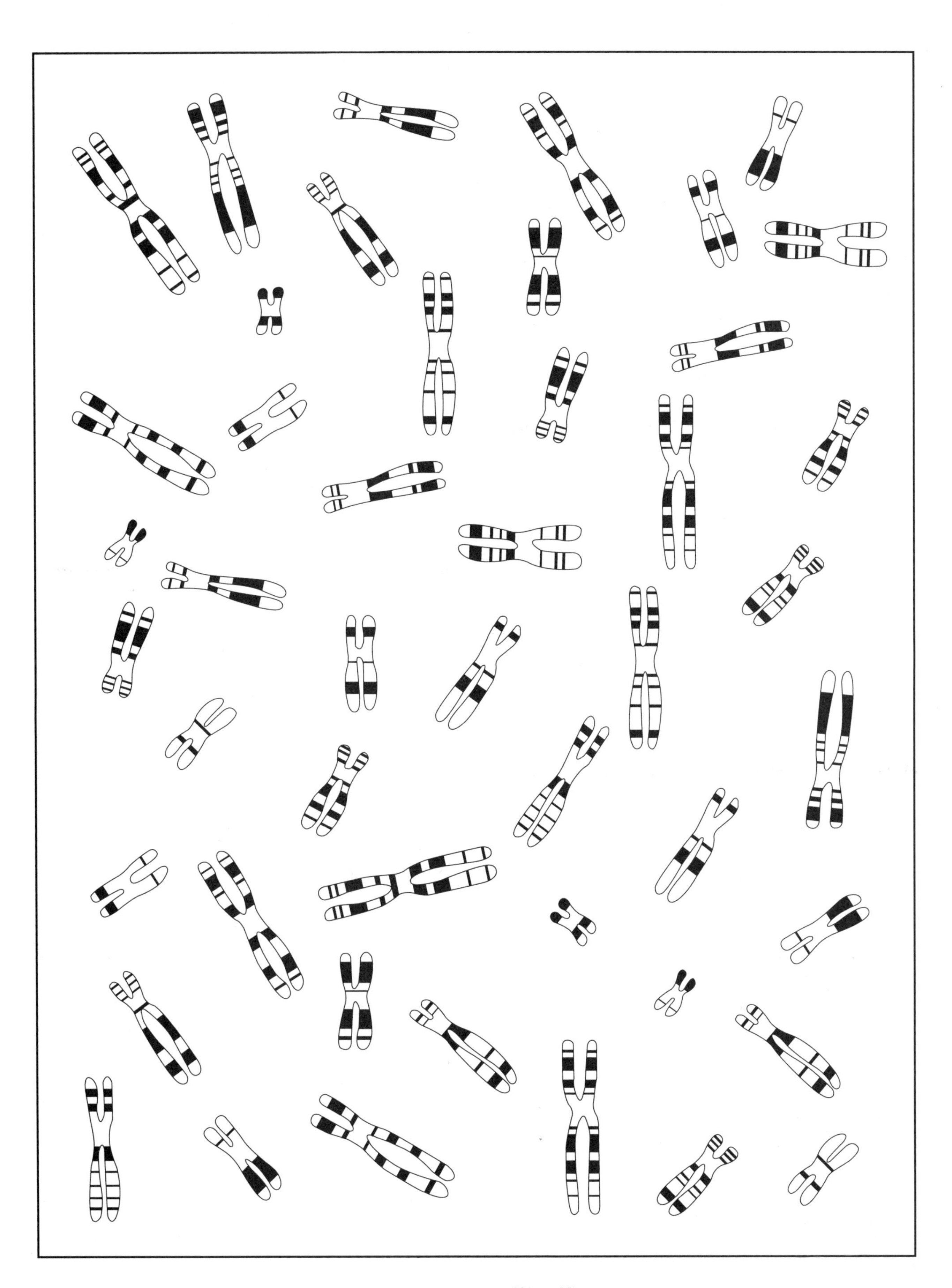

2 Is it possible for two people who have Down syndrome to give birth to a normal child? Explain your answer.

■ Yes, as the homologous chromosomes line up along the equatorial plate during meiosis I one cell will inherit two pairs, the other will inherit one of the pairs. If the normal cell combines with another normal cell from the other parent, the child will have the normal diploid chromosome number. The diagram below may help you explain the answer to your students.

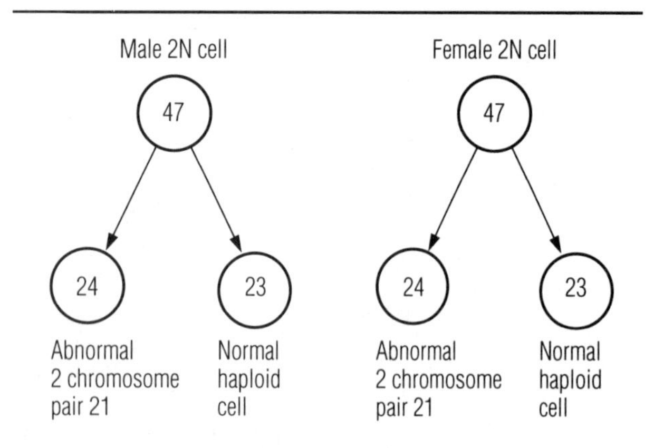

3 Would it ever be possible to produce a baby who has 48 chromosomes? Explain your answer.

■ Yes, it would be possible if either nondisjunction occurred at both stages of meiosis or if nondisjunction occurred in both male and female gametes. Students should not be expected to get both combinations.

4 More males than females suffer from color blindness. Speculate as to why females who have Turner syndrome have a similar incidence of color blindness as males.

■ The color blindness gene is found on the X chromosome. The female with Turner syndrome, like a male, has only one X chromosome.

APPLYING THE CONCEPTS

(page 465)

1 Explain why sexual reproduction promotes variation.

■ Homologous chromosomes segregate independently. Sperm cells and egg cells contain different combinations of chromosomes.

2 Predict what might happen if the polar body were fertilized by a sperm cell.

■ The cell would not develop because it would lack the required amount of cytoplasm to fuel all of the required cell divisions.

3 A microscopic water animal called *Daphnia* reproduces from an unfertilized egg. This form of reproduction is asexual because male gametes are not required. Indicate the sex of the offspring produced.

■ The offspring are always females, because they are clones of the mother.

4 Indicate which of the following body cells would be capable of meiosis. Provide a brief explanation for your answers.

a) brain cells
b) fat cells
c) cells of a zygote
d) sperm-producing cells of the testes

■ Meiosis only occurs in the cells of the sex organs. Specialized cells such as brain cells in humans are not totipotent.

5 Compare the second meiotic division with the second mitotic division.

■ The process is identical, except that meiosis II involves haploid cells.

6 King Henry VIII of England beheaded his wives when they did not produce sons. Indicate why a little knowledge of meiosis might have been important for Henry's wives.

■ The father determines the sex of the child. If a sperm cell carrying an X chromosome fertilizes the egg before a sperm cell carrying a Y chromosome, the child will be a girl.

7 Explain how it is possible to produce a trisomic female XXX.

■ Nondisjunction must occur. The diagram below provides an explanation for meiosis I nondisjunction.

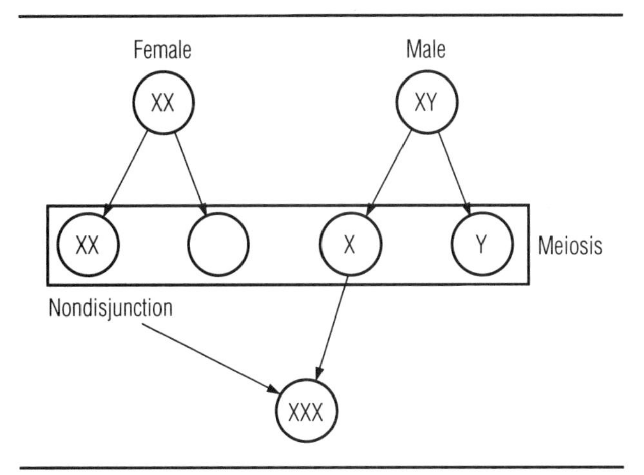

8 Abnormal cell division can produce an XYY condition for males. Diagram the nondisjunction that would cause a normal male and female to produce an XYY offspring.

■ Nondisjunction must occur at meiosis II. See the diagram below.

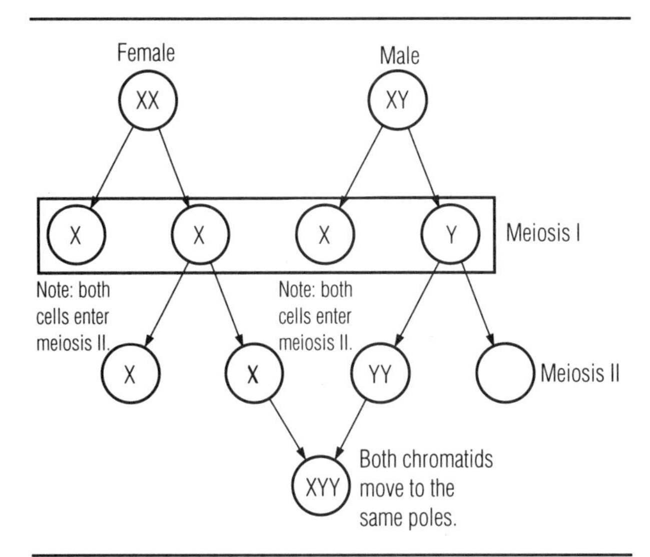

9 A number of important genes are found on the X chromosome. Explain why many biologists suggest that genetic differences account for the fact that more male than female babies die shortly after birth.

■ Females have two X chromosomes. A defect on one of the chromosomes will often be overruled by the normal gene located on the other chromosome. Because males only have one X chromosome any defect will be expressed immediately.

10 Why might a physician decide to perform amniocentesis on a pregnant woman who is 45 years of age?

■ The children of women over 45 have a higher incidence of Down syndrome.

CRITICAL-THINKING QUESTIONS

(page 465)

1 According to one report, the incidence of infertility appears to be increasing in countries like Canada. The sperm count in males has fallen more than 30% in the last half-century and is continuing to fall. Although thee is no explanation for this phenomenon at present, environmental pollution is suspected. Suggest other reasons for decreased fertility in males and females.

■ That parents on average are getting older, or a higher incidence of sexually transmitted diseases, which damage reproductive organs, may be suggested. Do not expect conclusive answers. Students should be encouraged to consider many alternatives.

2 A couple who are unable to have childen decide to hire another woman to carry the fetus. The procedure will cost them $10 000. They want absolute legal rights to the child, both while it is in the womb and when it is born. Discuss the ethical implications of hiring surrogate mothers.

■ Responses will vary. Students should recognize that technological change often initiates social and moral changes.

3 Advances in modern medicine have increased the number of people who carry genetic disorders. People who might have died at an early age because of their disorder are living longer, marrying and producing children, and thus passing on their defective genes. Nondisjunction disorders could be eliminated by screening prospective sperm and egg cells. Sperm and egg banks could all but eliminate many genetic disorders. Comment on the implications to society of the systematic elimination of genetic disorders in humans.

■ Responses will vary. Students should recognize that technological change often initiates social and moral changes.

4 A technique called egg fusion involves the union of one haploid egg cell with another. The zygote contains the full $2n$ chromosome number and is always a female. Discuss the implications for society if this technique were to be employed for humans.

■ Males would become obsolete. Females could reproduce without males.

U N 6 I T

Heredity

SUGGESTIONS FOR INTRODUCING UNIT 6

In this unit the hereditary basis for biological diversity is examined. Chapters 20 and 21 acquaint the reader with classical genetics. The significance of meiosis is explored through monohybrid and dihybrid crosses in chapter 20. Chapter 21 continues by examining the connections between genetics and cytology by investigating mechanisms that control sex linkages and chromosomal crossing-over. Chapters 22 and 23 delve even further and present the molecular basis for heredity. Chapter 22 examines the molecule of life, DNA. Chapter 23 takes a look at how DNA regulates protein synthesis in cells.

Ask students, in small brainstorming groups, to generate a list of genetic traits that they possess. Answers to questions similar to the following could be entered in their journals. Is it possible for two black Labrador dogs to have an offspring with yellow-colored fur? If cows are made of protein, why do you not look like a cow after eating cow protein? The list can be checked once again at the end of the unit. How many of the students' initial questions have been answered?

KEY SCIENCE CONCEPTS

Unit 6, Heredity, provides students with an opportunity to link the key concepts of **change** and **diversity.** Genetic structure and molecular biology provide the basis for understanding why things are different. The relationship between science and technology is vividly depicted in this unit as technological applications of selective breeding and gene manipulation are explored. Social aspects concerning the application of science are also fostered throughout this unit. This unit also develops some of the skills and knowledge of Unit 3 of the Alberta Biology 30 Program of Studies. ■

CHAPTER 20

Genes and Heredity

INITIATING THE STUDY OF CHAPTER TWENTY

Chapter 20 provides an introduction to heredity. The following suggestions provide examples for beginning the chapter:

1 Introduce the social issue at the beginning of the chapter. Ask students to write their initial reactions prior to discussing the social issue. After the chapter has been completed, they can go back to their journal and see if learning more about genetics altered their initial thoughts.
2 Challenge students to devise a list of human phenotypes. Ask them to predict which traits are dominant and recessive.

ADDRESSING ALTERNATE CONCEPTIONS

- Students beginning a study of genetics may confuse the relationship between gene and chromosome. Students might be asked to refer back to the laboratory on Human Karyotypes presented in chapter 19, Sexual Cell Reproduction, and indicate why the chromosomes have bands. The bands represent different genes. Genes are units of instruction, located on chromosomes, which produce or influence a specific trait in the offspring.
- The concepts of segregation and independent assortment can be confused. Segregation refers to the separation of paired genes during meiosis. A pair of factors, or genes, separate or segregate during the formation of sex cells. We recommend introducing segregation as monohybrid crosses are introduced. The law of independent assortment indicates that if two different genes are located on separate chromosomes, the genes are inherited independently of each other. Consider a pea plant (*YyRr*) in which yellow color (*Y*) is dominant to green (*y*) and round shape (*R*) is dominant to wrinkled (*r*). As the homologous chromosomes for (*YyRr*) move to opposite poles during meiosis, a gene for pea color will segregate with genes for pea shape in equal frequency. This means that the sex cells containing (*YR*) will equal the number of sex cells containing (*yR*). Similarly, the green genes will segregate with round and wrinkled genes in equal frequency. The number of gametes containing (*Yr*) will be equal to the number of gametes containing the (*yr*) alleles. Independent assortment involves dihybrid problems (or those containing even more gene combinations).
- Some students, believing that they look more like one parent, may have constructed an understanding of dominance which is determined by sex. Such theories, based on limited empirical evidence, seem to be logical and consistent. About 2000 years ago Aristotle suggested that heredity could be traced to the power of the male's semen. He believed that hereditary factors from the male outweighed those from the female. Other scientists speculated that the female determined the characteristics of the offspring and that the male gamete merely set events in motion. By 1865, the year that Mendel published his papers, many of the misconceptions had been cleared up. Nineteenth-century biologists knew that the egg and sperm unite to form a new individual, and it was generally accepted that a blending of factors from the egg and sperm were responsible for the characteristics of the offspring. However, Mendel knew nothing about the structure or location of the hereditary material. Mendel did not know

about meiosis, nor did he understand how the genetic code worked. Yet, in spite of the limited knowledge of the times, he set geneticists on the right pathway.

- Many times the terms gene and allele are inappropriately interchanged. The two terms are not identical. Alleles are two or more alternate forms of a gene. The pea plant, for example, may have a gene for tall-stem length, designated by *T*, and a gene for short-stem length, designated by *t*. *T* is one allele and *t* is a second allele. Each of the alleles is located at the same position on one of the pair of homologous chromosomes. The section of the chapter entitled, Multiple Alleles, may help them sort out this quandary. Students will recognize that some single genes have multiple alleles.
- Students may associate inbreeding with negative connotations. Purebred animals, many of which have desired traits, are products of inbreeding.
- The effect of the environment on phenotype is often undervalued. Many students will develop a determinist outlook of the world, in which they conclude that we are totally a product of our genes. Although genes have a major role to play in our identity, they are not the only factor. All genes interact with the environment, and some phenotypes are modified by the environment. At times it is difficult to identify how much of the phenotype is determined by the genes (nature) and how much is determined by the environment (nurture). Fish of the same species show variable numbers of vertebrae if they develop in water of different temperatures. Primrose plants are red if they are raised at room temperature, but become white when raised at temperatures above 30°C. Himalayan rabbits are black when raised at low temperatures, but white when raised at high temperatures.
- Students are rarely provided an opportunity to talk about the nature of science. Many students believe that scientific breakthroughs occur because of tremendous insight or luck. Most often they overlook the social context for the development of a scientific theory. Technological achievements, the development of new paradigms in related fields of inquiry, or social needs often guide scientific discovery. It is often said that scientific theories have their time. The fact that Sutton and Boveri came to the same conclusion in 1902 appears to support this notion. In chapter 4, Adaptation and Change, students read that Darwin and Wallace simultaneously published articles explaining evolution by natural selection.

MAKING CONNECTIONS

- Mendelian genetics provides a context for understanding variation, the basis of natural selection. Natural selection is presented in chapter 4, Adaptation and Change, and again in greater detail in chapter 24, Population Genetics. By understanding the principles of inheritance, students recognize why some gene pools are more successful than others. In turn, this section of the chapter acts as a springboard to developing an understanding of gene selection and changes in gene frequencies, introduced in chapter 24. The case study, Genetic Disorders as Models for Evolution, presented in chapter 24, establishes a bridge between meiosis (chapter 19), genetics (chapters 20–23), and evolution (chapter 24).
- The section, Research in Canada: The Plant Breeders, establishes additional links between Unit 1, Life in the Biosphere, and Unit 6, Heredity.

POSSIBLE JOURNAL ENTRIES

- Students might be asked to write a dialogue between the text and themselves, as they refute the idea that dominance is sex dependent for all traits. By acting as "devil's advocate," they can challenge their own learning and push understanding to a higher level.
- Students can be encouraged to develop a concept map that includes all key terms presented in the chapter.
- Students can be encouraged to express the difficulties they experienced in formulating an understanding of heredity. What things did they employ to aid them in constructing their knowledge? For example, some students may indicate that the laboratory, Genetics of Corn, aided their understanding of dihybrid crosses. Other students may have attempted a series of their own summary charts or constructed concept maps.

- Decision-making strategies may be recorded as each group prepares for the debate. Did the students change their minds during the preparation for the debate? Students may even be asked to record their initial positions about the social issue and to reflect upon these feelings after the debate has been completed. Did they change their minds after listening to opposing arguments?

USING THE VIDEODISC

Section title	Page	Code	Frame (side)	Description
Importance of Genetics	468 – 469		1640 (all)	Human chromosome and Drosophila chromosome
Importance of Genetics	468 – 469		1641 (all)	Human chromosome, low power magnification
Importance of Genetics	468 – 469		1642 (all)	Human chromosome, high power magnification
Pioneer of Genetics: Gregor Mendel	469 – 470		3014 (all)	Mendel's first law
Pioneer of Genetics: Gregor Mendel	469 – 470		3015 (all)	Mendel's second law
Mendel's Experiments	470 – 471		3130 (all)	Pollen
Mendel's Experiments	470 – 471		2926 (all)	Gametes
Mendel's Experiments	470 – 471		3131 (all)	Pollination
Genetic Terms	472 – 473		2716 (all)	Alleles
Genetic Terms	472 – 473		2955 (all)	Heterozygous
Genetic Terms	472 – 473		2931 (all)	Gene
Genetic Terms	472 – 473		2933 (all)	Genotype
Single-Trait Inheritance	473 – 474		2874 (all)	Dominance
Single-Trait Inheritance	473 – 474		3168 (all)	Recessive allele
Single-Trait Inheritance	473 – 474		2637 (all)	Three-frame sequence showing gametes
Single-Trait Inheritance	473 – 474		3160 (all)	Punnett square

Section title	Page	Code	Frame (side)	Description
Multiple Alleles	475 – 476		1721 (all)	Red-eyed fruit fly
Incomplete Dominance	476 – 477		2828 (all)	Codominance
Dihybrid Crosses	480 – 481		1713 (all)	General review of crosses up to dihybrids
Dihybrid Crosses	480 – 481		1716 (all)	Dihybrid crosses: purple—starchy; yellow—sweet corn
Probability	482 – 483		1717 (all)	Blood typing showing probability

IDEAS FOR INITIATING A DISCUSSION

Table 20.1: Dominance Hierarchy and Symbols for Eye Color in *Drosophila*

Challenge students to devise their own multiple-alleles question. All questions can be placed in a hat and drawn by individual students. The originator of the question should place his or her name on the question; this enables the person working on the question to check his or her answer with the originator. Any failure to reach consensus between the originator and the person who has selected the question should be arbitrated by the teacher or a group of other students.

Case Study: A Murder Mystery

Challenge students to devise their own murder mystery.

Figure 20.16: Photo of Lysenko

What lessons does Lysenko's story provide to young scientists?

Figure 20.21: Shape of ear lobes and hairline

Using the characteristics shown in the photographs, develop a pedigree chart for either your family or some other family. Ask students to try the same. Students who are adopted may choose their adopted family to determine how many similarities they share. Which traits appear to be dominant?

Figures 20.29 and 20.30: Environmental influences on phenotype

Brainstorm about other traits that are affected by environmental influences.

REVIEW QUESTIONS

(page 474)

1 For Labrador retrievers, black fur color is dominant to yellow. Explain how a homozygous black dog can have a different genotype than a heterozygous black dog. Could the heterozygous black dog have the same genotype as a yellow-haired dog?

■ The heterozygous black dog would have a *Bb* genotype. The homozygous black dog would have a *BB* genotype. Black is dominant, so both genotypes are possible. A yellow dog must be homozygous (*bb*), because yellow hair color is recessive.

2 A pea plant with round seeds is cross-pollinated with a pea plant that has wrinkled seeds. For the cross, indicate each of the following:

a) the genotypes of the parents if the round-seed plant were heterozygous

■ The round plant must be *Rr*, and the wrinkled one *rr*.

b) the gametes produced by round and wrinkled-seed parents

■ Round gametes are *R* and *r*. Wrinkled gametes are both *r*.

c) the genotypes and the phenotypes of the F_1 generation

d) the F_2 generation if two round plants from the F_1 generation were allowed to cross-pollinate

■ **c)**

	R	*r*
r	*Rr*	*rr*
r	*Rr*	*rr*

1/2 round, *Rr*
1/2 wrinkled, *rr*

■ **d)**

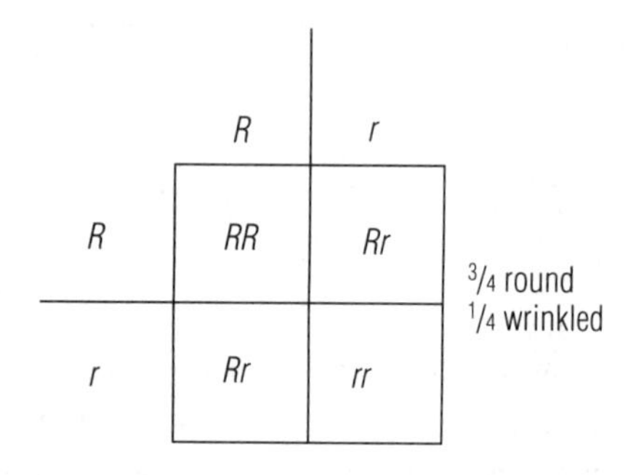

3 For Dalmatian dogs the spotted condition is dominant to non-spotted.

a) Using a Punnett square, show the cross between two heterozygous parents.

Parents: Spotted *(Ss)* x Spotted *(Ss)*

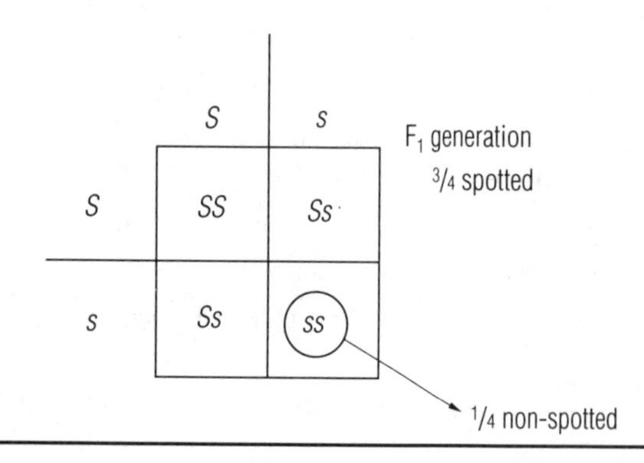

b) A spotted female Dalmatian dog mates with an unknown father. From the appearance of the pups, the owner concludes that the male was a Dalmatian. The owner notes that the female had six pups, three spotted and three non-spotted. What is the phenotype of the unknown male?

■ The mother must have at least one *S*, spotted gene. The father is unknown. Begin by writing the known genotype of the offspring. The non-spotted offspring must be *ss*. This indicates that each parent must have contributed at least one non-spotted gene.

Parents: Spotted *(Ss)* x (?)

	S	*s*
S	*S*	*ss*
s	*Ss*	*ss*

The mother must have at least one non-spotted gene.

1/2 spotted

F_1 generation

1/2 non-spotted

The father must have at least one non-spotted gene.

Now all that remains is to determine the final gene. If the final gene for the male were a *S* gene indicating spotted, 3/4 of the offspring would be spotted. However, if the remaining gene were *s* indicating non-spotted, the ratio would be 1/2 spotted and 1/2 non-spotted.

Parents: Spotted *(Ss)* x *(Ss)*

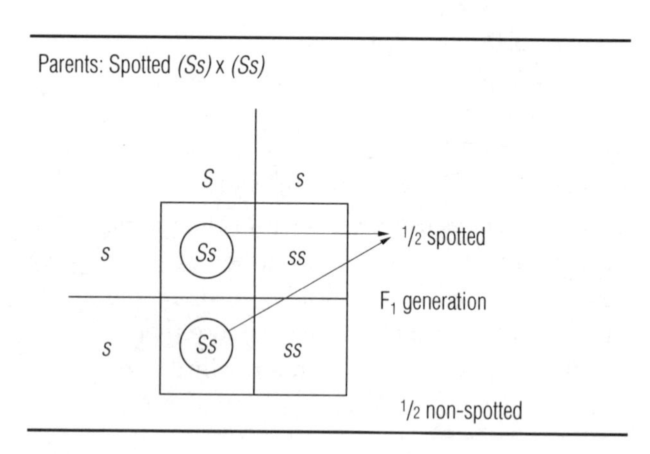

Note: Some students may point out that the 3 to 3 ratio does not rule out the mating of *Ss* and *Ss*. Although we might expect a ratio of 4 to 2 or 5 to 1, the 3 to 3 ratio is quite possible. This question provides the teacher with an opportunity to stress not jumping to conclusions. Although we have indicated the more likely genotype ($Ss \times ss$), we haven't eliminated the $Ss \times Ss$ possibility. The small sample size creates a variety of problems for geneticists.

4 For Mexican hairless dogs, the hairless condition is dominant to hairy. A litter of eight pups is found; six are hairless and two are hairy. What is the genotype of their parents?

■ The ratio is 3/4 hairless and 1/4 hairy. The 1/4 hairy indicates that each parent must contribute at least one hairy gene. Both parents are *Hh*.

(page 478)

5 Multiple alleles control the intensity of pigment in mice. The gene D^1 designates full color, D^2 designates dilute color, and D^3 is deadly when homozygous. The order of dominance is $D^1 > D^2 > D^3$. When a full-color male is mated to a dilute-color female, the offspring are produced in the following ratio: two full color to one dilute to one dead. Indicate the genotypes of the parents.

■ Because the F_1 generation contained a homozygous D^3 offspring (dead) both parents must carry the D^3 gene. Therefore, the full-color male must be D^1D^3, and the dilute-color female must be D^2D^3.

6 Multiple alleles control the coat color of rabbits. A gray color is produced by the dominant allele *C*. The C^{cb} allele produces a silver-gray color when present in the homozygous condition, $C^{cb}C^{cb}$, called chinchilla. When C^{cb} is present with a recessive gene, a light silver-gray color is produced. The allele C^b is recessive to both the full-color allele and the chinchilla allele. The C^b allele produces a white color with black extremities. This coloration pattern is referred to as Himalayan. An allele, C^a, is recessive to all genes. The C^a allele results in a lack of pigment, called albino. The dominance hierarchy is $C > C^{cb} > C^b > C^a$. The table below provides the possible genotypes and phenotypes for coat color in rabbits. Notice that four genotypes are possible for full color but only one for albino.

Phenotypes	Genotypes
Full color	CC, CC^{ch}, CC^h, CC^a
Chinchilla	$C^{ch}C^{ch}$
Light gray	$C^{ch}C^h$, $C^{ch}C^a$
Himalaya	C^hC^h, C^hC^a
Albino	C^aC^a

a) Indicate the genotypes and phenotypes of the F_1 generation from the mating of a heterozygous Himalayan-coat rabbit with an albino-coat rabbit.

■ For this cross ($C^hC^a \times C^aC^a$), the F_1 generation genotypes and phenotypes are: C^hC^a, Himalayan, and C^aC^a, albino.

b) The mating of a full-color rabbit with a light-gray rabbit produces two full-color offspring, one light-gray offspring, and one albino offspring. Indicate the genotypes of the parents.

■ Because the F_1 generation contains an albino offspring, both parents must carry the recessive C^a gene. Therefore, the full-color rabbit was genotype CC^a, and the light-gray rabbit was genotype $C^{ch}C^a$.

c) A chinchilla-color rabbit is mated with a light-gray rabbit. The breeder knows that the light-gray rabbit had an albino mother. Indicate the genotypes and phenotypes of the F_1 generation from this mating.

■ The chinchilla phenotype is produced by the genotype $C^{ch}C^{ch}$. Because the light-gray rabbit had an albino mother (C^aC^a), it must be genotype $C^{ch}C^a$. Therefore, the F_1 generation is: genotype $C^{ch}C^{ch}$, phenotype chinchilla, and genotype $C^{ch}C^a$, phenotype light-gray.

d) A test cross is performed with a light-gray rabbit and the following offspring are noted: five Himalayan-color rabbits and five light-gray rabbits. Indicate the genotype of the light-gray rabbit.

■ Test crosses are always performed with homozygous recessives, in this case, C^aC^a. Because the F1 generation contains equal numbers of Himalaya and light-gray rabbits, the genotype of the light-gray parent rabbit must be $C^{ch}C^h$.

7 A geneticist notes that crossing a round-shaped radish with a long-shaped radish produces oval-shaped radishes. If oval radishes are crossed with oval radishes, the following phenotypes are noted in the F_2 generation: 100 long, 200 oval, and 100 round radishes. Use symbols to explain the results obtained for the F_1 and F_2 generations.

■ Let *L* represent the long gene and *R* represent the round gene. The combination of *L* and *R* genes produced all oval phenotypes. This is then an example of incomplete dominance. The genotypes and their respective phenotypes are listed below:

Genotype	Phenotype
LL	long
RR	round
LR	oval
RL	oval

Therefore, crossing two oval-shaped parents, *LR* × *LR* will produce:

Genotype	Phenotype
1 *LL*	long
2 *LR*	oval
1 *RR*	round

8 Palomino horses are known to be caused by the interaction of two different genes. The allele C^r in the homozygous condition produces a chestnut, or reddish-color, horse. The allele C^m produces a very pale cream coat color, called cremello, in the homozygous condition. The palomino color is caused by the interaction of both the chestnut and cremello alleles. Indicate the expected ratios in the F_1 generation from mating a palomino with a cremello.

■ Parents $C^mC^m \times C^rC^m$

F_1 ratio	Genotype	Phenotype
2	C^mC^m	cremello
2	C^rC^m	palomino

(page 483)

9 In guinea pigs, black coat color (*B*) is dominant to white (*b*), and short hair length (*S*) is dominant to long (*s*). Indicate the genotypes and phenotypes from the following crosses:

a) Homozygous for black, heterozygous for short-hair guinea pig crossed with a white, long-hair guinea pig.

■ Parents *BBSs* × *bbss*

F_1 ratio	Genotype	Phenotype
1	*BbSs*	black, short
1	*Bbss*	black, long

b) Heterozygous for black and short-hair guinea pig crossed with a white, long-hair guinea pig.

■ Parents *BbSs* × *bbss*

F_1 ratio	Genotype	Phenotype
1	*BbSs*	black, short
1	*Bbss*	black, long
1	*bbSs*	white, short
1	*bbss*	white, long

c) Homozygous for black and long-hair crossed with a heterozygous black and short-hair guinea pig.

■ Parents *BBss* × *BbSs*

F_1 ratio	Genotype	Phenotype
1	*BBSs*	black, short
1	*BbSs*	black, short
1	*BBss*	black, long
1	*Bbss*	black, long

10 Black coat color (*B*) in cocker spaniels is dominant to white coat color (*b*). Solid coat pattern (*S*) is dominant to spotted pattern (*s*). The pattern arrangement is located on a different chromosome than the one for color, and its gene segregates independently of the color gene. A male that is black with a solid pattern mates with three females. The mating with female A, which is white, solid produces four pups: two are black, solid, and two white, solid. The mating with female B, which is black, solid produces a single pup, which is white, spotted. The mating with female C, which is white, spotted, produces four pups: one white, solid; one white, spotted; one black, solid; one black, spotted. Indicate the genotypes of the parents.

■ Female C is the key to this question as she is a homozygous recessive. Thus, the pairing of female C and the male is a test cross. The test cross produced white, solid; white, spotted; black, solid; and black, spotted. Thus, the male must be heterozygous for both traits and his genotype must be *BbSs*.
The mating with female A, white, solid, produced black, solid, and white, solid offspring. Therefore, female A does not have the recessive, spotted gene, and her genotype is *bbSS*.
The mating with female B, black, solid, produced a white, spotted offspring. Thus, female B must have been heterozygous for both traits, and her genotype is *BbSs*.

11 For human blood type, the alleles for types A and B are codominant, but both are dominant over the type O allele. The Rh factor is separate from the ABO blood group and is located on a separate chromosome. The Rh+ allele is dominant to Rh-. Indicate the possible phenotypes from the mating of a woman, type O, Rh-, with a man type A, Rh+.

■ The mother must be I^oI^o *Rh-Rh-*. The father, however, could be any one of four genotypes: I^aI^a *Rh+Rh+*, I^aI^a *Rh+Rh-*, I^aI^o *Rh+Rh+*, I^aI^o *Rh+Rh-*. Thus, the child could be phenotype A, Rh+; A, Rh-; O, Rh+; or O, Rh-.

(page 489)

12 In mice, the gene *C* causes pigment to be produced, while the recessive gene *c* makes it impossible to produce pigment. Individuals without pigment are albino. Another gene, *B*, located on a different chromosome, causes a chemical reaction with the pigment and produces a black coat color. The recessive gene, *b*, causes an incomplete breakdown of the pigment, and a tan, or light-brown, color is produced. The genes that produce black or tan coat color rely on the gene *C*, which produces pigment, but are independent of it. Indicate the phenotypes of the parents and provide the genotypic and phenotypic ratios of the F_1 generation from the following crosses.

a) *CCBB* × *Ccbb* **b)** *ccBB* × *CcBb*
c) *CcBb* × *ccbb* **d)** *CcBb* × *CcBb*

■ **a)** Parents *CCBB* (black) × *Ccbb* (tan)

F_1 ratio	Genotype	Phenotype
1	*CCBb*	black
1	*CcBb*	black

■ **b)** Parents *ccBB* (albino) × *CcBb* (black)

F_1 ratio	Genotype	Phenotype
1	*CcBB*	black
1	*CcBb*	black
1	*ccBB*	albino
1	*ccbb*	albino

Phenotypic ratio 1/2 black and 1/2 albino

■ **c)** Parents *CcBb* (black) × *ccbb* (albino)

F_1 ratio	Genotype	Phenotype
1	*CcBb*	black
1	*Ccbb*	tan
1	*ccBb*	albino
1	*ccbb*	albino

Phenotypic ratios 1/4 black, 1/4 tan, and 2/4 or 1/2 albino

■ **d)** Parents *CcBb* × *CcBb* (black)

F_1 ratio	Genotype	Phenotype
1	*CCBB*	black
2	*CCBb*	black
1	*CCbb*	tan
2	*CcBB*	black
4	*CcBb*	black
2	*Ccbb*	tan
1	*ccBB*	albino
2	*ccBb*	albino
1	*ccbb*	albino

Phenotypic ratios 9/16 black, 3/16 tan, and 4/16 albino

13 The mating of a tan mouse and a black mouse produces many different offspring. The geneticist notices that one of the offspring is albino. Indicate the genotype of the tan parent. How would you determine the genotype of the black parent?

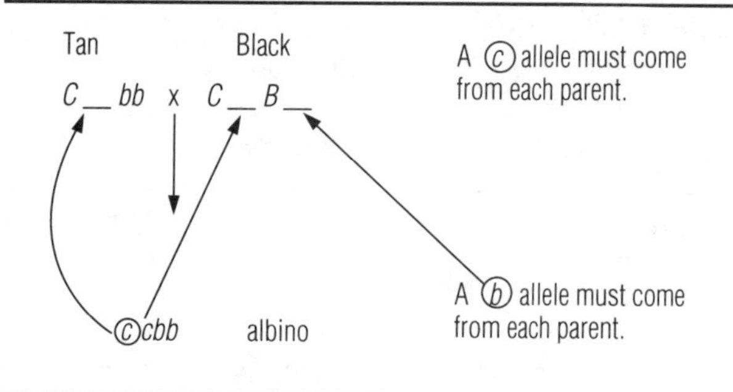

■ The tan parent must be *Ccbb*, while the black parent must be *CcBb*.

14 The gene *R* produces a rose comb in chickens. An independent gene, *P*, which is located on a different chromosome, produces a pea comb. The absence of the dominant rose comb gene and pea comb gene (*rrpp*) produces birds with single combs. However, when the rose and pea comb genes are present together, they interact to produce a walnut comb (*R_ P_*). Indicate the phenotypes of the parents and give the genotypic and phenotypic ratios of the F_1 generation from the following crosses.

a) *rrPP* × *RRpp* **b)** *RrPp* × *RRPP*
c) *RrPP* × *rrPP* **d)** *RrPp* × *RrPp*

■ **a)** Parents *rrPP* (pea) × *RRpp* (rose)

F_1 ratio	Genotype	Phenotype
all	*RrPp*	walnut

■ **b)** Parents *RrPp* × *RRPP* (both walnut)

F_1 ratio	Genotype	Phenotype
1	*RRPP*	walnut
1	*RRPp*	walnut
1	*RrPP*	walnut
1	*Rrpp*	rose

Phenotypic ratio: 3/4 walnut comb, 1/4 rose comb

■ **c)** Parents *RrPP* (walnut) × *rrPP* (pea)

F_1 ratio	Genotype	Phenotype
1	*RrPP*	walnut
	rrPP	pea

Phenotypic ratio: 1/2 walnutcomb, 1/2 pea comb

■ **d)** Parents *RrPp* × *RrPp* (both walnut)

F_1 ratio	Genotype	Phenotype
1	*RRPP*	walnut
2	*RRPp*	walnut
1	*RRpp*	rose
2	*RrPP*	walnut
4	*RrPp*	walnut
2	*Rrpp*	rose
1	*rrPP*	pea
2	*rrPp*	pea
1	*rrpp*	single

CASE STUDY
A MURDER MYSTERY

(page 479)

Time Requirements

Approximately 15 min are required, although some students will solve the question in less than 5 min.

The reasoning which gives the answer is this: The murderer must have had blood type O, Rh negative. The only possibilities are Henry, Tom, and Beth. Henry refused to give a blood sample, but he can be eliminated because neither of his parents had freckles. Because the nonfreckle condition is recessive it would be impossible for Henry to have freckles. Both of his parents do not have freckles. Like Henry, Tom does not have freckles, but he is not actually the son of Lord Hooke. Tom's daughter Beth must have murdered the Lord to ensure her future inheritance.

LABORATORY
GENETICS OF CORN

(page 484)

Time Requirements

Approximately 40 min are required to complete the laboratory.

Notes to the Teacher

- For sample A, order dihybrid purple starchy: purple, sweet: yellow, starchy: yellow, sweet, 9:3:3:1 ratio. See biological supply houses.
- For sample B order yellow, starchy: yellow, sweet: purple, starchy: purple, sweet, 1:1:1:1 ratio. See biological supply houses.

Observation Questions

a) Indicate the two different traits.

■ They are color and shape.

b) Predict the dominant phenotypes.

■ Purple and starchy (students may refer to it as wrinkled or shrunken) are dominant.

c) Predict the recessive phenotypes.

■ Yellow and sweet (students may refer to it as smooth) are recessive.

d) Indicate the phenotype of the *PPss* parent.

■ It is purple, sweet.

e) Indicate the phenotype of the *ppSS* parent.

■ It is yellow, sweet.

f) Indicate the expected genotypes and phenotypes of the F_1 generation.

■ All of the F_1 generation are genotype *PpSs*, and are purple and starchy.

g) Use a Punnett square to show the expected genotypes and the phenotypic ratio of the F_2 generation.

■ The phenotypic ratio is: 9 purple, starchy; 3 purple, sweet; 3 yellow, starchy; 1 yellow, sweet. See the Punnett square below.

	PS	*Ps*	*pS*	*ps*
PS	*PPSS*	*PPSs*	*PpSS*	*PpSs*
Ps	*PPSs*	*PPss*	*PpSs*	*Ppss*
pS	*PpSs*	*PpSs*	*ppSS*	*ppSs*
ps	*PpSs*	*Ppss*	*ppSS*	*ppss*

h) Indicate the phenotypic ratio of the F_1 generation (the kernel).

■ The phenotype ratio is: 1 purple, starchy; 1 purple, sweet; 1 yellow, starchy; 1 yellow, sweet.

i) Give the phenotype of the unknown parent.

■ The phenotype is purple, starchy.

Laboratory Application Questions

1 Why are test crosses important to plant breeders?

■ Test crosses allow the plant breeder to determine the genotypes of the plants. By knowing the genotypes of the parents they are able to predict the phenotypes of offspring.

2 A dihybrid cross can produce 16 different combinations of alleles. Explain why 100 seeds were counted rather than 16.

■ The larger the number, generally the more accurate the ratio. By using only 16 different kernels, there is a high probability that one of the gene combinations would not be represented.

3 A dominant allele *Su*, called starchy, produces kernels of corn that appear smooth. The recessive allele *su*, called sweet, produces kernels of corn that are wrinkled. The dominant allele *P* produces purple kernels, while the recessive *p* allele produces yellow kernels. A corn plant with starchy, yellow kernels is cross-pollinated with a corn plant with sweet, purple kernels. One hundred kernels from the hybrid are counted and the following results are obtained: 52 starchy, yellow kernels and 48 starchy, purple kernels. What are the genotypes of the parents and the F_1 generation?

■ The parents are *SuSupp* × *susuPp* (homozygous for starchy, and yellow X homozygous for sweet, heterozygous for purple).
The F_1 generation are *SusuPp* (starchy, purple) and *Susupp* (starchy, yellow).

APPLYING THE CONCEPTS

(page 492)

1 Explain why Mendel's choice of the garden pea was especially appropriate.

■ First, Mendel observed that garden peas have a number of different characteristics, each of which is expressed in one of two ways. For example, some garden peas produce green seed coats, while others produce yellow seed coats. Some plants are tall, while others are short. Mendel even noticed different flower positions and different flower colors.
A second reason for using garden peas is associated with the method in which the plant reproduces. Garden peas are both self-fertilizing and cross-fertilizing. Under normal circumstances the pollen produced from the male stamens attaches to the pistil inside of the flower. The process is called self-pollination. Because male and female sex cells usually come from the same plant, the parentage of the offspring rarely changes.
Mendel was able to cross-pollinate plants by removing the stamen from one plant and transferring the pollen to the pistil of another plant. Thus, the male and female sex cells of different plants were combined. Mendel was able to ensure that the seeds produced were cross-pollinated by first removing the stamen from the recipient plant. The pollen present must then have originated from the donor plant.

2 Long stems are dominant over short stems for pea plants. Determine the phenotypic and genotypic ratios of the F_1 offspring from the cross-pollination of a heterozygous long-stem plant with a short-stem plant.

■ *Ss* long-stem plant × *ss* short-stem plant, 1/2 tall-stem and 1/2 short-stem plants.

3 Cystic fibrosis is regulated by a recessive allele, *c*. Explain how two normal parents can produce a child with this disorder.

■ If each parent is heterozygous, neither one will show the cystic fibrosis phenotype, but 1/4 of their offspring will have two recessive genes. One half of their children will be heterozygous for the condition and only 1/4 will be homozygous (*CC*) for the normal condition.

4 In horses, the trotter characteristic is dominant to the pacer characteristic. A male trotter mates with three different females, and each female produces a foal. The first female, a pacer, gives birth to a foal that is a pacer. The second female , also a pacer, gives birth to a foal that is a trotter. The third female, a trotter, gives birth to a foal that is a pacer. Determine the genotypes of the male, all three females, and the three foals sired.

■ Male is a trotter (*Tt*), female 1 is a pacer (*tt*), foal 1 is a pacer (*tt*), female 2 is a pacer (*tt*), foal 2 is a trotter (*Tt*), female 3 is a trotter (*Tt*), and foal 3 is a trotter (*TT* or *Tt*).

5 For shorthorn cattle, the mating of a red bull and a white cow produces a calf that is described as roan. Roan is intermingled red and white hair. Many matings between roan bulls and roan cows produce cattle in the following ratio: 1 red, 2 roan, 1 white. Is this a problem of codominance or multiple alleles? Explain your answer.

■ This is a problem of codominance, because multiple alleles would demonstrate various levels of dominance. The *RW* alleles cause the roan condition.

6 For ABO blood groups, the A and B genes are codominant, but both A and B are dominant over type O. Indicate the blood types possible from the mating of a male who is blood type O with a female of blood type AB. Could a female with blood type AB ever produce a child with blood type AB? Could she ever have a child with blood type O?

■ No, not if she mated with the man with blood type O. She would not have a child with blood type O; only blood types A and B children would be produced.

7 Thalassemia is a serious human genetic disorder that causes severe anemia. The homozygous condition (T^mT^m) leads to severe anemia. People with thalessemia die before sexual maturity. The heterozygous condition (T^mT^n) causes a less serious form of anemia. The genotype T^nT^n causes no symptoms of the disease. Indicate the possible genotypes and phenotypes of the offspring if a male with the genotype T^mT^n marries a female of the same genotype.

■ The offspring would be: 1/4 T^mT^m (thalassemia), 1/2 T^nT^m (less serious anemia), and 1/4 T^nT^n (normal).

8 For guinea pigs, black fur is dominant to white fur color. Short hair is dominant to long hair. A guinea pig that is homozygous for white and homozygous for short hair is mated with a guinea pig that is homozygous for black and homozygous for long hair. Indicate the phenotype(s) of the F_1 generation. If two hybrids from the F_1 generation are mated, determine the phenotypic ratio of the F_2 generation.

■ All members of the F_1 generation are *BbSs*.
The F_2 generation is 9/16 black, long; 3/16 black, short; 3/16 white, long; and 1/16 white, short.

9 For chickens, the gene for rose comb (*R*) is dominant to that for single comb (*r*). The gene for feather-legged (*F*) is dominant to that for clean-legged (*f*). Four feather-legged, rose-combed birds mate. Rooster A and hen C produce offspring that are all feather-legged and mostly rose-combed. Rooster A and hen D produce offspring that are feathered and clean, but all have rose combs. Rooster B and hen C produce birds that are feathered and clean. Most of the offspring have rose combs, but some have single combs. Determine the genotypes of the parents.

■ Rooster A is *RRFf*; rooster B is *RrFf*; hen C is *RrFf*; and hen D is *RRFf* or *RrFf*.
Note: If rooster A and hen C continued to mate, some single-comb offspring would be eventually seen in the F_1 generation.

10 For mice, the allele *C* produces color. The allele *c* is an albino. Another allele, *B*, causes the activation of the pigment and produces black color. The recessive allele, *b*, causes the incomplete activation of pigment and produces brown color. The alleles *C* and *B* are located on separate chromosomes and segregate independently. Determine the F_1 generation from the cross $CcBb \times CcBb$.

■ The F_1 generation would be 9/16 black, 3/16 brown, and 4/16 albino.

CRITICAL-THINKING QUESTIONS

(page 493)

1 Baldness (H^B) is dominant in males but recessive in females. The normal gene (H^n) is dominant in females, but recessive in males. Explain how a bald offspring can be produced from the mating of a normal female with a normal male. Could these parents ever produce a bald girl? Explain your answer.

■ The male must be H^nH^n (normal) and the female must be H^nH^B (not bald for a female, but if the H^B allele is passed on to her sons, they will develop baldness). They could not produce a bald girl, since females must be H^BH^B to be bald.

2 Use the phenotype chart (below) to answer the following questions.

a) How many children do the parents A and B have?

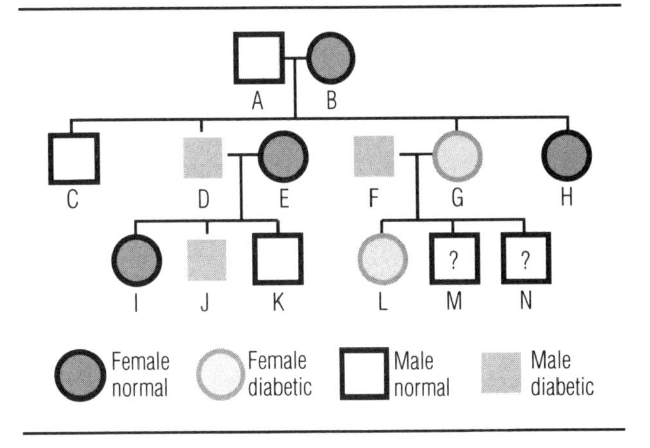

■ They have four children.

b) Indicate the genotypes of the parents.

■ The male A is *Nn* (heterozygous for normal); the female B is *Nn* (heterozygous for normal).

c) Give the genotypes of M and N.

■ Both will be *nn*.

3 In Canada, it is illegal to marry your immediate relatives. Using the principles of genetics, explain why inbreeding in humans is discouraged.

■ There are many possible answers, but students should consider the increased incidence of recessive genetic disorders occurring in the offspring from parents with similar genetic structures.

CHAPTER 21

The Source of Heredity

INITIATING THE STUDY OF CHAPTER TWENTY-ONE

Chapter 21 provides a union of genetics, cytology, and emergent technologies. The development and improvements in microscopy allowed genetics to progress. In this chapter and the next, students will be provided with cogent examples of how science and technology are intertwined. The light microscope, electron microscope, and biochemical analysis techniques such as X-ray diffraction, and gel electrophoresis, have provided a more complete picture of the mechanism of gene action.

- View the picture of Bobby and Brett Hull. Ask students to comment on the genetic basis for athletics. Are great athletes made or born? Consider the horses sired by Northern Dancer, the great Canadian race-horse and winner of the Kentucky Derby. Encourage discussion but avoid reinforcing any answers until the students begin addressing their ideas within the context of the chapter. The initial discussion merely sets a framework for learning.

ADDRESSING ALTERNATE CONCEPTIONS

- People question the relevence of some scientific research. Although studies on cancer, hereditary human diseases, and environmental toxins are self-evident, other studies seem less germane. Morgan's Experiments, as presented in chapter 21, stress the nature-of-science strand of an STS model. Students are often surprised to discover that a genetic difference between male and female had not been established prior to the twentieth century. Even more surprising was that Morgan's model for understanding gender came from fruit flies. The description of his experiments provides a compelling example of different ways of problem solving. There are many pathways to understanding.
- Students assume that sex differences are obvious; however, their basis for comparisons is most often restricted to anatomical characteristics, many of which are only evident after puberty. Most physical exams can readily identify sex. But, in some situations, a true distinction between male and female is not as easy as some would have you believe. Although the sex chromosomes help define sex, not all of the genes that code for sex are located on the sex chromosomes. The autosomes carry some traits that control the production of sex hormones. Males carry genes that code for female sex hormones. Most organisms contain genes for both sexes. Sex is determined by both the genetic makeup of the person and the development of the sex organs. For example, females with Turner syndrome, which is characterized by a single X chromosome, do not have Barr bodies. It is believed that the Barr body results from a second X chromosome, which is inactive. Should sex be determined on the basis of the presence of Barr bodies, individuals with Turner syndrome would not be classified as women. However, a karyotype would prove that they are not males. An extreme example of difficulty arises from testicular feminization syndrome. These XY individuals appear to be females, but show a positive test for male Y chromatin and a negative test for female chromatin. The syndrome is caused by a gene mutation on the X chromosome that acts only in XY zygotes. These individuals do not react to injections of male sex hormones, and therefore do not display the muscle development associated with males.

MAKING CONNECTIONS

- Chapter 21 forges a nexus between genetics and cell biology, as introduced in chapter 5, Development of the Cell Theory, and expanded upon in chapter 18, Asexual Cell Reproduction, and chapter 19, Sexual Cell Reproduction. The development and refinement of the microscope led to advances in cytology, and a union of two previously unrelated fields of study, cell biology and genetics. As students continue reading and exploring genetics, they will discover how other sciences, such as biochemistry and nuclear physics, have also become coupled with genetics.
- The concept of segregation links crossing-over, introduced in chapter 19, Sexual Cell Reproduction, with the sections Gene Linkage and Crossing-Over and Mapping Chromosomes, presented in chapter 21.
- The section on sex-linked traits provides a genetic framework for understanding sex differences, introduced in chapter 17, The Reproductive System. In chapter 17 students discovered that the testes and ovaries originate within the abdominal cavity from the same location. The fact that females produce both male and female hormones (and similarly, males produce male and female hormones) was also introduced within that chapter. Chapter 21 considers sex differences, by exploring X and Y chromosomes. The sections, Sex Determination, and Gender Verification at the Olympics, make the notion of male/female differences problematic. What most students assume is obvious is anything but obvious. Examples taken from nondisjunction conditions such as Turner syndrome help bridge concepts learned in chapter 19, Sexual Cell Reproduction, with chapter 21.

POSSIBLE JOURNAL ENTRIES

- Students might be asked to provide examples of how one discovery complements another. The work of Walter Fleming, who described the separation of threads within the nucleus during cell division, provided a theoretical framework for understanding the similarity of genetic information in all cells of an individual. Fleming called the process mitosis. Prior to this, many scientists speculated that the cells of the liver had different genetic instructions from those of the kidney, heart muscle, or brain. Other examples come from Edouard van Benden, who noticed that sperm and egg cells of roundworms had two chromosomes, but that the fertilized egg had four chromosomes. By 1887, August Weisman had proposed that a special division took place in sex cells. The reduction division is known as meiosis. Weisman had added an important piece to the puzzle of heredity, and in doing so, he provided a framework in which Mendel's work could be understood. With a cellular framework in place, Mendel's experiments were re-discovered in 1900, and the true significance of his work became apparent.
- Students might be asked to write a dialogue between the text and themselves, as they refute the idea that every cell of the body contains the same genetic information. They may ask why cells that have the same genetic information look so different. By acting as a "devil's advocate," the students can challenge their own learning and push understanding to a higher level.
- Students can be encouraged to develop a concept map that includes all key terms presented in the chapter.
- Students can be encouraged to express the difficulties they experienced in formulating an understanding of this challenging chapter. What things did they employ to aid them in constructing their knowledge? For example, some students may indicate that the laboratory, Human Sex-Linked Genes, aided their understanding of sex linkage. Other students may have used the problems provided within the chapter.
- Decision-making strategies may be recorded as each group prepares for the debate. Did the students change their minds during the preparation for the debate? Students may even be asked to record their initial positions about the social issue and to reflect upon these feelings after the debate has been completed. Did they change their minds after listening to opposing arguments? Should limits be placed on gene therapy?
- Students may comment on the importance of finding the cystic fibrosis gene, as described in the section, Research in Canada, The Cystic Fibrosis

Gene. The discovery of the gene is not a cure, but leads the way to finding a cure.

- Students may comment on the potentials of gene mapping techniques.
- The work of Barbara McClintock, although correct, was virtually ignored by the scientific community for nearly 40 years. Comment on McClintock's perseverance. See Gene Recombinations in Nature.

USING THE VIDEODISC

Section title	Page	Code	Frame (side)	Description
Importance of the Chromosomal Theory	494 – 495	[barcode]	1640 (all)	Human chromosome and Drosophila chromosome
Importance of the Chromosomal Theory	494 – 495	[barcode]	1641 (all)	Human chromosome under low power magnification
Importance of the Chromosomal Theory	494 – 495	[barcode]	1642 (all)	Human chromosome under high power magnification
Morgan's Experiments	496 – 498	[barcode]	1720 (all)	Red-eyed female Drosophila
Morgan's Experiments	496 – 498	[barcode]	1721 (all)	Red-eyed male
Morgan's Experiments	496 – 498	[barcode]	1722 (all)	White-eyed female
Morgan's Experiments	496 – 498	[barcode]	1723 (all)	White-eyed male
Morgan's Experiments	496 – 498	[barcode]	1724 (all)	Red-eyed male and female
Morgan's Experiments	496 – 498	[barcode]	1725 (all)	White-eyed male and female
Morgan's Experiments	496 – 498	[barcode]	1726 (all)	White-eyed female and red-eyed male
Morgan's Experiments	496 – 498	[barcode]	1727 (all)	Red-eyed female and white-eyed male
Sex Determination	498 – 500	[barcode]	3194 (all)	Sex chromosome
Sex Determination	498 – 500	[barcode]	3195 (all)	Sex-linked characteristics
Gene Linkage and Crossing-Over	503 – 504	[barcode]	2855 (all)	Crossing-over
Gene Linkage and Crossing-Over	503 – 504	[barcode]	3169 (all)	Recombinations
Mapping Chromosomes	504 – 505	[barcode]	2818 (all)	Chromosomes

Section title	Page	Code	Frame (side)	Description
Mapping Chromosomes	504 – 505	[barcode]	2819 (all)	Chromosome map
Mapping Chromosomes	504 – 505	[barcode]	2983 (all)	Karyotypes
Gene Recombinations in Nature	507 – 508	[barcode]	3125 (all)	Plasmids
Gene Recombinations in Nature	507 – 508	[barcode]	3268 (all)	Transposons
Gene Splicing Techniques and Gene Mapping	508 – 509	[barcode]	3243 (all)	Temperate bacteriophages
Gene Splicing Techniques and Gene Mapping	508 – 509	[barcode]	3266 (all)	Translocation

IDEAS FOR INITIATING A DISCUSSION

Figure 21.4: Cross between a homozygous red-eyed female and a white-eyed male

Ask students to consider whether or not other scientists working with *Drosophila* may have noticed that only males had white eyes in the F_2 generation. Was Morgan the first to ever notice that male and female *Drosophila* may have different traits (for the cross of a white-eyed male and red-eyed female parent)? Is it possible that others may have ignored results that they were unable to explain? A great number of laboratories were engaged in work with fruit flies at the time when Morgan made his announcement.

Figure 21.8: Gender verification

Students generally think of the difference between male and female as being clear. In reality, the difference is much more obscure. Many will argue that nondisjunction occurrences, such as XXY conditions, make it difficult to determine maleness and femaleness. Is the difference dependent on who defines what a female is?

Figure 21.13: Photo of Barbara McClintock

Ask students to speculate about why Barbara McClintock's work was ignored for nearly 40 years, despite substantial data which supported her conclusions. She opposed the ideas of Watson and Crick, who then believed that genes were in fixed positions.

REVIEW QUESTIONS

(page 500)

1 A recessive sex-linked gene (*h*) located on the X chromosome increases blood-clotting time.This causes the genetic disease, hemophilia.

a) Explain how a hemophilic offspring can be born to two normal parents.

■ Because the gene for hemophilia is a sex-linked recessive gene, hemophilic offspring can be produced by two normal parents. If the mother is heterozygous, and contributes the *h* gene, 1/2 of the male children will be hemophilic.

b) Can any of the female offspring develop hemophilia?

■ No female offspring of normal parents can be hemophilic because the father can only contribute a dominant gene to the female offspring.

2 A mutant sex-linked trait called "notched" (*N*) is deadly in *Drosophila* when homozygous in females. Males who have a single (N) allele will also die. The heterozygous condition (Nn) causes small notches on the wing. The normal condition in both males and females is represented by the allele n.

a) Indicate the phenotypes of the F_1 generation from the following cross: $X^nX^N \times X^nY$.

F_1 ratio	Genotype	Phenotype
1	$X^n X^n$	Normal, female
1	X^nX^N	Small notches, female
1	X^nY	Normal, male
1	X^NY	Dead, male

b) Explain why dead females are never found in the F_1 generation, no matter which parents are crossed.

■ The researcher never finds dead females in the F_1 generation because no male with the N trait can mate, all such males are dead.

c) Explain why the mating of female X^nX^N and a male X^NY is unlikely.

■ The male would be dead.

3 In *Drosophila*, eye color is determined by two different genes located on different chromosomes. A recessive gene (b), found on the second chromosome, produces brown eye color. The second chromosome is an autosomal chromosome. A recessive sex-linked gene (v) causes vermilion eye color to be produced. The presence of the dominant genes (B) and (V) results in a wild-type color. The presence of both the brown and vermilion alleles in the homozygous condition results in an individual with white eyes. Indicate the genotypes and phenotypes produced from the following crosses:

a) BB,VV (wild-type female) × bb,vY (white-eyed male)

■ Parents $BB,Vv \times bb,Vy$

F_1 ratio	Genotype	Phenotype
1	$BbVv$	Wild-type, female
1	$BbVY$	Wild-type, male

b) Bb, Vv (wild-type female) × Bb, VY (wild-type male)

■ Parents $BbVv \times BbVY$

F_1 ratio	Genotype	Phenotype
1	$BBVV$	Wild-type, female
1	$BBVv$	Wild-type, female
2	$BbVV$	Wild-type, female
2	$BbVv$	Wild-type, female
1	$bbVV$	Brown-eyed, female
1	$bbVv$	Brown-eyed, female
1	$BBVY$	Wild-type, male
1	$BBvY$	Vermilion-eyed, male
2	$BbVY$	Wild-type, male
2	$BbvY$	Vermilion-eyed, male
1	$bbVY$	Brown-eyed, male
1	$bbvY$	White-eyed, male

c) bb, VV (brown-eyed female) × Bb,vY (vermilion-eyed male)

■ Parents $bbVV \times BbvY$

F_1 ratio	Genotype	Phenotype
1	$BbVv$	Wild-type, female
1	$bbVv$	Brown-eyed, female
1	$BbVY$	Wild-type, male
1	$bbVY$	Brown-eyed, male

(page 512)

4 List three difficulties that arise when genes are studied in human populations.

■ It isn't always clear whether the phenotype is caused by a single genotype or whether or not the trait resides on an autosome or sex chromosome. With humans, test crosses cannot be done to determine a genotype. Ethical considerations prevent mating to determine gene action. Humans also produce very few offspring when compared with *Drosophila.* This also prevents us from determining crossover ratios and inhibits us in determining dominance.

5 What are gene crossovers? How does this process affect segregation?

■ Gene crossovers are the joining of corresponding strands of DNA from homologous chromosomes because of strand breakage and rejoining.

6 What are linked genes?

■ Linked genes are genes that are located on the same chromosome.

7 How can gene crossover frequencies be used to construct gene maps?

■ It is known that the rate of crossing-over between two genes is a function of the distance separating them on the chromosome; the further apart they are, the greater the rate of crossing-over. The percentage of crossover, of map units, is additive. This means that the linear order and distance along the chromosome can be calculated for groups of three or more genes.

8 What are some possible benefits of gene therapy?

■ Gene therapy has too many potential benefits to attempt an exhaustive list. Two of the positive foreseeable aspects are the prevention of hereditary genetic disorders and the curing of somatic cell gene disorders. Gene therapy also has many potential uses which may prove to be beneficial, although it may also be used for unethical purposes. These potential uses include the altering of DNA to produce "better" offspring, the designing of people for specific tasks such as underwater living, and the altering of the genome to slow aging and increase life span. In fact, the potential of gene therapy is limitless. In time, it may be possible to accomplish in the span of a few years, what evolution has spent eons on.

LABORATORY
HUMAN SEX-LINKED GENES

(page 502)

Time Requirements

Approximately 10 min are required to complete the procedure.

Observation Questions

a) Answers will vary.

b) Answers will vary but expect more males to be color-blind.

Laboratory Application Questions

1 How would your laboratory results differ if color blindness were not sex-linked?

■ Males and females would show equal frequency. (This is also assuming a large sample size.)

2 Explain why a woman who is not color-blind, but whose father was color-blind, can give birth to a son who is color-blind.

■ The son inherits the color-blind gene from the mother. The mother is X^CX^c heterozygous. The father contributes the Y chromosome, which determines maleness.

3 Diabetes is caused by a recessive gene located on an autosomal chromosome. You already know that color blindness is caused by a recessive sex-linked trait. Explain why the ratio of women to men who have diabetes is much closer than the ratio of women to men who have color blindness.

■ For males to acquire diabetes, they must receive two *dd* alleles; however, for males to get color blindness, they only need a single color-blind allele.

4 Hemophilia A, another sex-linked disorder, is very rare in females, yet color blindness is not rare. Explain why the color-blindness gene is more common.

■ For females to get hemophilia, the father would have to carry a hemophilic gene and the mother must carry at least one hemophilic gene. Hemophilia, for females, would be especially serious after puberty. The female reproductive cycle presents many problems which ultimately could result in death. This ensures that the homozygous recessive disorder becomes very rare—these genes will not survive. However, color blindness presents no such detrimental condition. People who are color blind can go on to lead what can be described as a normal lifestyle.

CASE STUDY
MAPPING CHROMOSOMES

(pages 506–507)

Time Requirements

Approximately 40 min are required.

Notes to the Teacher

The following points should be reviewed prior to the case study:

- Genes are located in a linear series along a chromosome, much like beads on a string.
- Genes that are closer together will be separated less frequently than those that are far apart.
- Crossover frequencies can be used to construct gene maps.

Observation Questions

a) Indicate the areas of the chromatids that show crossing-over.

■ The lower section of the chromatid would show crossing-over, involving the *FG* section from one chromosome and *fg* section from the other chromosome pair.

b) According to the diagram, which genes appear farthest apart? (Choose from *EF*, *FG*, or *EG*.)

■ *E* and *G* are farthest apart.

c) Which alleles have been exchanged?

■ The *fg* and *FG* alleles have been exchanged.

d) Using the following data, determine the distance between genes *E* and *F*.

■ The distance is 6 units apart.

e) What is the distance between genes *E* and *G*?

■ The distance between *E* and *G* is 10 units.

f) What is the distance between genes *F* and *G*?

■ The distance between *F* and *G* is 4 units.

Case-Study Application Questions

1 What mathematical evidence indicates that gene *F* must be found between genes *E* and *G*?

■ *E* to *F* = 6 units and *E* to *G* = 10 units, therefore *F* to *G* must equal the difference between the two, or 4 units.

2 Draw the gene map to scale. (1 cm = 1 unit)

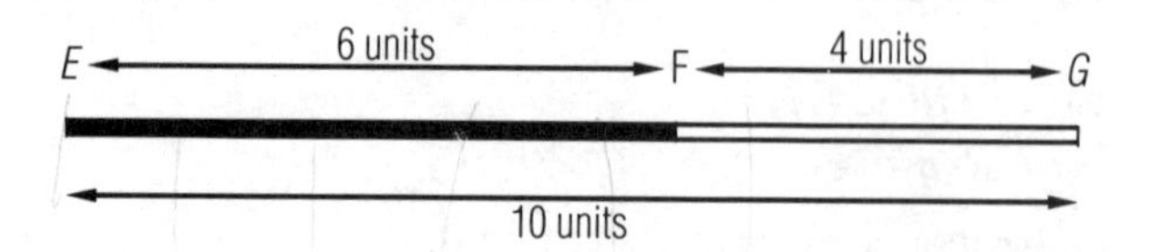

3 For a series of breeding experiments, a linkage group composed of genes *W*, *X*, *Y*, and *Z* was found to show the following gene combinations. (All recombinations are expressed per 100 fertilized eggs.)

Genes	W	X	Y	Z
W	—	5	7	8
X	5	—	2	1
Y	7	2	—	1
Z	8	3	1	—

Construct a gene map. Show the relative positions of each of the genes along the chromosome and indicate distances in gene units.

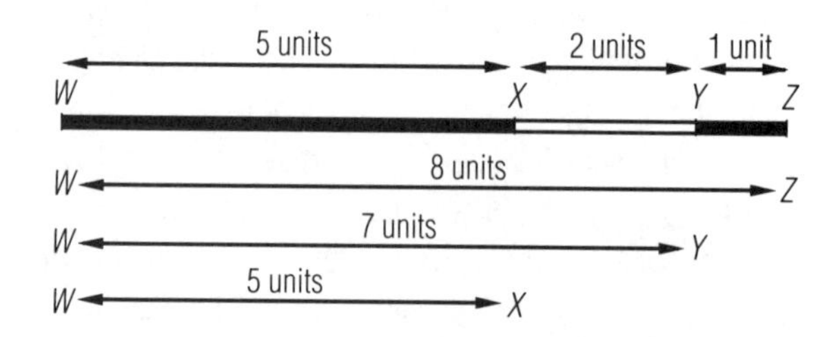

4 For a series of breeding experiments, a linkage group composed of genes, *A*, *B*, *C*, and *D* was found to show the following gene combinations. (All recombinations are expressed per 100 fertilized eggs.)

Genes	A	B	C	D
A	—	12	15	4
B	12	—	3	8
C	15	3	—	11
D	4	8	11	—

Construct a gene map. Show the relative positions of each of the genes along the chromosome and indicate distances in gene units.

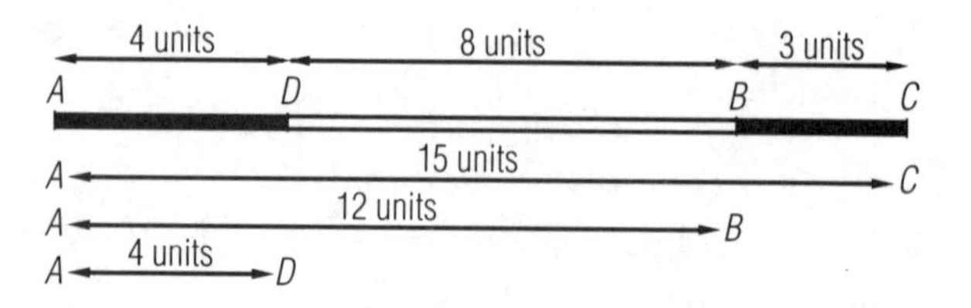

APPLYING THE CONCEPTS

(page 514)

1 In what ways was the development of the chromosomal theory linked with improvements in microscopy?

■ About the same time as Mendel was doing his experiments with garden peas, new techniques in lens grinding were providing better microscopes. The nucleus was discovered in 1831, just 34 years before Mendel published his results. In 1882, Walter Fleming described the separation of threads within the nucleus during cell division. He called the process mitosis. In the same year Edouard van Benden noticed that sperm and egg cells of roundworms had two chromosomes, but the fertilized egg had four chromosomes. By 1887 August Weisman had proposed that a special division took place in sex cells. The reduction division is known as meiosis. Weisman had added an important piece to the puzzle of heredity, and in doing so, he provided a framework in which Mendel's work could be understood. With a cellular framework in place, Mendel's experiments were re-discovered in 1900, and the true significance of his work became apparent.

2 Discuss the contributions made by Walter Sutton, Theodor Boveri, Thomas Morgan, and Barbara McClintock to the development of the modern chromosomal theory of genetics.

■ Sutton and Boveri observed that chromosomes come in pairs which segregate during meiosis. The chromosomes form new pairs when the egg and sperm unite. The paired chromosomes or homologous chromosomes supported Mendel's two-factor explanation of inheritance.

Sutton and Boveri knew that the egg was much larger than the sperm, but that the expression of a trait was not tied to it being located in a male or female sex cell. Therefore, some structure in both the sperm cell and the egg cell must determine heredity. Sutton and Boveri deduced that Mendel's factors (genes) must be located on the chromosomes. The fact that humans have 46 chromosomes, but thousands of different traits, led Sutton to hypothesize that each chromosome contains many different genes.

Thomas Hunt Morgan discovered that some genes are located on sex chromosomes. From experiments on *Drosophila,* he discovered that females have an XX chromosome pair and males have an XY chromosome pair. Morgan also discovered various mutations in *Drosophila.* He noted that some of the mutations seemed to be linked to other traits. Morgan concluded that the two genes responsible for the traits must be located on the same chromosome. This added support to the theory that the genes were located on chromosomes.

Barbara McClintock believed that genes could exchange position on chromosomes. With the exception of a few new combinations that might occur because of crossing-over, chromosome structure was thought to be fixed. Barbara McClintock interpreted her results of experiments with Indian corn and came to a conclusion that would shatter the traditional view of gene arrangement on chromosomes. McClintock suggested that genes can move to a new position. Her theory was dubbed the "jumping gene theory."

3 The gene for wild-type eye color is dominant and sex-linked in *Drosophila.* White eyes are recessive. The mating of a male with wild-type eye color with a female of the same phenotype produces offspring that are 3/4 wild-type eye color and 1/4 white-eyed. Indicate the genotypes of the P_1 and F_1 generations.

■ X^WX^w (wild-type color, female) × X^WY (wild-type color, male) F_1 generation

	X^W	X^w
X^W	X^WX^W	X^WX^w
Y	X^WY	X^wY

2 wild-type eye color females
1 wild-type eye color male
1 white-eyed male

4 Use the information from the pedigree chart to answer the following questions.

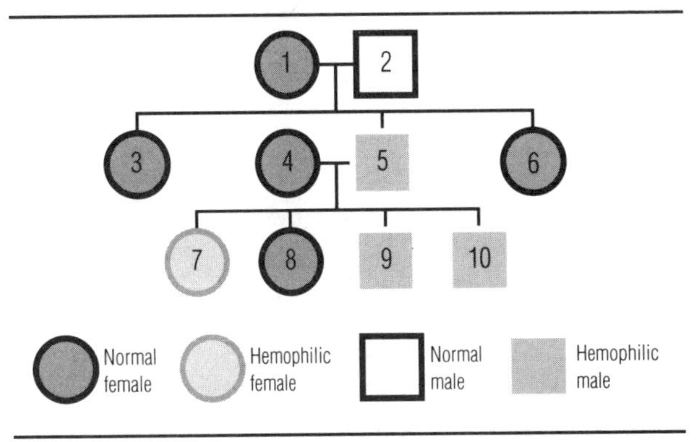

a) State the phenotypes of the P_1 generation.

■ Number 1 is a normal female; number 2 is a normal male.

b) If parents 1 and 2 were to have a fourth child, indicate the probability that the child would have hemophilia.

■ There is a 3/4 probability that the child would be normal, and a 1/4 probability that it would be hemophilic.

c) If parents 1 and 2 were to have a second male child, indicate the probability that the boy would have hemophilia.

■ There is a 1/2 probability that the boy would have hemophilia.

d) State the genotypes of 4 and 5.

■ 4 = X^HX^h; 5 = X^HY. Note the female must carry a hemophilic gene because female child 7 carries two hemophilic genes, one from each parent.

5 The autosomal recessive gene *tra* transforms a female into a phenotypic male when it occurs in the homozygous condition. The females who are transformed into males are sterile. The *tra* gene has no effect in XY males. Determine the F_1 and F_2 generations from the following cross: XX, +/*tra* crossed with XY, *tra*/*tra*. (Note: the + indicates the normal dominant gene.)

■ For F_1 generation:

	+	*tra*
tra	+/*tra*	*tra*/*tra*
tra	+/*tra*	*tra*/*tra*

The *tra*/*tra* gene will cause sterile females; *tra*/*tra* males are normal.

■ For F_2 generation:

+/*tra* × +/*tra*
female male

	X +	X *tra*
X +	+/+	+/*tra*
Y *tra*	+/*tra*	*tra*/*tra*

All females and males are normal.

+/*tra* × *tra*/*tra*
female male

	X +	X *tra*
X *tra*	+/*tra*	*tra*/*tra*
Y *tra*	+/*tra*	*tra*/*tra*

1/2 of the females are sterile.

6 Edward Lambert, an Englishman, was born in 1717. Lambert had a skin disorder characterized by very thick skin that was shed periodically. The hairs on his skin were very coarse, like quills, giving him the name "porcupine man." Lambert had six sons, all of whom exhibited the same traits. The trait never appeared in his daughters. In fact the trait has never been recorded in females. Provide an explanation for the inheritance of the "porcupine trait."

■ The allele is most likely sex-linked. The male only needs one mutated gene to express the phenotype, but the female requires two recessive genes, which should occur in much lower frequency.

7 A science student postulates that dominant genes occur with greater frequency in human populations than recessive genes. Using the information that you have gathered in this chapter, either support or refute the hypothesis.

■ This is not always so. Consider blood types; type O is recessive but occurs more frequently. Other examples such as Huntington's chorea may be cited.

8 In 1911, Thomas Morgan collected the following crossover gene frequencies while studying *Drosophila*. Bar-shaped eyes are indicated by the *B* allele, and carnation eyes are indicated by the allele *C*. Fused veins on wings (*FV*) and scalloped wings (*S*) are located on the same chromosome.

Gene combinations	Frequencies of recombinations
FV/B	2.5%
FV/C	3.0%
B/C	5.5%
B/S	5.5%
FV/S	8.0%
C/S	11.0%

Use the crossover frequencies to plot a gene map.

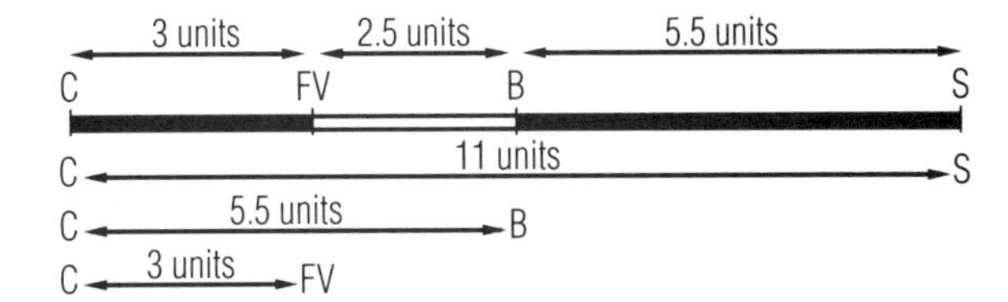

9 Huntington's chorea is a dominant neurological disorder that usually appears when a person is between 35 and 45 years of age. Many people with Huntington's chorea, however, do not show symptoms until they are well into their sixties. Explain why the slow development of the disease has led to increased frequencies in the population.

■ The disorder does not affect people until after they have reached reproductive age. The genes have already been passed on.

10 Explain the significance of locating the cystic fibrosis gene.

■ Many potentials can be mentioned. The most dramatic is with gene replacement, but others such as isolating enzymes to understand gene action may also be mentioned.

CRITICAL-THINKING QUESTIONS

(page 515)

1 Aristotle suggested that heredity could be linked to male semen. Other early scientists suggested that the female determined the traits of the offspring. Based on the knowledge that you have gathered about genetics, provide evidence that would refute both of these theories.

■ Mendel discovered that the dominant trait was not found exclusively in either the pollen or egg cells. If the allele was the dominant one, it did not matter which sex carried the allele. Many other genetic crosses, including those of Morgan, indicate that genes usually are not dominant or recessive because of their location on sex chromosomes.

2 Despite meeting all the criteria for good scientific research, the work of Barbara McClintock was ignored for almost 20 years. The significance of her work was only acknowledged when other scientists, working with bacteria, confirmed the fact that genes move along chromosomes. Why do you think her work was not readily accepted?

■ Many answers are possible. McClintock herself felt that her ideas opposed the popular theories of the day.

3 Discuss ways in which gene mapping technology can be used for beneficial purposes.

■ Students might indicate how gene markers can be used to diagnose diseases before they become acute. Students might also speculate on how gene mapping could be used as a forerunner of gene therapy.

4 Discuss ways in which gene mapping technology might be applied for harmful purposes.

■ Once a person's genotype is known, who will have access to the information? Will it be used to limit employment, restrict marriages, or limit someone's education? Students might speculate about how it could eventually lead parents to gene selection.

5 The gene for Huntington's chorea has been located near another gene that can be used as a genetic marker. The marker travels with the gene for Huntington's chorea. When segments of DNA are chopped up with restriction enzymes, scientists test for the Huntington's gene by locating the marker. If the marker is present, there is a 96% chance that the patient also carries the Huntington's gene. However, despite the success of the test, some people who are at risk have decided not to take the test. Why would some people be reluctant to take the test?

■ Many answers are possible. For example, some individuals may not want to know their fate in terms of illness. Others may fear that the information may prejudice their lifestyle.

6 The ability to fix genes of people who have genetic diseases lies on the horizon of genetic research. Gene therapy has been employed for some blood diseases, such as sickle-cell anemia. This disease is characterized by a single defective gene that forms abnormal hemoglobin. As you read earlier in the chapter, the hemoglobin is unable to carry oxygen efficiently. In fetal development, two different genes control the production of hemoglobin. In infancy, one gene is turned off, and the other, called the beta gene, is turned on. Unfortunately, the defective beta gene produces the abnormal hemoglobin, and sickle-shaped cells result.

One technique of gene therapy involves manufacturing a drug that turns off the defective beta gene and switches the infant gene back on. Although results have so far proved inconclusive, the technique shows promise. Should scientists attempt to develop drugs that turn genes on and off? Do scientists have the right to alter human genes? Support your opinion.

■ Many possible viewpoints can be discussed. Most importantly, students should recognize that the question cannot be answered from a scientific paradigm.

CHAPTER 22

DNA: The Molecule of Life

INITIATING THE STUDY OF CHAPTER TWENTY-TWO

Chapter 22 provides the biochemical basis for the understanding of heredity. The new era of genetics is often described in terms of molecular biology. Not only does this chapter provide many answers for questions that address the replication of life, but it creates a bridge between science and technology. Nowhere is the promise of biotechnology more grand, nor the fears more warranted than in technologies that use recombinant DNA biology.

- Introduce the social issue at the beginning of the chapter. Ask students to record their initial thinking at the beginning of the chapter. Their initial reactions can be reflected upon once they have gathered more information. Has the thinking of your students changed as information is gathered?
- Students can be asked if they believe that the same genetic information exists in a cell of their eye and a muscle cell found in their toe. Why do the two cells look so different?

ADDRESSING ALTERNATE CONCEPTIONS

- Some students will confuse the fact that DNA is double-stranded with the chromosomes containing two chromatids. Each chromatid is composed of a double helix of DNA. The demonstration, Comparing Meiosis and Mitosis, introduced in chapter 19, Sexual Cell Reproduction, might be re-visited during this chapter.
- Students may be challenged to explain how scientists came to understand that DNA and not proteins were the basis of heredity. Many good arguments can be introduced supporting the protein as a source of heredity. The protein molecule within the nucleus was thought to act as a master molecule, directing the arrangement of amino acids in the cytoplasm. Nucleic acids were thought to be monotonous repetitions of a single sugar molecule, identical phosphates, and four nitrogen bases. The same sugar, phosphate, and nitrogen bases were found in all living things. How could a language be constructed upon such a limited alphabet? The key to the genetic code was reasoned to

lie in the proteins. The case study, Evidence of Hereditary Material, will help students follow Avery and MacLeod's famous experiment, which helped confirm DNA as the source of heredity.

- Some texts, by presenting a limited profile of the discovery of DNA, leave students with the idea that Watson and Crick were solely responsible for our understanding of the biochemical basis for heredity. The work of Avery, MacLeod, McLyn, Wilkins, and Franklin helps show students clues that Watson and Crick followed. Most importantly, the monumental discovery by Watson and Crick was in providing a model that stimulated even more research. In many ways Watson and Crick opened a door to greater mysteries.
- Some students develop a view of scientists working in isolation, unaffected by material possessions, and consumed by the pursuit of knowledge. This view, popularized by television, movies, and some books, greatly distorts how the scientific enterprise is carried out. Watson and Crick might not have been the co-discoverers of DNA except for politics. The X-ray diffraction technique developed in England was used by Maurice Wilkins and Rosalind Franklin to view the DNA molecule. At that time, the leading investigator, Linus Pauling, an American citizen, was refused a visa to England to study the X-ray photographs. Pauling, along with others, had been identified by Senator Joe McCarthy as a communist sympathizer because of his support of the anti-nuclear movement. Many scientists believe that the United States passport office may have determined the winners in the race for discovery of the double-helix model of DNA.

MAKING CONNECTIONS

- The section, Replication of DNA, as presented in chapter 22, establishes a bridge between cell division, introduced in chapter 18, Asexual Cell Reproduction, and molecular biology. Replication helps explain how one cell can divide into two cells. Each of the daughter cells requires a complete set of genetic information that can only be obtained if the DNA molecule makes an exact duplicate of itself.
- The section on replication of DNA makes a connection between chapter 9, Digestion, and genetics. Students may not be able to explain why they do not begin to look like a cow after consuming and digesting a steak. If they take in cow DNA, why don't they look like a cow? A steak dinner supplies the muscle cells from a cow. The cells contain the nucleus and the genetic information from the cow. Specialized enzymes in their digestive tract break apart the cow DNA into nucleotides, which they use to make human DNA. Free nucleotides are floating around in the cells. They act as the building blocks for future DNA molecules.
- The section, Gene Recombinations in Nature, introduced in chapter 21, is complemented by Gene Recombinations in the Laboratory, presented in chapter 22. In turn, recombinant DNA technology serves as a bridge for the section on monoclonal antibodies, introduced in chapter 11, Blood and Immunity, and the work done on producing human insulin, presented in chapter 14, The Endocrine System and Homeostasis.
- The Frontiers of Technology section, entitled DNA Fingerprinting, uses a forensic theme and can make a connection with the section The Myth of Criminal Chromosomes, presented in chapter 19, Sexual Cell Reproduction.

POSSIBLE JOURNAL ENTRIES

- Students might be asked to provide examples of how politics influences science. The example provided by Linus Pauling's work on DNA and Lysenko's theory of genetics, presented in chapter 20, provide contrasting examples. Students may express how Lysenko's theories were structured to conform to Marxist philosophy or why Pauling was considered to be a government security risk.
- Students might be asked to write a dialogue between the text and themselves, as they refute the idea that you do not acquire genetic characteristics of other animals by consuming them. They may ask why they do not express characteristics of a cow by consuming steak. By acting as a

"devil's advocate," the students can challenge their own learning and push understanding to a higher level.

- Students can be encouraged to develop a concept map that includes all key terms presented in the chapter.
- Students can be encouraged to express the difficulties they experienced in formulating an understanding of this challenging chapter. What things did they employ to aid them in constructing their knowledge? For example, some students may indicate that the laboratory, Exploring DNA Replication, aided their understanding of cell reproduction. Other students may have used the diagrams within the chapter to organize an image of DNA.
- Group thinking and decision-making strategies may be recorded as each student prepares for the debate. Did the students change their minds during the preparation for the debate? Students may even be asked to record their initial positions about the social issue and to reflect upon these feelings after the debate has been completed. Did they change their minds after listening to opposing arguments? Should patents be placed on non-human life forms?

USING THE VIDEODISC

Section title	Page	Code	Frame (side)	Description
Importance of DNA	516 – 517		1663 (all)	Feulgen staining of broad bean root tip
Importance of DNA	516 – 517		1664 (all)	Root tips in fixative
Importance of DNA	516 – 517		1665 (all)	Acid hydrolysis of DNA
Importance of DNA	516 – 517		1666 (all)	Staining using the Feulgen's reagent
Importance of DNA	516 – 517		1667 (all)	Root tip showing staining
Structure of DNA	521 – 522		2865 (all)	DNA molecule
Structure of DNA	521 – 522		3076 (all)	Nucleotide
Structure of DNA	521 – 522		2762 (all)	Base pairings
Replication of DNA	523		25971 (1)	Movie sequence showing DNA replication and its enzyme controls
Gene Recombinations in the Laboratory	527 – 528		3169 (all)	Recombinations
Gene Recombinations in the Laboratory	527 – 528		3175 (all)	Restriction enzymes
Gene Recombinations in the Laboratory	527 – 528		3125 (all)	Plasmids
Gene Recombinations in the Laboratory	527 – 528		27255 (1)	Movie sequence showing graphic model of DNA and insertion of ethidium

IDEAS FOR INITIATING A DISCUSSION

(page 521)

Politics and Science: Watson and Crick

Students may have heard about the McCarthy era. It greatly affected the performing arts. Few people are aware that it also affected the scientific community. One of the Committee on Un-American Activities' other victims was Julius Robert Oppenheimer, the coordinator of the Manhattan project. Ironically, Linus Pauling went on to receive a Nobel Peace Prize in 1962.

Figure 22.6: Representation of DNA molecule

Ask your students if the strand on the right has been drawn upside-down. (The diagram is correct.) Some chains of DNA spiral to the right, while others spiral to the left.

Figure 22.8: DNA profile

DNA fingerprinting has been used to convict criminals in Canada. Allan Legere, a man who committed murder in New Brunswick, was Canada's first murderer to have DNA fingerprinting evidence presented in a murder trial.

Review Questions
(page 522)

1 Name two ways in which the DNA molecule is important in the life of a cell.

■ DNA is important to the life of a cell in that it provides the directions that guide the repair of worn or damaged cell parts, and contains the information that allows the cell to be duplicated, allowing the continuity of life.

2 What chemicals make up chromosomes?

■ Chromosomes are composed of roughly equal proportions of proteins and nucleic acids.

3 What are nucleotides?

■ Nucleotides are the basic components of nucleic acids.

4 What chemicals are found in a nucleotide?

■ Nucleotides are composed of a ribose sugar, phosphate, and nitrogen bases.

5 On what basis did some scientists conclude that proteins provided the key to the genetic code?

■ Some scientists erroneously concluded that proteins provided the key to the genetic code because of their complexity when compared with nucleic acids. Because proteins and nucleic acids were found in roughly equal proportions in the chromosomes, it was reasoned that the complicated proteins were logical master molecules. Nucleic acids were thought to be too simple to contain the vast amount of information required to produce a cell.

6 Who were the co-discovers of the double-helix model of DNA?

■ The co-discoverers of the double-helix model of DNA were James Watson and Francis Crick. They presented their model to the scientific world in 1953.

7 How did the X-ray diffraction technique provide a clue to the structure of DNA?

■ The X-ray diffraction provided an outline of the molecule of DNA.

8 Which nitrogen base pairs with guanine? Which base pairs with adenine?

■ Adenine bonds with thymine; cytosine bonds with guanine.

(page 527)

9 What is the significance of DNA replication for your body?

■ Because DNA is capable of self-replication, it is possible to pass the information contained in each of your cells to the cell's descendants. If DNA were not capable of self-replication, the information would have to be divided and reorganized each time a cell reproduced.

10 Name the nitrogen bases in the DNA molecule, and state which normally pairs with which.

■ The nitrogen bases of DNA and their complementary pairs are: guanine-cytosine; adenine-thymine.

11 Briefly describe the events of DNA replication.

■ DNA replication begins with the breaking of the hydrogen bonds between complementary nitrogen bases. This produces two parent strands of DNA. Each parent strand acts as a template for a new DNA molecule. Free nucleotides in the cell attach to their complementary bases on the parent strand. A set of enzymes, called polymerases, fuse the nucleotides together. In this way, two identical strands of DNA are formed from the original strand.

12 Proofreading enzymes scan the strands of DNA to check the nitrogen base pairings. Explain why these enzymes are important.

■ The proofreading enzymes that scan the DNA strands to check for errors in the nitrogen base pairings are very important. Errors are occasionally introduced by environmental factors such as radiation or chemicals. If left alone, these errors could be harmful or even fatal. The proofreading enzymes identify the damaged areas, which are then repaired or replaced by other specialized enzymes.

13 When you eat fish, you take in fish protein and fish DNA; however, you do not assume the characteristics of a fish. Explain why the nucleic acids of a fish do not change your appearance.

■ Even though we take in DNA of fish and other organisms, we do not incorporate this DNA into our cells. Rather, the DNA is broken down into its component nucleotides by enzymes in the digestive tract. These nucleotides are then used by our cells to make human DNA.

14 What is DNA fingerprinting and how does it work?

■ DNA fingerprinting is a method of identifying whether or not a sample of DNA comes from a specific person. Each person in the world (with the exception of identical twins, triplets, etc.) has segments of DNA that are specific to her or him. DNA fingerprinting entails comparing the target DNA with a sample from the suspected owner. Both samples are radioactively labeled and placed against X-ray film. The two samples are compared, and if they are identical, the DNA is proven to have come from the suspect.

CASE STUDY

EVIDENCE OF HEREDITARY MATERIAL

(pages 518–519)

Time Requirements

Approximately 40 min are required. It is recommended that students work in groups of three or four.

Notes to the Teacher

The following suggestions may help groups function:

- Organizer: will assume responsibility for assigning roles, organizing desks, and beginning the discussion
- Recorder: will assume responsibility for keeping a record of answers and checks for group agreement
- Scrutinizer: will seek other solutions and suggest other pathways
- Presenter: will ensure that all group members have had an opportunity to provide an opinion and will make a presentation

  Note: one student may assume more than one role.

Observation Questions

a) What conclusion can you derive from the experimental results?

■ The capsuled cells are more virulent, and they cause death.

b) Why might a scientist decide to repeat this experimental procedure on other mice?

■ The strain of mice may have been particularly sensitive to contaminants in one of the culture plates. The bacteria may not have caused death. By repeating the experiments, many variables can be checked.

c) Explain these experimental results.

■ Heating destroyed the bacteria.

d) Predict what would have happened to the mouse if the noncapsuled cells had been heated and then injected.

■ The mouse would have lived. Even untreated noncapsuled bacteria do not cause death.

e) Would you have predicted this observation? Explain why or why not.

- Students may answer yes or no to prediction. Predictions cannot be incorrect. They should be able to give an answer for their prediction. Consider any logical answer that supports their prediction to be correct—students are not expected to know the correct answer at this point. However, the heat-destroyed cells seem to have made the living uncapsuled cells capable of causing the mouse to die of pneumonia.

Case-Study Application Questions

1 A microscopic examination of the dead and live cell mixture revealed cells with and without capsules. What influence did the heat-destroyed cells have on the noncapsuled cells?

- Once heated, these bacteria were no longer capable of killing the mice.

2 Avery, MacCleod, and McCarty hypothesized that a chemical in the dead, heat-treated, capsuled cells must have altered the genetic material of the living noncapsuled cells. In similar previous experiments, scientists had dubbed this chemical the *transforming principle.* Avery and his associates believed that the transforming principle was DNA. What must have happened to the DNA when the cells divided?

- The genetic information of the heat-treated cells was incorporated into the noncapsuled living cells. The genetic material transformed the non-virulent bacteria into virulent bacteria, which also had capsules.

3 To test whether or not DNA was the transforming principle, Avery and his associates crushed capsuled cells to release their contents. The DNA was extracted. Next they added the DNA extract to a cell culture that contained noncapsuled pneumonia cells. These cells were later found to contain some cells with capsules. Did this confirm or disprove their hypothesis? Give your reasons.

- This supported their idea that the DNA entered the noncapsuled cells and transformed them into virulent, capsuled bacteria. As the noncapsuled bacteria reproduce, their offspring begin to show characteristics not found in the noncapsuled parents.

LABORATORY
EXPLORING DNA REPLICATION

(pages 525–526)

Time Requirements

Approximately 30 min are required.

Notes to the Teacher

A full page of the molecules that make up DNA is provided for duplication and student use on the following page.

Observation Questions

a) Why are the adenine and guanine molecules represented by larger shapes than the other two nitrogen bases?

- They are double-ring structures; see Figure 22.5.

b) What is this structure called?

- This structure is call a nucleotide.

c) Record the genetic code by indicating the letters of the nitrogen bases, beginning from the top of the page.

- Answers will vary.

d) Record the genetic code of the complementary strand.

- Answers will vary, but complementary base pairs will be A-T and G-C.

e) What do you notice about the two strands of DNA?

- The two strands of DNA are identical.

Laboratory Application Questions

1 Thymine always bonds with adenine. Explain why the thymine used in your model does not bind with guanine.

- The geometry will not match. The shapes do not fit together.

2 What determines the nitrogen base sequence of DNA in a new strand of DNA?

- The original strand of DNA provides the blueprint.

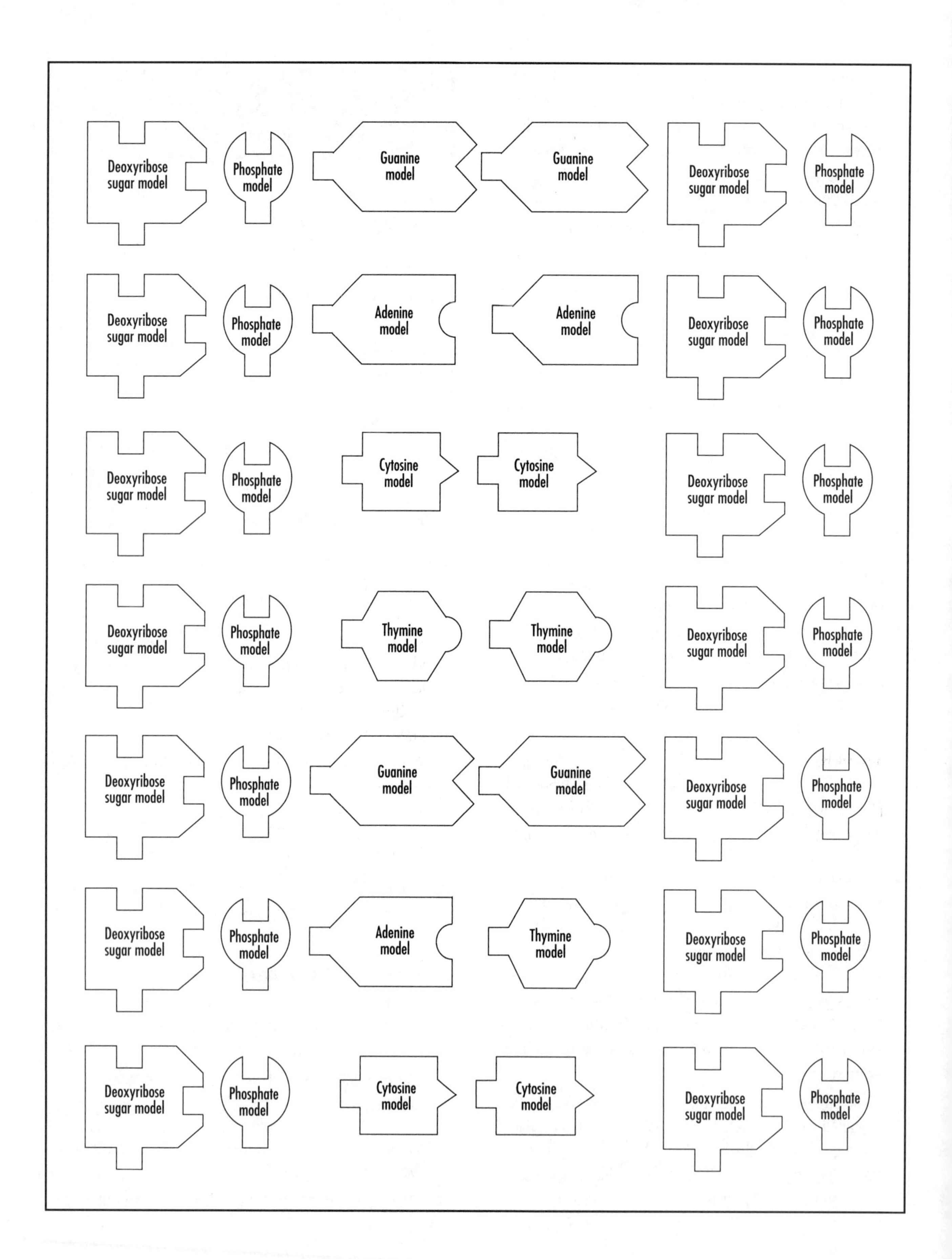
Deoxyribose sugar model
Phosphate model
Guanine model
Guanine model
Deoxyribose sugar model
Phosphate model
Deoxyribose sugar model
Phosphate model
Adenine model
Adenine model
Deoxyribose sugar model
Phosphate model
Deoxyribose sugar model
Phosphate model
Cytosine model
Cytosine model
Deoxyribose sugar model
Phosphate model
Deoxyribose sugar model
Phosphate model
Thymine model
Thymine model
Deoxyribose sugar model
Phosphate model
Deoxyribose sugar model
Phosphate model
Guanine model
Guanine model
Deoxyribose sugar model
Phosphate model
Deoxyribose sugar model
Phosphate model
Adenine model
Thymine model
Deoxyribose sugar model
Phosphate model
Deoxyribose sugar model
Phosphate model
Cytosine model
Cytosine model
Deoxyribose sugar model
Phosphate model

3 Adenine is a double-chain purine molecule. Purines always bond with pyrimidines. What special problems would be created if the purine adenine joined with guanine? with another purine?

■ The joining of the two large molecules (purine) would cause gaps along the ladder. The rungs joined by the smaller molecules (pyrimidines) would be too short to touch.

4 A special enzyme scans the DNA helix in search of improper nitrogen base pairings. Using your model, indicate how the enzyme would be able to detect incorrect nitrogen base pairings.

■ The geometry of the paired bases would prevent bonding, or two bases of inappropriate size would cause the strand to pull apart.

5 Radiation can cleave the bonds between phosphate and sugar molecules, thereby cutting a section of the DNA out of the ladder. Special enzymes have the responsibility of gluing the spliced segment of DNA back into place; however, the segment is not always glued back in the correct place. Explain why it is important to place the broken section of DNA back in its original position.

■ A genetic code that was once ATA CGC would be altered by removing a sequence of bases. In the next chapter students will be able to relate these changes to protein synthesis. This question is designed to get them started thinking about the importance of a genetic code.

APPLYING THE CONCEPTS

(pages 530–531)

1 What led scientists to speculate that proteins were the hereditary material?

■ Proteins exhibit much greater diversity than do nucleic acids.

2 Explain how Avery, MacLeod, and McCarty's experiment pointed scientists toward nucleic acids as the chemical of heredity.

■ The transformation of non-virulent bacteria without capsules to a more virulent form by combining their DNA with a virulent form suggested that DNA controls the expression of traits. In this case, the DNA from the virulent bacteria was incorporated into the harmless bacteria, which in turn were transformed into the capsulated forms, characteristic of the dangerous bacteria.

3 Science and technology are often referred to as *synergists*. This means that not only do scientific breakthroughs provide information for technological applications, but technological advances also spur scientific progress. Explain how X-ray diffraction techniques led to the discovery of DNA.

■ X-ray diffraction is a technique used to determine the shape of a molecule. The X ray is directed at the crystallized form of the molecule. The diffracted rays are then trapped by film. The pattern produced reveals the 3-D shape of the molecule. This allowed Watson and Crick to determine that DNA was a double helix, not a single helix, as others, like Linus Pauling had suggested.

4 Using DNA as an example, explain why scientists use models.

■ By studying the replica much can be learned about how the thing works. Molecules are already too small to see. Molecules are made larger to learn how the different atoms interact. X-ray diffraction techniques provide a picture that indicates how different chemical bonds interact with one another. Scientists use models as visual devices that help them understand how things look, and to see the relationship of different parts.

5 Compare the amount of DNA found inside one of your muscle cells with the DNA found in one of your brain cells.

■ They are the same; all came from the same initial cell. (Students often find it surprising to discover that cells of the eye have all the genetic information to be a muscle cell or a liver cell.)

6 Why are organ transplants more successful between identical twins than between other individuals?

■ The DNA of the identical twins should be identical or near identical. (Some changes caused by environmental mutagens and crossing over may have occurred.)

7 A drug holds complementary nitrogen bases with such strength that the DNA molecule is permanently fused in the shape of a double helix. Predict whether or not this drug might prove harmful. Provide your reasons.

■ DNA will not replicate, or the strand of DNA must be broken for that cell to replicate. Should a number of cells of the body absorb that drug, cell division in all of those cells would be impaired.

8 DNA fingerprinting has been used to identify rapists. Suggest at least two other applications for the DNA matching technique.

■ A great many could be identified. DNA fingerprinting has been used to locate lost or missing children in Argentina. In Canada, DNA fingerprinting has been used to identify poachers or individuals accused of catching illegal fish stocks. DNA fingerprinting can be used in paternity lawsuits.

9 What follows is a hypothetical situation. Genes that produce chlorophyll in plants are inserted into the chromosomes of cattle. Indicate some of the possible advantages of this procedure.

■ This would provide an animal that is capable of making its own food. The problem does not hint at the many difficulties associated with such gene implants, because the gene-regulating mechanism would also have to be included. However, at an introductory level, encourage students to speculate about the potentials of recombinant DNA research.

10 Analysis of chloroplasts and mitochondria reveals that DNA is located within these organelles. Explain how this discovery helps support the theory that these organelles might actually be descendants of individual living creatures.

■ DNA regulates cell activity. Organelles that were once independent would need their own genetic information. No organism survives without genetic information. If these structures were always parts of eukaryotic cells, they would need their own DNA. See chapter 24 for further information.

11 Explain the significance of the following statement to the search for the structure of the genetic material: Structure provides many clues about function.

■ The double-helix structure provides clues about DNA replication.

CRITICAL-THINKING QUESTIONS

(page 531)

1 The European Economic Community (EEC) has ruled that organisms created by biotechnology can be patented. Such patents have been awarded in the United States since 1984. The Supreme Court of Canada ruled that a patent application for a hybrid soybean could not be granted. Although this does not prevent another company from applying for patents in Canada, it does raise an interesting question: Will successful biotechnology companies choose not to locate in Canada unless they are given the same protection as companies in Europe and the United States? Give your opinions on this matter.

■ Many opinions can be explored. Students should be encouraged to listen to other viewpoints. Most importantly, students should recognize that the answer to the question is found not within the discipline of science but that of economics. The question permits students to express their own opinions and justify their conclusions.

2 DNA fingerprinting has had a tremendous impact on law enforcement. Speculate about how DNA fingerprinting will affect criminal trials. Should DNA fingerprinting ever be used to convict a criminal? Why or why not?

■ Be willing to accept a variety of answers. Some students may indicate that the DNA fingerprinting is not 100% accurate, but that the evidence may be accepted as such by jurors because it is scientific. Although the possibility is extremely remote, two people could have the same genetic structure. Could the technician make a mistake? Others will indicate that DNA testing will help convict criminals who may escape prosecution because of a lack of other evidence. DNA testing will not convict, it is just another form of evidence, similar to fingerprints.

3 In December 1988, researchers at Toronto's Hospital for Sick Children transplanted human bone marrow into mice. The mice allow scientists to study human blood cells, immune responses, and genetic blood disorders in a living system. Previously, bone marrow could only be studied in a living culture, not in a functional body. Many supporters of animal rights are appalled by this procedure. They believe that the mice are being exploited. Should animal modeling experiments, like the ones performed at the Hospital for Sick Children, be continued? Give reasons for your answer.

■ Expect at least two divergent types of arguments. Each opinion can be supported. Individuals who do not believe in animal testing will indicate that humans do not hold a superior position in the biosphere. Humans have no right to exploit nature. Individuals who support animal testing will indicate that direct human testing also creates problems. Animal testing is preferred to human testing.

4 Recombinant DNA technology has been described as the 20th century's most powerful technique since the splitting of the atom. Do you agree or disagree with this comparison?

■ Students may indicate the potential of recombinant DNA in biological warfare. Genes, such as the ones that produce botulism toxins, can be placed in *E. coli,* the microbe that lives in our guts. Students might also indicate limitations of biological warfare. It has never been used successfully. Students might also express how recombinant DNA will create economic domination. Biotechnology is a billion dollar industry that has spin-offs in agriculture, the chemical industry, and manufacturing.

CHAPTER 23

Protein Synthesis

INITIATING THE STUDY OF CHAPTER TWENTY-THREE

Proteins are the structural basis for life. Diversity and continuity of life can better be comprehended if we understand the manner in which DNA directs protein synthesis. Chapter 23 provides the biochemical basis for the understanding of protein synthesis. The expression of genetic mutations is presented as an extension of the understanding of protein synthesis. The HIV case study is also an extension of protein synthesis, but for this occurrence RNA from the HIV directs the synthesis of DNA which becomes incorporated into human genetic information. Eventually the DNA from the HIV virus directs human protein synthesis.

- Introduce the social issue at the beginning of the unit. Ask students to record their thinking at the beginning of the chapter. Their initial reactions can be reflected upon once they have gathered more information. Has the thinking of your students changed as information is gathered?
- Introduce the unit by asking students to read the section on biological warfare. Ask them to consider how protein synthesis through gene manipulation causes even greater concerns. The topic can be addressed once again near the end of the chapter. The initial discussion is not designed to provide answers, but merely frame a context for learning.

ADDRESSING ALTERNATE CONCEPTIONS

- Students may have difficulties explaining why offspring display characteristics not visible in their parents. The uniqueness of descendants can often be explained by new combinations of genes and/or mutations. In order to understand how genes affect the expression of an organism's traits, students will have to learn something about how DNA regulates the production of cell protein. Proteins are the structural components of cells. DNA, therefore, not only provides for a continuity of life, but it also accounts for a diversity of life forms.

- Students often consider biology to be composed of many subdisciplines. Ecology is not often related to genetics or physiology; however, such fragmentation restricts an authentic view of the nature of science. Advancements in one field of biology often lead to advancements in others, even though the two fields are not even closely linked. Students are often surprised to discover that a hypothesis that attempted to explain symptoms of hereditary diseases in people would receive support from research done on bread mold. Beadle and Tatum's one gene, one protein hypothesis provides an excellent opportunity to discuss the nature of science.
- The terms translation and transcription are often confused or interchanged. Transcription is the process by which the genetic information is transferred from DNA to mRNA. Translation is the process by which proteins are synthesized using DNA instructions encoded in mRNA. The tRNA plays a significant role in translation by bringing amino acids to the ribosome. Students may be encouraged to develop three-dimensional models of protein synthesis. Model kits can be purchased from a number of biological supply houses. Figures 23.4 and 23.5 are useful in developing models.
- Students often have difficulties developing a conceptual model for understanding how cells mutate or how mutations can lead to cancer. The concept of protein synthesis provides a framework for understanding pathological conditions. Mutations are inheritable changes in the genetic material that arise from mistakes in DNA replication. Cosmic rays, X rays, ultraviolet radiation, and chemicals that alter DNA are referred to as mutagenic agents. By changing the arrangement of the nucleotides in the double helix, the mutagen changes the genetic code. The ribosome will read the new code and assemble amino acids according to the new instructions provided. Unfortunately, the shift of a single amino acid will produce a new protein. The new protein has a different chemical structure, and in most cases is incapable of carrying out the function of the required protein. Without the required protein, cell function is impaired if not completely destroyed. Although some mutations can, by chance, improve the functioning of the cell, the vast majority of mutations produce adverse effects. Most importantly, the text points out that there is no one single reason for mutations. Some occur because of nitrogen base shortage, others because of nitrogen base substitution, still others because of chemical destruction.
- An understanding of protein synthesis helps students understand why mutation in a germ cell (or sex cells) is more dangerous than those which may take place in somatic cells. It also provides a basis for understanding why environmental mutagens are a greater peril for a fetus or young child than an adult. When mutations occur, they are repeated each time the cell divides. A mutation in an egg or sperm cell will lead to permanent change in the characteristics of an offspring. In the adult, many genes code for characteristics that have already developed. It is also important to note that every gene is not used in a particular cell. For example, a mutation in the segment of a chromosome of a cell in the cornea which codes for the production of insulin will not alter insulin production. Insulin is only made in the pancreas.

MAKING CONNECTIONS

- The first question presented in the chapter, "Why do organ transplants present medical problems?", establishes a link between the structure of cell membranes, presented in chapter 5, Development of the Cell Theory, and antibody formation, presented in chapter 11, Blood and Immunity. Diversity and individuality can be approached by examining the almost infinite variety of possibilities for proteins.
- Chapter 6, Chemistry of Life, first introduces the chemical structure of amino acids and proteins. Chapter 23 presents the function of proteins with-

in a context of gene expression. The production of each protein is controlled by one gene. The genetic message is imprinted in the DNA molecule. Allosteric enzyme reactions, presented in chapter 7, Energy within the Cell, are clarified by examining how the production of these protein catalysts is regulated by DNA.

- The expression of dominance, multiple alleles, and gene interaction, presented in chapter 20, Genes and Heredity, are explained within a molecular context in chapter 23. Students can begin linking the expression of mutated genes with alterations of DNA, which in turn regulate sequencing of the amino acid.
- The case study, Human Immunodeficiency Virus, links the study of immunity, as presented in chapter 11, with the expression of genetic information, as presented in chapter 23. An understanding of glycoprotein receptor molecules, first introduced in chapter 5 and again in chapter 11, is consolidated within the context of protein synthesis by the highly relevant topic of AIDS.

POSSIBLE JOURNAL ENTRIES

- Students might be asked to provide examples of their fear of AIDS. How did the case study answer some of their concerns. What other questions do they have?
- Students can be encouraged to develop a concept map which includes all key terms presented in the chapter.
- Students can be encouraged to express the difficulties they experienced in formulating an understanding of this challenging chapter. What things did they employ to aid them in constructing their knowledge? For example, some students may indicate that the laboratory, Antibiotics and Protein Synthesis, aided their understanding of transcription and translation. Other students may have used the diagrams within the chapter to organize a visual image of protein synthesis.
- Group thinking and decision-making strategies may be recorded as students prepare for the debate. Did the students change their minds during the preparation for the debate? Students may even be asked to record their initial positions about the social issue and to reflect upon these feelings after the debate has been completed. Did they change their minds after listening to opposing arguments? Should research to be used for biological warfare be conducted?

USING THE VIDEODISC

Section title	Page	Code	Frame (side)	Description
Importance of Proteins	532 – 533		3141 (all)	Primary structure of proteins
Importance of Proteins	532 – 533		3187 (all)	Secondary structure of proteins
Importance of Proteins	532 – 533		3247 (all)	Tertiary structure of proteins
Importance of Proteins	532 – 533		3165 (all)	Quaternary structure of proteins
One Gene, One Protein	533 – 534		2897 (all)	Enzymes
The Role of DNA in Protein Synthesis	534 – 535		2625 (all)	DNA molecule

Section title	Page	Code	Frame (side)	Description
The Role of DNA in Protein Synthesis	534 – 535		2634 (all)	Segment of DNA molecule
The Role of DNA in Protein Synthesis	534 – 535		2829 (all)	Codon
The Role of DNA in Protein Synthesis	534 – 535		3021 (all)	Messenger RNA
Protein Synthesis: Transcription	535		3261 (all)	Transcription
Protein Synthesis: Translation	535		3263 (all)	Transfer RNA
Protein Synthesis: Translation	535		2737 (all)	Anticodon
Protein Synthesis: Translation	535		3265 (all)	Translation
Protein Synthesis: Translation	535		22695 (1)	Movie sequence showing transcription and translation; amino acids assembled to make proteins
Laboratory: Antibiotics and Protein Synthesis	538 – 539		2735 (all)	Antibiotics
DNA and Mutations	539 – 540		2673 (all)	Three-frame sequence showing physical interactions between methylcations and various DNA sequences
Oncogenes: Gene Regulation and Cancer	544 – 546		3230 (all)	Structural genes
Oncogenes: Gene Regulation and Cancer	544 – 546		3172 (all)	Repressor proteins

IDEAS FOR INITIATING A DISCUSSION

Figure 23.6: Mutations resulting from breaking apart DNA

UV radiation (sunlight) is known to break apart DNA molecules.

Figure 23.8: Model of action of regulator genes

This figure provides clues for the molecular basis of cell division. Although we still don't know what causes cells to divide normally, we have a better understanding of the mechanism.

Figure 23.10: Ames test

The Ames test has become big business in screening for environmental mutagens. Some substances may be harmless by themselves, but when combined with other chemicals they display tremendous mutagenic capacity.

REVIEW QUESTIONS

(page 537)

1 In what ways is the structure of mRNA similar to DNA? How does mRNA differ from DNA?

- mRNA is similar to DNA in that it is composed of a sugar, phosphate, and nitrogen bases. However, mRNA contains ribose sugar, as opposed to deoxyribose sugar in DNA. mRNA also contains the nitrogen base uracil, which replaces the thymine base found in DNA.

2 Provide an example of a type of message that mRNA might carry.

- mRNA might carry a message from the DNA to start synthesizing proteins to fight a bacterial infection, for example, mRNA might carry the message UUU which codes for the amino acid phenylalanine.

3 What is the function of tRNA?

- tRNA is the molecule responsible for identifying and collecting amino acids circulating in the cytoplasm. The tRNA binds with the amino acid for which it is specific and delivers it to the mRNA.

4 What is a codon? an anticodon?

- A codon is a three-nitrogen base code for an amino acid. For example, AAG is the codon that specifies the amino acid lysine. An anticodon is a three-nitrogen base code found in tRNA that pairs with the corresponding codon of mRNA.

5 Differentiate between transcription and translation.

- Transcription and translation are two stages in the production of proteins from the instructions contained in DNA. Transcription is the process by which the sequence of bases, or genetic code, is copied from DNA to a mRNA molecule. Translation is the process by which the instructions on the mRNA molecule are encoded by ribosomes to synthesize proteins.

6 What anticodon on the tRNA molecule would gain access to the codon UUG?

- To gain access to the codon UUG, the anticodon would have to be AAC. Adenine pairs only with uracil, and cytosine pairs only with guanine.

7 Why is protein synthesis essential for life?

- Proteins are the building blocks of cells. Cell reactions also depend on enzymes which are a class of proteins. Life could not continue if we were unable to make new cells, or if our metabolism was slowed or stopped because of a lack of enzymes.

(page 547)

8 What are mutations?

- Mutations are inheritable changes in the genetic code of an organism.

9 Indicate three factors that can produce gene mutations.

- Cosmic rays, X rays, ultraviolet radiation, and chemicals that alter DNA are referred to as mutagenic agents. Even a shortage of a specific type of amino acid can cause mutations.

10 Explain why a food dye that has been identified as a chemical mutagen poses greater dangers for a developing fetus than for an adult.

- In the adult, many genes code for characteristics that have already developed. For example, your hands, feet, and toes are already in place. If the gene that controls the production of hair color is altered in a single cell, it will have little effect upon your body. However, if this same gene were altered in a fertilized egg, the mutation would be coded into the DNA of every succeeding cell of the body. The single mistake can be repeated billions of times. This explains why mutagenic agents are especially dangerous for pregnant women, particularly during the first trimester of pregnancy.

11 What are oncogenes?

- In 1982, molecular biologists were able to provide evidence that supported the hypothesis that cancer could be traced to genetic mutations. Segments of chromosomes extracted from cancerous mice transformed normal mouse cells growing in tissue cultures into cancerous cells. The cancer-causing genes, called oncogenes, seemed to turn on cell division.

12 How does the Ames test identify cancer-causing agents?

- The technique involves testing the potential of a chemical to alter the DNA of an organism. It is generally accepted that cells can become cancerous if their genetic information is altered. Although all gene mutations do not lead to cancer, it is assumed that a chemical's ability to cause a mutation is a measure of its potential to cause cancer. Any chemical that rapidly alters the DNA molecule could produce a mutation that affects a cell's ability to regulate its own rate of reproduction.

The Ames method is performed on a bacterium; *Salmonella typhimurium* is most commonly used. The bacteria have undergone a mutation that has led to their inability to produce an essential chemical called histidine. In order for the bacteria to grow, histidine must be supplied. A *Salmonella* culture is placed on a petri dish. The test chemical is then added to the petri dish. Because the growth medium in the petri dish does not contain histidine, no growth would be expected. If bacteria colonies are found, the conclusion must be that the microbe has mutated, and is now producing its own histidine. The more colonies, the more mutagenic is the test chemical. The tested chemical is often added to liver enzymes. Although the chemical itself may not be toxic, it may be broken down by the liver into toxic metabolites. Like most scientific models, the mutation of a microbe does not signify that the substance will cause cancer in all humans. It is possible that a toxic chemical could cause many genetic changes in a person, yet the changes would never manifest themselves as cancer. On the other hand, this same chemical could cause cancer in another individual. Although our DNA is composed of essentially the same chemicals, the arrangement of the nitrogen bases is unique. Some sequences seem more susceptible to change by chemical agents than others. The Ames test not only allows carcinogens to be identified in a less costly manner, but it also provides researchers with information on how chemicals alter DNA.

13 How do regulator genes turn off structural genes?

■ A gene, referred to as the regulator gene, acts like a switch that turns "off" segments of the DNA molecule. Structural genes are genes that direct the synthesis of proteins. Regulator genes control the production of repressor proteins which switch off structural genes.

LABORATORY
ANTIBIOTICS AND PROTEIN SYNTHESIS

(pages 538–539)

Time Requirements

Two class periods are needed to complete the laboratory. Approximately 20 min are required from each class.

Notes to the Teacher

- Antibiotic disks can be purchased from a number of biological supply companies.
- We recommend stressing sterile technique. Students must be cautioned against opening the petri dishes to measure the growth ring around each disk.
- Dispose of the disks immediately after the laboratory. Autoclave glass petri dishes only.

Observation Questions

a) Record which antibiotic was placed in each section of the petri dish.

■ Answers will vary.

b) Record your data in tabular form as shown.

Antibiotic	Measurement (mm)				
	#1	#2	#3	#4	Average
1					
2					
3					

■ Answers will vary.

Laboratory Application Questions

1 On the basis of the experimental evidence, indicate which antibiotic was the most effective.

■ The disk with the greatest clear zone around it is most effective at inhibiting the growth of the bacteria.

2 Would the antibiotic that best controlled the bacteria found on your face necessarily be the most effective at controlling the bacteria that cause strep throat? How would you go about testing your hypothesis?

■ No, different microbes grow in different environments. Even slight variations can cause different colonies to grow. A similar test to the one conducted in the laboratory would likely be repeated.

3 What potential problems might be created by placing low dosages of antibiotics in skin creams?

■ The microbe might develop a resistance to the antibiotic, or the long-term exposure to low levels of antibiotic could cause an allergic reaction to develop. The antibiotic might also kill a harmless microbe, thereby eliminating competition and allowing a more dangerous microbe to take up residence.

4 Some antibiotics prevent protein synthesis in more advanced cells (eukaryotic cells). Indicate some of the advantages of these antibiotics, and explain why their dosages must be carefully administered.

■ Antibiotics, like cycloheximide, block protein synthesis in eukaryotic cells. These antibiotics are capable of inhibiting the growth of fungi. Some types of fungi, like ringworm, are disease causing. The dosage must be carefully controlled so as not to interfere with protein synthesis of the host. The antibiotics that affect eukaryotic cells usually have a greater effect on parasites than the host, because the parasite usually carries out protein synthesis at an accelerated rate.

CASE STUDY
HUMAN IMMUNODEFICIENCY VIRUS

(pages 541–544)
Time Requirements
Approximately 40 min are required.

Notes to the Teacher

The following suggestion may help groups function:

- Organizer: will assume responsibility for assigning roles, organizing desks, and beginning the discussion
- Recorder: will assume responsibility for keeping a record of answers and checks for group agreement
- Scrutinizer: will seek other solutions and suggest other pathways
- Presenter: will ensure that all group members have had an opportunity to provide an opinion and will make a presentation
  Note: one student may assume more than one role.
- A review of the chapter entitled Blood and Immunity is recommended prior to the case study.

Observation Questions

a) Can HIV attach itself to a muscle cell or a cell from the skin?

■ No, the geometry of the HIV is not compatible with that of the binding sites on muscle and skin cells.

b) Explain why you cannot get AIDS by shaking hands. (Use the information that you have gained about binding sites.)

■ The binding sites of the skin will not allow the attachment of HIV.

c) Most viruses leave their coat on the membrane of the infected cell. Indicate why these viruses are much more easily identified than HIV.

■ T-4 lymphocytes (helper T cells) get an imprint of the virus coat and send a message to B cells to begin antibody production. Without the outer coat exposed, antibody production will not begin. Be willing to accept a wide variety of answers if your students have had the chapter on Blood and Immunity prior to this one.

d) Why is the enzyme referred to as reverse transcriptase?

■ It reverses the message; HIV RNA acts as a template for the synthesis of DNA.

e) What happens to the viral DNA if the T cell divides ?

■ The viral DNA is also duplicated.

f) Explain why it is possible for a human to be infected with HIV and not exhibit any of the symptoms of AIDS.

■ The segment containing the HIV has not been activated for protein synthesis. Once it is activated, the segment of DNA that carries HIV genes directs cell ribosomes to construct HIV protein coats. The cell becomes a HIV protein coat factory.

g) Normally, the killer T cells would destroy a cell infected with a virus long before it could become a virus factory. Why are the infected helper T cells not held in check ?

■ The helper T cells hide the virus.

h) Indicate why people infected with HIV most often die of another infection such as pneumonia?

■ Without a proper functioning immune system, the body becomes susceptible to many other infections.

i) David, "the boy in the plastic bubble," suffered from a disorder called "severe combined immunodeficiency syndrome." How does this disorder differ from acquired immune deficiency syndrome?

■ The severe combined immunodefiency syndrome is not acquired, but due to a genetic mistake prior to birth. HIV is a viral infection.

j) Why is it so difficult to destroy a virus that changes shape?

■ Each different shape eliminates antibodies with a specific geometry.

Case-Study Application Questions

1 Why does the Canadian Red Cross inquire about a person's travel before they accept blood donations?

■ Certain parts of the world have a much higher incidence of HIV.

2 How does transcription for HIV differ from normal cell transcription?

■ RNA transcribes to DNA, a reversal of normal transcription.

3 Can AIDS be transmitted through either food or beverages? Explain your answer.

■ No, to date no such indication exists. Infection requires appropriate binding sites.

4 Can AIDS be contracted by casual contact such as shaking hands, or using the same telephone or toilet seat? Explain your answer.

■ No, to date no such indication exists. Infection requires appropriate binding sites.

5 Do tattoos and pierced ears pose any potential risks for infection with AIDS?

■ Yes, blood is often collected from the end of a needle. If the needle is not sterilized, the infection could be spread. Dentists face the same problems, but it should be noted that strict guidelines for sterilization are mandated.

6 How is it possible to catch the HIV virus from a person who shows no symptoms associated with AIDS?

■ The symptoms are only manifest once many cells have become infected and the helper T cells actively begin making HIV protein coats.

7 Should people with AIDS be quarantined? Justify your answer.

■ Consider a diversity of answers. Students should recognize that this question requires scientific information, but that it cannot be solved by using scientific thinking.

8 Should health-care workers such as doctors, dentists, and nurses be screened for HIV? Justify your answer.

■ Consider a diversity of answers. Students should recognize that this question requires scientific information, but that it cannot be solved by using scientific thinking.

APPLYING THE CONCEPTS

(page 552)

1 Why are somatic cell mutations less harmful than germ cell mutations?

■ When mutations occur, they are repeated each time the cell divides. A mutation in an egg or sperm cell will lead to permanent change in the characteristics of an offspring. In the adult, many genes code for characteristics that have already developed. For example, your hands, feet, and toes are already in place. If the gene that controls the production of hair color is altered in a single cell, it will have little effect upon your body. However, if this same gene were altered in a fertilized egg, the mutation would be coded into the DNA of every succeeding cell of the body. The single mistake can be repeated billions of times. This explains why mutagenic agents are especially dangerous for pregnant women, particularly during the first trimester of pregnancy.

2 Explain how Beadle and Tatum's experiments with the bread mold *Neurospora* helped explain protein synthesis.

■ Working with a variety of bread mold called *Neurospora crasa*, Beadle and Tatum examined metabolic pathways.

Neurospora will grow on minimal nutrient medium which contains only sugar, salts, and one of the B vitamins. All other materials can be synthesized by the bread mold. If a gene synthesizes one of the enzymes, what would happen to the enzyme should the gene mutate?

Beadle and Tatum found that a mutated strain of Neurospora would only grow when vitamin B_6 was added to the minimal nutrient medium. Other strains only grew if vitamin B_1 was added. Mutated strains of Neurospora that could not synthesize a specific vitamin would have to be supplied with the vitamin in order to grow. But why could the Neurospora not synthesize the vitamins? Analysis of cell extracts showed that each gene mutation could be associated with alterations of a specific enzyme. The scientists concluded that defective enzymes must be linked to genetic mutations. Genes regulate the production of proteins. The explanation provided by Beadle and Tatum has been referred to as the "one gene, one enzyme hypothesis."

3 In what ways does mRNA differ from DNA?

RNA	DNA
Ribose sugar	Deoxyribose sugar
Single-stranded	Double-stranded
Uracil	Thymine

4 Suppose that during protein synthesis, nucleotides containing uracil are in poor supply. The uracil is substituted with another nitrogen base to complete the genetic code. How will the protein be affected by the substitution of nitrogen bases?

■ A new protein would be produced. The genetic code would be altered and different amino acids would be sequenced along the ribosomes.

5 In what ways do the codons and anticodons differ?

■ Codons come from mRNA; anticodons are found on tRNA. The anticodon has the same genetic code as DNA, but uracil is substituted for thymine. mRNA has a complemenatry code to that of the DNA. Uracil for mRNA, replaces thymine for DNA. An anticodon pairs with a codon.

6 A scientist discovers a drug that ties up the site at which mRNA attaches to the ribosome. Although mRNA is transcribed in the nucleus, it has no way of attaching itself to the ribosome. Speculate about how the drug will affect the functioning of a cell. If the drug is introduced into a single brain cell, will the organism be destroyed?

■ The drug will prevent protein synthesis. It is highly unlikely that the death of one single cell could cause death.

7 Outline the advantages of Dr. Ram Mehta's genetic tests using yeast cells over those that use animals.

■ Dr. Mehta's techniques are cheaper and faster. Some individuals may also indicate that they prefer yeast testing to animal testing for humane reasons.

8 Compare protein synthesis in cells to an automobile assembly plant. Match the parts of the auto assembly plant with the correct part of the cell. Provide reasons for each of the matches.

■ **1** Corporate headquarters resemble the nucleus, the command center.
2 The master blueprint for the car is the DNA, the genetic code of life.
3 The entire shop area is the cytoplasm, the area of the cell in which work occurs.
4 The supervisor who carries the blueprints is the mRNA which carries the genetic code to the ribosomes.
5 The stockperson is the tRNA which carries amino acids to the ribosomes.
6 The assembly worker is the ribosome, the organelle that assembles proteins.
7 The parts of the automobile are the amino acids, the building blocks of proteins.

9 Thalidomide was a drug given to pregnant women in the early 1960s to reduce morning sickness. Unfortunately, the drug caused irreparable damage to the developing fetus. Thalidomide inhibited proper limb formation. Many children were born without arms and legs. Explain how the Ames test could have prevented the tragedy.

■ The Ames test may have identified the drug as a mutagen. Although the exact type of mutation would not have been known, the drug may have never been given to pregnant women without much greater testing.

10 Recombinant DNA provides a valuable technique for those engaged in biological warfare. The gene that produces the botulism toxin, a deadly food poison, can be extracted from its resident bacteria and placed into a harmless bacterium, called *Escherichia coli*, which is a natural inhabitant of the large intestine of humans. Why would the transfer of the botulism gene to the bacteria found in your gut be so dangerous?

■ Your immune system would not detect the *E. coli* as a foreign invader; it is a natural inhabitant of the large intestine. The toxins produced from the bacterium your body identifies as harmless would kill you.

CRITICAL-THINKING QUESTIONS

(pages 552–553)

1 Recombinant DNA has produced human insulin in bacteria. Because millions of people suffer from diabetes, the market for human insulin is enormous, and so are the profits. Because of economic pressures, the nature of scientific research has changed from one of sharing information through publishing to one of patents and secretiveness. Companies are unwilling to release any breakthrough for fear that their ideas might be stolen. Scientific research, at least in some fields, is no longer controlled by researchers, but by investors, who seek the advice of accountants and lawyers. Should government remove biotechnology from private enterprise? Support your opinion.

■ Many opinions can be expressed, but caution students against viewing all individuals who are interested in profit as being potentially evil, while viewing scientists as inherently good. Some scientists, like some people from other professions, are motivated by profit. Encourage students to avoid stereotypes in presenting their rationale. Students might also be asked to consider how the secretive world of higher finance is altering the more social world of sharing. Prior to large profits, scientists were much more willing to share. This also allows students to view science as a social enterprise.

2 John Moore suffered from a rare form of leukemia. A team of doctors from UCLA removed his spleen and and he went on to live a fruitful life. However, years after his operation, Moore discovered that the surgeons were cloning cells from his extracted spleen for cancer research. The cells produced significant amounts of a substance called interferon, which has tremendous commercial value. Moore now wants what he believes is his fair share of the money, and he is suing the people who are cloning the cells. Should Moore be entitled to any money? Is the spleen Moore's property? Support your conclusions.

■ Many opinions can be explored. Students should be encouraged to listen to other viewpoints. Most importantly, students should recognize that the answer to the question is found not within the discipline of science, but that of ethics. The question permits students to express their own opinions and justify their conclusions. This question also provides an example of the limits of the scientific paradigm for answering questions.

3 *Pseudomonas syringae* is a bacterium found in raindrops and in most ice crystals. Researchers have been able to snip the frost gene from its genetic code, thereby preventing the bacteria from forming ice crystals. The *Pseudomonas* has been aptly named "frost negative." By spraying the bacteria on tomato plants, scientists have been able to reduce frost damage. The bacteria can extend growing seasons, thus increasing crop yields, especially in cold climates. A second version, called "frost positive" has also been developed. The frost-positive strain promotes the development of ice and, not surprisingly, has been eagerly accepted by some ski resorts. When this microbe is sprayed on ski slopes, a longer season and better snow base can be assured. However, environmental groups have raised serious concerns about releasing genetically engineered bacteria into the environment. Could these new microbes gain an unfair advantage over the naturally occurring species? What might happen if the genetically engineered microbes mutate? Could the mutated microbe become a super microbe? Do you think genetically engineered microbes should be introduced into the environment? Support your conclusions.

■ Many opinions can be explored. Students should be encouraged to listen to other viewpoints. Most importantly, students should recognize that the answer to the question is found not within the discipline of science, but that of ethics. Many people currently wonder if the genetically altered species or strains might not have an unfair advantage. Only certain gene combinations are possible in nature. Recombinant DNA technology, by placing plant and animal genes in bacteria, or vice versa, has far exceeded the realm of nature's capabilities.

4 The gene for the growth hormone has been extracted from human chromosomes and implanted into bacteria. The bacteria produce human growth hormone, which can be harvested in relatively large quantities. The production of human growth hormone is invaluable to people with dwarfism. Prior to the development of this hormone people with dwarfism relied on costly pituitary extracts. Although the prospect of curing dwarfism has met with approval from a majority of the scientific community, some concerns about the potentially vast supply of growth hormone have been raised. How can scientists ensure that the growth hormone produced by these genetically engineered bacteria will not be used by normal individuals who wish to grow a few more centimeters? Do people have the right to choose their own height? Give your opinion and support your conclusions.

■ Many opinions can be explored. Students should be encouraged to listen to other viewpoints. Most importantly, students should recognize that the answer to the question is found not within the discipline of science, but that of ethics. The question permits students to express their own opinions and justify their conclusions. This question also provides an example of the limits of the scientific paradigm for answering questions. Growth hormone is also used to prevent aging.

U N 7 I T

Change in Populations and Communities

SUGGESTIONS FOR INTRODUCING UNIT 7

Chapters 24 and 25 examine how genetic principles and environmental factors can be applied to the study of populations. Populations in equilibrium and population growth strategies are examined within the context of communities. Variation provides the basis for change, while environmental fluctuations provide a necessity for change.

This section of the text is written in a manner that attempts to engage students in problem-solving strategies. Quantitative aspects of biology are stressed, as students are expected to use laws of probability to determine whether or not changes in a population are due to chance or some environmental factor. The approach also acknowledges that prior learning has occurred. Previous formalized learning of genetics and ecology are woven into a fabric of understanding.

Ask students to identify a population of organisms which has changed over time. Encourage small brainstorming groups to speculate about why those changes have occurred, but avoid coming to a single conclusion or identifying a single factor. Students may consider changes in population size, such as those represented by whooping cranes, or changes in morphology, such as those represented by the Galápagos finches (e.g., Why are people getting larger? Have you ever noticed the size armor or the length of beds from the middle ages?).

Begin the study with the social issue, Interbreeding of Plains and Woodland Bison, presented on page 574 of chapter 24, Population Genetics. Students can revisit this debate once the unit has been completed.

KEY SCIENCE CONCEPTS

Unit 7 links the key concepts of **equilibrium** and **change** by returning to a more global perspective. Human populations are studied from both a perspective of genetics and communities. This unit reflects the skills, attitudes, and knowledge developed in Unit 4 of the Alberta Biology 30 Program of Studies. ■

CHAPTER 24

Population Genetics

INITIATING THE STUDY OF CHAPTER TWENTY-FOUR

Chapter 24 links genetics and reproduction with a study of populations and communities. The Hardy-Weinberg principle examines equilibrium in natural populations. Chi-square tests allow predictions of outcomes based on laws of probability. The chapter leads from populations in equilibrium to a study of changing populations—speciation. Geographic isolation and reproductive isolation are factors that influence speciation.

The following suggestions provide examples for beginning the chapter:

1. Introduce the debate about the bison at the beginning of the chapter, with directions that students will prepare to debate the topic at the end of the chapter. Ask the students to provide a short journal entry, expressing their initial views. Examine their views once the chapter has been completed. Have any of the views changed?
2. View the tongue-rolling photograph on page 557. Can other variations be identified in the human population?

ADDRESSING ALTERNATE CONCEPTIONS

- Students often think of survival of the fittest as meaning the strongest organism wins out. Although the fittest individual within a species is often the strongest, equating fittest with strongest creates misconceptions about interspecific competitions. Certain species might be better able to obtain necessary resources such as food and water, be better protected against predators, or have higher reproductive potentials. Accordingly, they would be better able to survive and reproduce in greater numbers. Any factor that increases gene frequencies disrupts equilibrium.
- Some students will find the mathematical representation of the Hardy-Weinberg principle difficult to interpret. Whenever possible, encourage these students to describe the problem before trying to establish mathematical relationships. The Hardy-Weinberg principle indicates conditions under which allele and gene frequencies will remain constant for succeeding generations. Only if this stability, called genetic equilibrium, is upset can a population evolve.

MAKING CONNECTIONS

- The case study, Tracing the Hemophilia Gene, enables students to connect the knowledge of sex-linked genes learned in chapter 21, The Source of Heredity, with the principles of population genetics, presented in chapter 24.
- The case study, Genetic Disorders as Models for Evolution, gives students an opportunity to link information that they gained about gene frequencies and equilibrium with the concept of mutations, as presented in chapter 23, Protein Synthesis. The concept also provides a bridge to Unit 6, Heredity, as students learn how genetic variation provides a basis for evolution. Sickle-cell anemia is also presented in chapter 11, Blood and Immunity.
- Physiological and structural adaptations presented in chapter 4 provide important background for establishing a context for understanding why gene pools in populations change.

POSSIBLE JOURNAL ENTRIES

- Students might be asked to write a dialogue between the text and themselves, as they refute the idea that gene frequencies will change over time. By acting as a "devil's advocate," they can challenge their own learning and push understanding to a higher level.
- Students can be encouraged to develop a concept map which includes all key terms presented in the chapter.
- Students can be encouraged to express the difficulties they experienced in formulating an understanding of how gene frequencies change. What things did they employ to aid them in constructing their knowledge? For example, some students may indicate that the case studies aided their understanding of gene selection. Other students may have attempted a series of their own summary charts or used computer simulations to work through the mathematics of calculating chi-square.

USING THE VIDEODISC

Section title	Page	Code	Frame (side)	Description
Importance of Variation	556 – 557		1720 (all)	Red-eyed drosophila (female)
Importance of Variation	556 – 557		1722 (all)	White-eyed drosophila (female)
Population Equilibrium	560 – 561		2949 (all)	Hardy-Weinberg principle
Population Equilibrium	560 – 561		1717 (all)	Blood typing, anti-A and anti-B
Population Equilibrium	560 – 561		1718 (all)	Blood typing Rh factors
Speciation	571 – 573		2717 (all)	Allopatric speciation
Speciation	571 – 573		3235 (all)	Sympatric speciation

IDEAS FOR INITIATING A DISCUSSION

Case study: Tracing the Hemophilia Gene

Ask students to identify members of the Commonwealth's royal family on the chart. Are the royal families of Europe related?

Figure 24.4: Mitochondrion

Ask students to check older biology textbooks to see if the mitochondria were identified as organelles containing DNA. Is it possible that other organelles might have DNA? This provides an excellent opportunity to begin a discussion about the tentative nature of scientific knowledge.

REVIEW QUESTIONS

(page 563)

1 What is the main source of variation in living organisms?

■ The main source of variation in living organisms is the differences in the genes carried by the chromosomes.

These genes determine an organism's appearance, and can be passed on to offspring.

2 How would you define a gene pool?

■ A gene pool is all the genes that occur within a specific population.

3 Why is population sampling a useful technique in studying population genetics?

■ Population sampling allows researchers to determine trends or frequencies of genes and characteristics without measuring the entire population.

4 What purpose does a pedigree chart serve in tracing particular genetic traits? Give an example.

■ A pedigree chart allows a trait to be traced from parents through successive generations. For example, a pedigree chart can be used to trace the inheritance of the hemophilia gene in the British royal family.

5 What prediction does the Hardy-Weinberg principle make with respect to populations?

■ The Hardy-Weinberg principle predicts that, if there are no external influencing factors, the population gene pool will maintain the same composition generation after generation.

6 Why does the Hardy-Weinberg principle apply mainly to large populations?

■ The Hardy-Weinberg principle applies mainly to large populations because only a large population can ensure that changes in gene frequencies are not the result of chance.

7 What are three factors that may bring about evolutionary change?

■ Three factors that may bring about evolutionary change are mutations, genetic drift, and migration or gene flow. Mutations are changes in the genetic makeup of an organism. Genetic drift is the random loss or decrease of genes in small populations. Migration or gene flow is the addition or subtraction of genes from a population through the physical movement of individual organisms.

8 Describe the two main categories of mutation.

■ There are two main types of mutations: chromosome mutations in which the organism either lacks or has additional chromosomes, and gene mutations which are the result of a chemical change within a gene. Down syndrome is an example of a chromosome mutation, while sickle-cell anemia is an example of a gene mutation.

(page 573)

9 What does the concept of probability enable scientists to predict?

■ The concept of probability allows scientists to predict what should happen, and the likelihood of an event occurring.

10 What do scientists mean when they say a chi-square value is "significant"?

■ When a chi-square value is found to be significant, it indicates that another factor may have influenced the outcome of the experiment. The outcome is not due to chance alone.

11 Why are human genetic disorders useful models for studying evolution?

■ Human genetic disorders are useful models for studying evolution because they can often be easily identified and they show changes in gene frequencies.

12 What common features are shared by the two genetic disorders in the case study?

■ Both sickle-cell anemia and Tay-Sachs disease are the result of genetic mutations. They are also uncommon except in a specific population, African blacks for sickle-cell anemia, and eastern European Jews for Tay-Sachs.

13 Why did the discovery that mitochondria contain their own DNA surprise many scientists?

■ Scientists had thought that all of a cell's structure and activities were under the control of the DNA inside the nucleus. The discovery that mitochondria contained their own DNA led to speculation that mitochondria may have evolved as separate organisms, before combining with other cells in a symbiotic relationship.

14 What features of the mitochondrion make it useful in the study of evolutionary relationships?

■ Since mt-DNA only comes from the mother new combinations do not occur. Therefore it provides a direct link to maternal ancestors.

15 What is speciation?

■ Speciation is the process by which species originate. The currently accepted definition of a species is a group of similar organisms that can interbreed and produce fertile offspring in their natural environment.

16 Explain two ways in which new species may arise.

■ Two ways in which a new species can arise are through geographic isolation and reproductive isolation. Geographic isolation is caused by physical barriers, such as mountains or bodies of water. Gene flow between populations is stopped, and eventually the two populations become so different that they are unable to interbreed. Reproductive isolation occurs when organisms in a population can no longer interbreed.

CASE STUDY
TRACING THE HEMOPHILIA GENE

(page 558)

Time Requirements

Approximately 40 min are required to complete the case study.

Observation Questions

a) Who was Queen Victoria's father?

■ Edward Duke of Kent was Queen Victoria's father.

b) How many children did Queen Victoria and Prince Albert have?

■ They had nine children.

c) Using the legend, provide the genotypes of both Alice of Hesse and Leopold.

■ X^HX^h for Alice and X^hY for Leopold.

d) Explain why Alexis was the only child to have hemophilia.

■ Only the male child has hemophilia when he inherits a single gene.

Case-Study Application Questions

1 Is it possible for a female to be hemophilic? If not, explain why not. If so, identify a male and female from the pedigree chart who would be capable of producing a hemophilic, female offspring.

■ Female hemophiliacs are possible, but the mother must be a carrier (or homozygous for the condition), and the father must have hemophilia. The mating of Victoria Eugenia with Frederick William could produce a hemophilic female offspring. (Note: other combinations are possible; however, none of these combinations ever married.)

2 On the basis of probability, calculate the number of Victoria's and Albert's children who would be carriers of the hemophilic trait?

■ One half of the females would carry the hemophilic gene. This is not what actually happened. Only two of five daughters were carriers of the hemophilic gene.

CASE STUDY
GENETIC DISORDERS AS MODELS FOR EVOLUTION

(pages 567–569)

Time Requirements

Approximately 40 min are required.

Notes to the Teacher

Small group work is recommended. The following roles might be considered for group work:

- Organizer: will assume responsibility for assigning roles, organizing desks, and beginning the discussion
- Recorder: will assume responsibility for keeping a record of answers and check for group agreement
- Scrutinizer: will seek other solutions and suggest other pathways
- Presenter: will ensure that all group members have had an opportunity to provide an opinion and will make an oral presentation to the class if called upon

Observation Questions

a) Write the genotypes of the offspring from a cross between two carriers of the sickle-cell gene. What is the probability that the couple will have **(i)** a normal child? **(ii)** a child with the sickle-cell disorder?

- The genotypes of the offspring are $Hb^A Hb^A$, $2Hb^A Hb^S$, $Hb^S Hb^S$. **(i)** The probability is 3/4. **(ii)** The probability is 1/4.

b) What is the probability that a child born of a normal parent and a carrier parent would have a normal child? A child with the sickle-cell disorder?

- Normal parent ($Hb^A Hb^A$) × carrier parent ($Hb^A Hb^S$) = all offspring are normal.

c) Why would there be no concern about the offspring of two homozygous individuals having the sickle-cell disorder?

- All offspring will be homozygous for the dominant condition. There is also little chance that they would have had offspring. Most would have died prior to reproductive age.

d) How does Pauling's discovery provide an answer to the observation that carriers experience mild "sickling" during strenuous exercise?

- It demonstrated that the heterozygous phenotyope was related to the sickling condition which was expressed by the homozygous recessive genotype.

e) What selective advantage is afforded a heterozygous individual ($Hb^A Hb^S$) in certain African populations?

- They are resistant to malaria, a parasitic infection.

f) What might happen to the frequency of the sickle-cell gene if malaria were eliminated in Africa? Why?

- The sickle-cell gene would eventually disappear, since there would no longer be any advantage in being a carrier. However, this would take a number of generations.

g) Estimates suggest that the frequency of the sickle-cell gene in North American blacks has decreased from an original 22% in the early slavery period to a current value of 10% or less. What might explain this difference?

- The mosquito that carries the disorder is being controlled. Also many North American blacks may have migrated to areas where the mosquito which carries malaria is not found.

h) Are the odds in this example like those of Mendel's F_2 generation? Under what circumstances would the next generation yield the same odds?

- No, the ratio is much lower than 1/4. The next generation should yield the same results unless the *tt* condition which causes Tay-Sachs provides some environmental disadvantage or some advantage.

i) Why can scientists be certain that each parent must have had at least one good copy of the gene along with the defective copy?

- Otherwise the parent would have the disorder. The phenotype is expressed as the homozygous genotype.

j) What similarities exist between the biochemical abnormalities that cause sickle-cell anemia and those that cause Tay-Sachs disorder?

- Both are caused by an abnormal protein (enzyme). In both cases the DNA sequence has been altered.

k) Explain how scientists recognize that the selection pressure against Tay-Sachs (*tt*) is high?

- Individuals with *tt* have motor coordination problems and often convulsions.

l) What are the two selection pressures acting on the eastern European Jewish population?

- The *tt* genotype often caused death or severe deterioration of the nervous system, while the *Tt* condition may have provided resistance against tuberculosis.

m) The Tay-Sachs recessive gene has survival value in areas where there is a high incidence of TB. How does natural selection operate in its favor in these areas?

- Individuals with the heterozygous genotype have an increased survival rate.

n) What significance can be attached to the fact that eastern European Jews are susceptible to 10 other genetic disorders that are not found in other Jewish and non-Jewish populations?

- It may indicate that migration from this population was restricted or that some other environmental factor may have occurred in this area. This group's genotype may have been susceptible to these environmental influences. Although no answer is known, the question provides students with an opportunity to engage in presenting hypotheses; however, no conclusions can be drawn without additional data. The hypotheses provide students with an opportunity to construct scientific questions.

o) How does this case illustrate that evolution results from the interaction between an organism's genetic makeup and its environment?

■ The environment likely induces gene mutations—mutations are not universal.

Case-Study Application Questions

1 How are the factors in the evolutionary process illustrated in this case study? In your answer consider the ideas of mutation, natural selection, and survival value.

■ The environment selects certain genotypes. Genotypes that are deleterious in the homozygous condition continue to remain part of the gene pool because their heterozygous condition provides some selective advantage.

2 What advice should a genetics counsellor give to carriers who are contemplating giving birth to a child?

■ The couple should be informed about the chances that a child would have the homozygous recessive condition. Treatment for the homozygous recessive condition could also be explored.

3 What is the meaning of the statement, "Recessive genetic disorders can be both a blessing and a curse"?

■ The recessive disorders provide a selective advantage in the heterozygous condition. The heterozygous condition requires that some genes remain in the gene pool. The mating of heterozygous individuals, by chance, will produce some homozygous recessive individuals. You cannot produce a 100% heterozygous population.

APPLYING THE CONCEPTS

(page 575)

1 Would it be more correct to say "an organism evolves" or "a species evolves"? Explain.

■ Only a population evolves, individual organisms change, but they do not evolve. Evolution refers to changes in the gene pool of a population. A species can evolve, because a species refers to the genes in a population of interbreeding organisms.

2 The five conditions of the Hardy-Weinberg principle are rarely met in nature. Yet the theory is still useful for studying "real" populations. How can you account for this apparent contradiction?

■ The principle makes it possible to make predictions about populations that are *not* evolving. From this reference point, we can consider the special conditions which serve as a measure of the rates of evolutionary change.

3 In a given population of organisms, the dominant allele (p) has a frequency of 0.7, and the recessive allele (q) has a frequency of 0.3. Use the Hardy-Weinberg formula to determine the genotype frequencies within the population.

■ $p^2 + 2pq + q^2$, $(0.7)^2 \times 2(0.7 \times 0.3) + (0.3)^2 = 49\%$ *AA* + 42% *Aa* + 9% *aa*

4 In Tanzania, 4% (0.04) of the population are homozygous sickle-cell anemics (*ss*) and 32% (0.32) are heterozygous (*Ss*). From these data, calculate the proportion of alleles that are *s* or *S*.

■ 4% *ss* + 32% *Ss* + 64% *SS* = 0.2 *s* and 0.8 *S*

5 Mutation rates are usually quite low in sexually reproducing organisms, yet mutations are known to be the raw material for evolution. Explain how this is so.

■ Most mutations are not successful; however, occasionally a mutation occurs that provides some selective advantage for an organism. Mutations provide for variation.

6 A cross between two pea plants, in which tall (T) is dominant to short (t), yields 1000 seeds. Of this number, 550 produce plants that are tall, while 450 produce short plants. Use the chi-square test to determine whether the deviation is the result of chance or some other complicating factor.

■ If we assume $Tt \times tt$, then 50% of the offspring should be tall and 50% of the offspring should be short.

$$\chi^2 = \Sigma \frac{(O - E)^2}{d},$$

$$\chi^2 = \Sigma \frac{(450 - 500)^2}{500} + \frac{(550 - 500)^2}{500}$$

$= 5.0 + 5.0 = 10.0$

It was not due to chance alone.

If we assume $Tt \times Tt$, then 75% of the offspring should be tall and 25% of the offspring should be short.

$$\chi^2 = \Sigma \frac{(O - E)^2}{d},$$

$$\chi^2 = \Sigma \frac{(550 - 750)^2}{750} + \frac{(450 - 250)^2}{250}$$

$$= \frac{(-200)^2}{750} + \frac{(200)^2}{250}$$

53.3 + 160 = 213.3, clearly indicating that this would not have happened because of chance.

7 Describe why long periods of geographic isolation of a small group from other members of a population favors speciation.

■ The isolated group is often subjected to different environmental pressures. The isolated group may also have different mutations which cannot be transferred to other members of the gene pool beyond the geographic boundary.

CRITICAL-THINKING QUESTIONS

(page 575)

1 How do the genetic disorders discussed in this chapter illustrate the point that evolution involves interactions between an organism's genetic makeup and its environment?

■ Genes are selected by the environment. Both sickle-cell anemia and Tay-Sachs disorder have deleterious effects when in the homozygous recessive condition, but provide an advantage when in the heterozygous condition. Although one environmental factor works against the selection of the *aa* genotype, another selects for the *Aa* genotype.

2 The Hardy-Weinberg formula, $p^2 + 2pq + q^2 = 1$, is said to represent all possible genotypes in a population. Verify this statement by determining (a) the frequencies of dominant and recessive alleles, and (b) the number of heterozygotes, in a population of 200 pigs in which 72 have the recessive trait. If natural selection removed all of the individuals with the recessive trait, what would be the gene frequencies in the next generation?

■ $72/200 = aa$, $0.36 = a^2$, then $\sqrt{0.36} = a$, or $0.6 = a$, then $0.4 = A$.

To find the number of heterozygous individuals in the population use the following formula: $p^2 + 2pq + q^2$, or $(0.6)^2 + 2(0.6 \times 0.4) + (0.4)^2$. The heterozygotes have a frequency of 0.48 from 200 = 96.

CHAPTER 25

Populations and Communities

INITIATING THE STUDY OF CHAPTER TWENTY-FIVE

Chapter 25 provides a quantitative approach for studying change. Population changes within a community and environmental changes, referred to as succession, are explored.

The following suggestions provide examples for beginning the chapter:

1 Ask the students to view Figure 25.1. The following questions may serve as a brainstorming session: Have there always been so many geese? Why do they return to Cap Tourmente? Are the geese endangered? If so, from whom and do they require special protection? If they are feeding so intensively, is there enough food to support the geese and other species of waterfowl living in the same area?

2 If there is a river valley in your locale, ask students to make a journal entry which describes the different appearance of the north and south slopes of the river. Once the journal entry has been completed, ask students to refer to Figure 25.5. Does their description agree with the diagram? (Some textbooks refer to the south-facing slope of the river, which is actually the north bank. The north-facing slope of the river is actually the south bank of the river. We have attempted to avoid confusion, by referring only to the north slope, meaning north bank, and south slope, meaning south bank, of the river.)

ADDRESSING ALTERNATE CONCEPTIONS

- Students often indicate difficulties differentiating terms like community and ecosystem. Communities, like ecosystems, have their boundaries defined by the study. The study of a community involves only the organisms, whereas the ecosystem includes both the biotic and abiotic components of a specific area. In spite of this difference between community and ecosystem, it is virtually impossible to examine the structure and activities of any community without some reference to the abiotic factors that may influence its populations.
- The chaos theory provides students with a new paradigm for scientific investigation. The theory has origins in mathematics and physics, but it applies equally well in biology. Classical opinion has always held that to understand the behavior of a system it is only necessary to study each of its individual components. By identifying generalizations or tendencies based on the study of each of the parts, scientists attempted to predict events. Predictions were based upon identifying repeating ordered events. Conversely, chaos assumes that since randomness is a basic feature of complex systems, long-term predictions may well be extremely difficult. In its simplest form, chaos suggests that small uncertainties in short-term prediction may be magnified to such an extent over the long term that the expected behaviors become quite unpredictable. If the data appear to be random and without connections, chaos suggests that there may be an orderly system producing the observations. Chaos in this situation may actually allow us to discover order.

MAKING CONNECTIONS

- In mature ecosystems, populations tend to remain relatively stable over the long term. In many ways dynamic equilibrium, a term used by ecologists, can be compared with homeostasis, the term used by physiologists, described in Unit 4, Coordination and Regulation in Humans. Dynamic equilibrium describes how populations adjust to changes in the environment to maintain equilibrium, while homeostasis describes how an organism tends to maintain a relatively constant internal environment despite a changing external environment.

- The study of populations and communities occurs within ecosystems, introduced in chapter 2, Energy and Ecosystems. Many of the concepts introduced in chapter 2 are expanded upon with greater emphasis placed on quantification in chapter 25.

POSSIBLE JOURNAL ENTRIES

- Students might be asked to write a dialogue between the text and themselves, as they refute the idea that populations will undergo predictable changes following a forest fire. They may argue that succession is random. By acting as a "devil's advocate," the students can challenge their own learning and push understanding to a higher level.
- Students can be encouraged to develop a concept map which includes all key terms presented in the chapter.
- Students can be encouraged to express the difficulties they experienced in formulating an understanding of this challenging chapter. What things did they employ to aid them in constructing their knowledge? For example, some students may indicate that the case study, Calculating the Size of a Mammal Population, aided their understanding of growth rates or population density. Other students may have used the problems provided within the chapter or developed their own histograms to clarify population changes.
- Thinking and decision-making strategies may be recorded as students prepare for the debate. Did the students change their minds during the preparation for the debate? Students may even be asked to record their initial positions about the social issue and to reflect upon these feelings after the debate has been completed. Did they change their minds after listening to opposing arguments? Should forest fires in national parks be fought?
- Students may comment on the importance of the work of Austin Reed.
- Students might use histograms shown in Figure 25.21 to predict the fate of Mexico in the next few years. What disadvantages are presented by a young population? Does an aging population, as shown by Sweden, create problems?

USING THE VIDEODISC

Section title	Page	Code	Frame (side)	Description
Habitats, Geographic Range, and the Ecological Niche	577 – 578		9794 (4)	Movie sequence of blue whale showing surfacing behavior
Habitats, Geographic Range, and the Ecological Niche	577 – 578		4398 (4)	Movie sequence of sand lizard showing burrowing behavior
Habitats, Geographic Range, and the Ecological Niche	577 – 578		4845 (4)	Movie sequence showing swan feeding behavior
Distribution of Populations	578 – 579		5289 (3)	Movie sequence showing bread mold: growth and sporulation. Time-lapse photography of hyphae growth across the surface of bread and the development of sporangia
Distribution of Populations	578 – 579		7861 (4)	Movie sequence showing mother and young feeding
Distribution of Populations	578 – 579		5371 (4)	Movie sequence of ostrich and young

Section title	Page	Code	Frame (side)	Description
Intraspecies and Interspecies Competition	594 – 595		5813 (4)	Movie sequence of Emperor and Adelie penguins swimming behavior and interaction at nesting sites
Intraspecies and Interspecies Competition	594 – 595		7041 (4)	Bighorn sheep rutting
Intraspecies and Interspecies Competition	594 – 595		7229 (4)	Movie sequence of giraffes: competition behavior of males
Intraspecies and Interspecies Competition	594 – 595		9363 (4)	Elephant seals: males fighting
Symbiotic Relationships	596 – 598		31833 (4)	Movie sequence of hermit crab, worm, and sea anemone
Symbiotic Relationships	596 – 598		14516 (4)	Movie sequence showing parasitic life cycle of fluke
Symbiotic Relationships	596 – 598		33725 (4)	Movie sequence of trapdoor spider and sowbug
Symbiotic Relationships	596 – 598		34008 (4)	Pangolin and lion cubs
Symbiotic Relationships	596 – 598		34768 (4)	Movie sequence of rhinoceros and tick bird
Symbiotic Relationships	596 – 598		34994 (4)	Movie sequence of ghost crab and land crab, showing territorial relationship

IDEAS FOR INITIATING A DISCUSSION

Figure 25.4: Population patterns

Challenge students to come up with examples of clumped, uniform, and random population patterns.

Figure 25.8: Graph showing population of snow geese

The population growth curve indicates that the snow goose population is increasing. Ask students whether or not population increases are desirable. We often talk about economic growth as being positive, and students begin thinking that any type of growth is always beneficial. Encourage them to consider both positive and negative aspects of increases of the snow goose population.

Figure 25.10: Population growth curve

Ask your students what phase of growth they believe that the human population is in. Will the human population ever enter a stationary phase? a death phase?

Figure 25.21: Population histograms of Mexico and Sweden

Ask students to draw conclusions about Mexico and Sweden from the population histogram. Encourage them to be cautious about the number and extent of the conclusions that they are willing to infer.

REVIEW QUESTIONS

(pages 581–582)

1 List three characteristics of a population.

■ Three characteristics of a population are its geographic range, its habitat, and its size. Other characteristics include its distribution and geographic niche.

2 Distinguish between a population and a community.

■ A population is a group of individuals of the same species living in the same area at the same time. A community is the interactions of two or more populations living in the same place at the same time.

3 How does an animal's habitat differ from its geographic range?

■ An animal's geographic range is a region where organisms of the species have been spotted. By contrast, its habitat is the area where organisms of the species live. For example, bears are found all across Canada, however, their habitat is forests. They are not found in cities or in the middle of lakes.

4 Explain the meaning of the term ecological niche.

■ An ecological niche is a term that refers to a population's role within its community. This role includes all the biotic and abiotic factors that the population influences, and that influence the population.

5 Distinguish among clumped, random, and uniform population distribution patterns and describe the factors that may be responsible for each pattern.

■ A population can be distributed in several ways. A clumped distribution is a distribution pattern where a population is concentrated in several distinct areas throughout its habitat. This uneven distribution is caused by abiotic factors. Random distribution of a population is uncommon, but occurs when members of the population have neither attraction nor repulsion between each other. Uniform distribution occurs when the habitat is uniform, and there is competition among individuals for raw materials.

6 Identify examples of clumped and uniform distributions around your home or school. Explain the environmental factors responsible for each type of distribution.

■ Answers will vary.

7 Calculate the density of a meadow vole population if 78 animals were observed in a 20 ha area.

■ The density of the meadow vole population is 3.9 meadow voles per hectare.

8 Define rate of change in density of a population.

■ The rate of change in density of a population is defined as the amount by which a population changes over a stated period of time.

9 The following are data on the density of lemmings in a defined area of the tundra:

September 1981: 15 animals per hectare
September 1991: 3 animals per hectare

Calculate the rate of change in density of the lemming population. Is the population increasing, decreasing, or remaining stable? Explain.

■ The rate of change in density of the lemming population is −1.2 lemmings per hectare per year. Thus, the lemming population is decreasing.

(page 589)

10 Puffins are small marine birds found off the coast of Atlantic Canada. Calculate the population growth rate of a puffin colony based on the following:

1990 Data (original population 200 000)

Natality	Mortality	Immigration	Emigration
15 000	10 000	175 000	160 000

■ The population growth rate for the puffin colony is +10%.

11 Define dynamic equilibrium.

■ Dynamic equilibrium is a condition where a population size fluctuates between certain limits. The average population over time tends to remain stable.

12 How does an open population differ from a closed population?

■ An open population is a population in which density is controlled by internal and external factors. These include natality and mortality, as well as immigration and emigration. A closed population is one in which the density of the population is controlled only by internal factors. Natality and mortality are the only interactions with neither wastes nor food being imported or exported.

13 For growth curves of a closed population, describe the characteristic features of the lag, growth, stationary, and death phases.

■ The growth curve of a closed population is characterized by four distinct phases. These are the lag phase, a period of inactivity, during which time the population is thought to adjust to the environment before beginning reproduction. The next phase is the growth phase, during which the population increases exponentially. At some point, the growth rate slows, and the population enters the stationary phase where birth and death rates are in equilibrium. As wastes accumulate and nutrients become scarce, the population enters into the death phase. During this phase, the organisms die at a constant rate. Quite often, the entire population dies out.

14 Define carrying capacity, biotic potential, and environmental resistance.

■ The carrying capacity of an environment is the maximum number of organisms of a species that the environment can support. The term biotic potential describes the maximum number of offspring that can be produced by a species under ideal circumstances. Environmental resistance is a term used to describe all of the environmental factors that negatively affect population numbers.

15 Name four factors that affect environmental resistance.

■ Four factors that affect environmental resistance include climate, predation, availability of space, and disease.

16 Differentiate between S-shaped and J-shaped growth curves.

■ An S-shaped growth curve is the usual result of a population over time plot for a closed population that has had additional nutrients added to it. The bottom of the S represents the lag phase, the middle represents the growth phase, and the top of the curve represents the new equilibrium population. A J-shaped population growth curve is representative of a population that grows very quickly, and exceeds its environment's carrying capacity. The population then experiences a very rapid drop in population.

17 Differentiate between r and K population strategies. Give at least two examples of each.

■ r and K population strategies are dependent on environmental conditions. r-selected populations are extremely variable, and are characterized by high birth rates and short life spans. Most insects are examples of r-selected populations. K-selected populations are very stable, and tend to maintain their populations at or close to the carrying capacity of their habitat. Most large mammals such as bears and humans are examples of K-selected populations.

18 Define the law of the minimum and Shelford's law of tolerance.

■ The law of the minimum states that of all substances that are essential for growth, the one with the minimum concentration will be the limiting factor. Similarly, essential nutrients in high concentrations can also be harmful. Organisms function only when essential nutrients are in the range between the upper and lower concentration limits. This theory is known as Shelford's law of tolerance.

19 Distinguish between density-dependent and density-independent factors. Provide examples of each.

■ Density-dependent factors are things that affect a population differently based upon its population density. For example, disease, predation, and starvation rates are all density dependent. Density-independent factors are things that affect a population, and are not dependent on the number of organisms. For example, temperature, rainfall, and fires affect populations, but are not dependent on the density of the population.

(page 598)

20 What information is provided by a population histogram?

■ Population histograms show a population in terms of its age structure. They also show the proportion of males and females at a specific moment in time.

21 Name the basic factors that have resulted in the rapid growth in the human population during the past 150 years. Explain.

■ The three major factors which have fueled the explosive growth in human population in the past 150 years are the increased mechanization of agriculture, resulting in a much greater supply of food, the improvement in transportation, allowing better distribution of food, and the improvements in medicine, particularly in the area of infant mortality, which allowed greater numbers of people to reach reproductive ages.

22 What is the difference between interspecific and intraspecific competition?

■ Interspecific competition is competition between organisms of different species. For example, the predator-prey relationship is an example of interspecific competition. Intraspecific competition is competition between members of the same population. For example, members of the same species compete for food, space, and reproduction.

23 Why is the parasite-host relationship considered to be symbiotic?

■ Symbiosis is a relationship where two organisms live in close association with each other. Thus, the parasite-host relationship is considered to be symbiotic because both organisms are closely associated.

24 Distinguish between mutualism and commensalism.

■ Two other types of symbiotic relationships are mutualism and commensalism. Mutualistic relationships are beneficial to both organisms involved. Pollination is an example of a mutualistic relationship. Bees get access to a food source, and flowers have the pollen transferred to other flowers, allowing them to reproduce. Commensalism is a type of symbiotic relationship that one organism benefits from, but the other is not harmed in any way. The example of sharks and remoras is a good illustration of commensalism.

25 How does a predator differ from a parasite?

■ A predator differs from a parasite in that it kills its source of food. A parasite harms its host when it gets nutrients from it, however, it does not kill it immediately.

26 The symbiotic relationship between two organisms is described as +/0. Classify the different relationships. Support your answer.

■ If a symbiotic relationship is classified as +/0, then one organism benefits from the relationship, while the other is neither harmed nor helped. This relationship is thus an example of commensalism.

(page 603)

27 Define succession.

■ Succession is the gradual replacement of the dominant species of an area by other species. This replacement occurs while the vegetation of the area is developing.

28 Distinguish between primary and secondary succession.

■ There are two modes of succession. Primary succession refers to the occupation of a previously sterile environment by vegetation, such as a newly formed volcanic island. Secondary succession refers to the occupation of an area which previously contained vegetation, such as the land left after a forest fire.

29 What is meant by a climax community? How would you recognize a climax forest community?

■ A climax community is the relatively stable community reached after a period of succession. The climax community is dominated by the climax vegetation, but also contains extensive areas of vegetation representing every stage of development.

30 Name two human activities that can result in secondary succession.

■ Two human activities that can result in secondary succession are deliberately set forest fires, and harvesting of old growth forest for wood.

31 Why does succession proceed in a series of stages?

■ Succession proceeds in stages because the different stages require different soil and shelter conditions. These varying conditions are provided by the successive species of plants that establish themselves.

CASE STUDY
CALCULATING THE SIZE OF A SMALL MAMMAL POPULATION

(pages 591-592)

Time Requirements

Approximately 50 min are required.

Notes to the Teacher

- A review of graphing skills may be necessary for some students.
- Small group work is recommended.

Observation Questions

a) Could a population of masked shrews have evolved in Newfoundland?

- The population is not naturally occurring in Newfoundland, and therefore, it can be argued that the population will not be under many of the selective pressures of a naturally occurring population. However, the shrew may compete with some indigenous small mammals. It might change over time.

b) Would you expect there to be natural predators of the shrew living in Newfoundland? Explain.

- No, because the rodent is not indigenous to Newfoundland.

c) What is the advantage of sampling the population rather than trapping the entire area?

- Trapping the entire area is not only time consuming, but the trapping procedure also places the small mammals at risk.

d) Why were the quadrats the same size, and the traps set at the same time and location each year?

- To avoid over-sampling a smaller area. Variables were controlled as much as possible.

e) Why did the biologist mark each trapped shrew?

- To avoid counting the same shrew twice and to provide some data to monitor migration the biologist marked each trapped shrew.

f) Why were four trapping stations used rather than one?

- To obtain a larger sample size four trapping stations were used. Anomalies from one site could be reduced by increasing the population.

g) Record the total number of shrews trapped each year, then calculate the average number caught per year. Tabulate your results.

- Quadrat #1 = 53 total, 5.3 average; Quadrat #2 = 30 total, 3.0 average; Quadrat #3 = 17 total, 1.7 average; Quadrat #4 = 75 total, 7.5 average.

h) Are there any indications of a preferred habitat for the shrews? What factors might account for these differences?

- Quadrat #4 has the greatest number of shrews. This area may have the greatest density of larch and spruce trees, and hence the greatest number of sawflies. The sawflies provide food for the shrew. Students might also consider preferred reproductive sites or an area that has minimal competition from other small mammals.

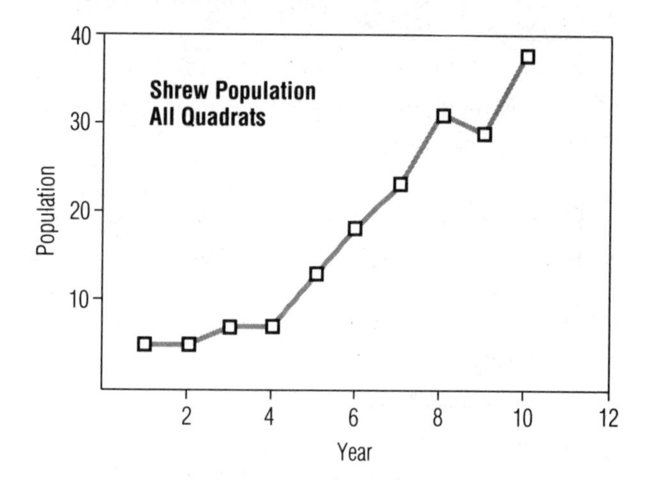

i) Does the graph show continuous growth in the population? Explain any trends.

- If a best fit curve is used, the general trend would show continuous increases in population; however, two years did not produce an increased growth.

j) Compare your graph with the theoretical population growth curves described in the text.

- Years 1 to 4 appear to be the lag phase, while years 5 to 10 appear to be the growth phase.

k) Predict what might happen to the population over the next two years.

- Accept growth phase continuing or the beginning of the stationary phase.

l) What may have happened to the population in year 9?

- The drop in population may be related to food sources declining, interspecific competition, intraspecific competition, increased parasitism, or predation. Many other variables should be considered.

m) If other species of shrew were already living in the area, how might this have affected the population being studied? Explain.

- Interspecific competition may have occurred. The more similar the animals, the greater is the competition.

Case-Study Application Questions

(For each of the following questions use the appropriate formula when necessary and show your work.)

1 Calculate the following:

a) How many square meters are there in a trapping quadrat?

■ 20 m X 20 m = 400 m^2

b) What is the size of a quadrat in hectares?

■ 10 000 m^2 = 1 ha, 400 m^2/10 000 = 0.04 ha

c) How many hectares are there in the total study area?

■ Release area = 10.0 km^2 or 1000 ha

2 Using the average number of shrews per quadrat for year 10 calculate the population density in shrews per hectare.

■ Each quadrat is 0.04 ha, therefore to find the average: (12 + 7 + 4 + 15)/4 = 9.5 shrews per 0.04 ha or 237.5 shrews per ha.

3 From your answer to question 2, calculate the number of shrews for the total study area.

■ 237.5 shrews/ha, total area = 1000 ha, or 237 500 shrews

4 Explain why the answer to question 3 is usually called a population "estimate."

■ Each shrew is not counted; the population is determined by a random sample. The sample assumes that the animals are equally distributed in each of the quadrats.

5 Calculate the rate of change in shrew density between years 1 and 5 and years 6 and 10. Which four-year period shows the greater population change?

■ Density (*D*) can be calculated by the average number of shrews per quadrat.
For year 1, 5 shrews/4 quadrats or 1.25 shrews/quadrat
For year 5, 13 shrews/4 quadrats or 3.25 shrews/quadrat
For year 6, 18 shrews/4 quadrats or 4.5 shrews/quadrat
For year 10, 38 shrews/4 quadrats or 9.5 shrews/quadrat
The density per quadrat can be changed to hectares (each quadrat is 0.04 ha).

$$D_{\text{year 1}} \quad \frac{1.25 \text{ shrews/quadrat}}{0.04 \text{ ha/quadrat}} = 31.25 \text{ shrews/ha}$$

$$D_{\text{year 5}} \quad \frac{3.25 \text{ shrews/quadrat}}{0.04 \text{ ha/quadrat}} = 81.25 \text{ shrews/ha}$$

$$D_{\text{year 6}} \quad \frac{4.5 \text{ shrews/quadrat}}{0.04 \text{ ha/quadrat}} = 112.5 \text{ shrews/ha}$$

$$D_{\text{year 10}} \quad \frac{9.5 \text{ shrews/quadrat}}{0.04 \text{ ha/quadrat}} = 237.5 \text{ shrews/ha}$$

Calculate the rate of density change between years 1 and 5.

$$R = \frac{\Delta D}{\Delta t} \text{ or } \frac{D_{\text{year 5}} - D_{\text{year 1}}}{5-1}$$

$$\frac{81.25 - 31.25}{5-1}$$

$$= \frac{50.0}{4} \text{ or } 12.5$$

The rate of density change is 1.5 shrews/ha/yr.

Calculate the rate of density change between years 6 and 10

$$R = \frac{D_{\text{year 10}} - D_{\text{year 6}}}{10-6}$$

$$= \frac{237.5 - 112.5}{10-6}$$

$$= \frac{125.0}{4} \text{ or } 31.25$$

The rate of density change is 31.25 shrews/ha/yr

The greater population change occurred between years 6 and 10.

6 Although there is a wide range of shrews and other small mammals all across Canada, only the meadow vole, a mouse-like animal, naturally inhabits Newfoundland. Suggest reasons why so few rodents live in Newfoundland.

■ Newfoundland is geographically isolated from other land masses.

CASE STUDY
SUCCESSION IN GLACIER BAY

(pages 604–605)
Time Requirements
Approximately 30 to 40 min are required.

Observation Questions

a) Algae are often found in the snow and ice. Indicate what function they would serve in this ecosystem. If so, where?

■ The algae are producers. They are concentrated on the surfaces that receive most sunlight.

b) Provide a possible explanation for the uneven distribution of the pioneer plants seen in diagram (iii).

■ The melting of the glacier carries minerals and dead algal matter into the lower depressions between the rocks. These low-lying areas make an ideal location for the accumulation of soil.

c) The mountain avens contain nitrogen-fixing bacteria in their roots. What special advantage do the mountain avens gain from this mutualistic relationship?

■ Because the microbes can fix nitrogen from the air, these plants can live in soils that contain little nitrogen. The nitrogen-fixing microbes and the plant exhibit a mutualistic relationship. The plant is supplied with nitrogen and the bacterium is provided a place to live and a source of carbohydrates.

d) Describe how the landscape has changed between diagrams (iv) and (v).

■ As the mountain avens die more soil accumulates and larger plants can be supported. The nitrogen added to the soil also helps larger plants which have greater nutrient requirements. The alders and shrubs become more dense and population density increases. The alders, like the mountain avens, have a mutualistic relationship with nitrogen-fixing bacteria.

e) Referring to diagram (vi), speculate why the cottonwood shrubs only begin to colonize the area after the mountain avens are established.

■ The mountain avens die and decompose, adding organic matter to the soil. After a number of years, more organic matter adds nutrient and the amount of soil found in the area increases. The cottonwood shrubs do make their own nitrogen, and therefore they require a higher nutrient soil before they can colonize. The thicker richer soil established by the mountain avens also provides an anchor for the roots of the shrubs.

f) Compare diagram (iii) with diagram (vii). Which plant provides the greatest biomass?

■ The later stages of succession have the greatest biomass. As soil quality improves, larger plants begin to grow. These plants can support more primary consumers, and hence more secondary and tertiary consumers.

Case-Study Application Questions

1 Compare the pioneer vegetation shown in diagram (iii), the intermediary vegetation shown in diagram (iv), and the vegetation of the climax community shown in diagram (vii).

a) Which community would support the greatest number of organisms? Give your reasons.

■ In community (vii) there are more producers to support a greater number of consumers. This community has a greater available food source.

b) Which community would support the greatest diversity of organisms? Give your reasons.

■ Community (vii) will have the greatest diversity of organisms. A mature forest can support consumers at ground level and in the trees. The taller and larger the tree, the more diverse the habitat.

2 Provide sample food webs for diagrams (iii) and (vii).

■ Answers will vary, but expect a more complicated food web for diagram (vii).

3 Briefly explain how the abiotic factors within the community change because of the succession of vegetation.

■ As trees grow less sunlight reaches the forest floor. Decomposing leaves form a blanket which reduces water loss from the soil by evaporation and adds organic matter which improves soil quality.

APPLYING THE CONCEPTS

(page 608)

1 Calculate the rate of change in a moose population using the following data:
1978: 25 moose; area surveyed = 40 ha
1984: 11 moose; area surveyed = 30 ha

■ $\frac{25 \text{ moose/40 ha} - 11 \text{ moose/30 ha}}{1984 - 1978}$, which is

$\frac{0.625 - 0.367 \text{ moose/ha}}{6 \text{ years}} =$

0.043 moose/ha/yr increase.

2 Plot a population curve from the following data, which were obtained after the introduction of deer mice on an isolated hillside.

Date	Numbers	Date	Numbers	Date	Numbers
1979	20	1983	25	1987	130
1980	20	1984	28	1988	128
1981	22	1985	40	1989	133
1982	26	1986	80	1990	132

a) Is the curve characteristic of an open or closed population? Why?

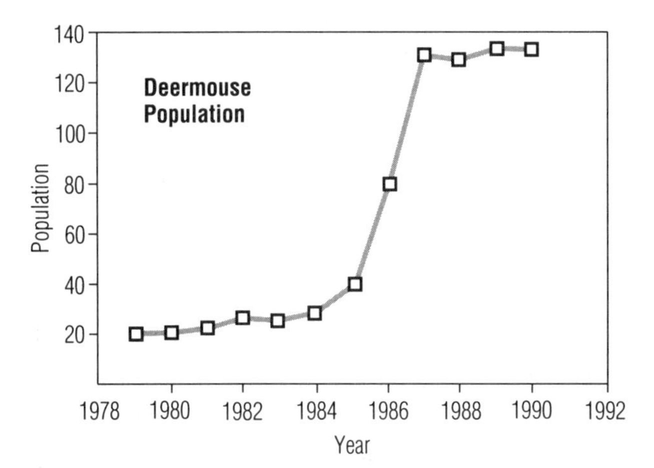

■ The graph appears to show the lag phase, growth phase, and stationary phase of a closed population; however, no death phase is shown. Students may suggest that because the death phase is not shown, the population may actually represent a S-shaped curve from an open population. Both answers make an assumption. Before any conclusive agreement can be reached, the plateau phase should be monitored for a longer period of time.

b) Using a line graph, show changes in the population between 1979 and 1990.

■ See graph above.

c) Use a red line to indicate the probable position of the carrying capacity of the ecosystem. Why have you placed it where you did?

■ The carrying capacity should be at or near the new plateau for the period 1987 to 1990.

3 Using the data provided above, calculate the growth rate between 1986 and 1987. Identify factors that might account for the accelerated growth rate.

■ $R = \Delta D/\Delta T$ or 130 - 80 / 1 year = 50 deer mice per year.

4 Explain how the carrying capacity of an ecosystem is related to the biotic potential of a species and its environmental resistance.

■ The term biotic potential (R_{max}) is used to refer to this theoretical maximum birth rate. Biotic potential is regulated by four important factors as shown in the chart that follows. Biotic potential is the number of offspring which could be produced by a species under ideal conditions.

Biotic potential

Factor	Description
Offspring	Maximum number of offspring/birth
Capacity for survival	Chances the organism's offspring will reach reproductive age
Procreation	Number of times the organism reproduces each year
Maturity	Age at which reproduction begins

5 Predict what might happen to a population of moose if their numbers exceeded the carrying capacity of their environment. Provide the reasons for your prediction.

■ The population would drop quickly and readjust at a lower plateau.

6 Design a research study that would supply you with data on the effect of latitude on the population cycle of the common deer mouse.

■ Sampling sites must be established at different times of the year. Habitat preference and food preference throughout the year must be determined. The study should also obtain a profile of predators, abundance of food source, and parasite levels. Many possible variations of studies can be indicated.

7 In what ways is Shelford's law a more appropriate principle in population studies than the law of the minimum?

■ If any nutrient is reduced, the development of an organism will be affected regardless of the amounts of the other substances. Later this became known as the law of the minimum. Victor Shelford added to the work by noting that not only a deficiency of nutrients, but an excess can be detrimental to an organism's survival. Subsequently it was determined that an organism will grow only within a range between the upper and lower limits of abiotic factors. The greater the range of tolerance the greater the survival ability of an organism. There is an optimum range of conditions for maximum population size. If one could plot tolerance curves for each environmental factor on a single graph, few would fit exactly on each other. Therefore the overall optimum range for a population is restricted. The organisms may respond to changes in any one of the abiotic factors by having their population reduced or increased.

8 Draw a population histogram from the following data on the white-tailed deer.

Age	Males	Females
1	72	75
2	35	33
3	24	25
4	17	15
5	14	11
6	8	9
7	7	6
8	5	5
9	4	3
10	2	3

What information is provided by this histogram? (Note: The few animals over 10 years of age are not included.)

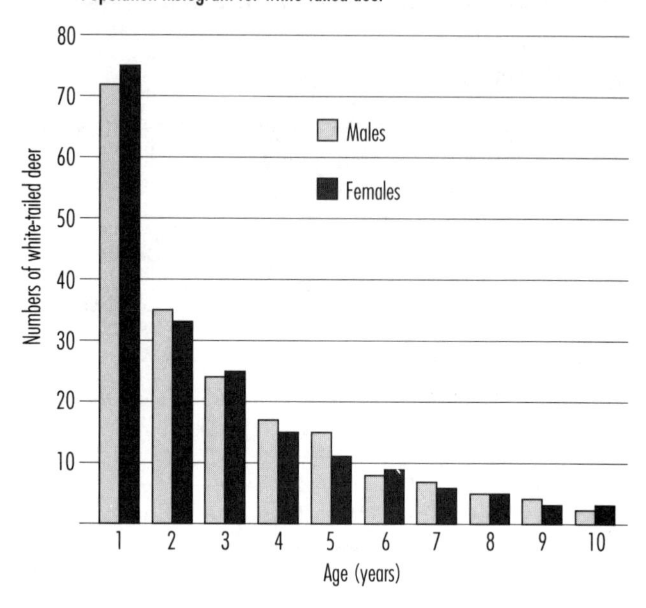

CRITICAL-THINKING QUESTIONS

(page 608)

1 How could habitat preference influence the results of a population density study? Is it possible to account for this factor when designing a population study? Explain clearly.

■ Habitat preference, especially for things such as nesting sites, has a profound influence on population density. The more desirable the area, the greater is the population density. This can be accounted for by selecting samples that represent densely populated communities and ones that represent sparsely populated areas. Expanding the study area may also help provide a better profile.

2 Compare the concept of dynamic equilibrium in an ecosystem with homeostasis in the body of an advanced animal.

■ In many ways dynamic equilibrium, a term used by ecologists, can be compared with homeostasis, the term used by physiologists. Dynamic equilibrium describes how populations adjust to changes in the environment to maintain equilibrium, while homeostasis describes how an organism tends to maintain a relatively constant internal environment despite a changing external environment. Dynamic equilibrium refers to any condition within the biosphere that remains stable within fluctuating limits.